UCLA Symposia on Molecular and Cellular Biology, New Series

Series Editor, C. Fred Fox

RECENT TITLES

Volume 84
Technological Advances in Vaccine Development, Laurence Lasky, *Editor*

Volume 85
B Cell Development, Owen N. Witte, Norman R. Klinman, and Maureen C. Howard, *Editors*

Volume 86
Synthetic Peptides: Approaches to Biological Problems, James P. Tam and Emil Thomas Kaiser, *Editors*

Volume 87
Gene Transfer and Gene Therapy, Arthur L. Beaudet, Richard Mulligan, and Inder M. Verma, *Editors*

Volume 88
Molecular Biology of the Eye: Genes, Vision, and Ocular Disease, Joram Piatigorsky, Toshimichi Shinohara, and Peggy S. Zelenka, *Editors*

Volume 89
Liposomes in the Therapy of Infectious Diseases and Cancer, Gabriel Lopez-Berestein and Isaiah J. Fidler, *Editors*

Volume 90
Cell Biology of Virus Entry, Replication, and Pathogenesis, Richard W. Compans, Ari Helenius, and Michael B.A. Oldstone, *Editors*

Volume 91
Bone Marrow Transplantation: Current Controversies, Robert Peter Gale and Richard E. Champlin, *Editors*

Volume 92
The Molecular Basis of Plant Development, Robert Goldberg, *Editor*

Volume 93
Cellular and Molecular Biology of Muscle Development, Laurence H. Kedes and Frank E. Stockdale, *Editors*

Volume 94
Molecular Biology of RNA, Thomas R. Cech, *Editor*

Volume 95
DNA-Protein Interactions in Transcription, Jay D. Gralla, *Editor*

Volume 96
Stress-Induced Proteins, Mary Lou Pardue, James R. Feramisco, and Susan Lindquist, *Editors*

Volume 97
Molecular Biology of Stress, Shlomo Breznitz and Oren Zinder, *Editors*

Volume 98
Metal Ion Homeostasis: Molecular Biology and Chemistry, Dean H. Hamer and Dennis R. Winge, *Editors*

Volume 99
Human Tumor Antigens and Specific Tumor Therapy, Richard S. Metzgar and Malcolm S. Mitchell, *Editors*

Volume 100
Mechanisms of Action and Therapeutic Applications of Biologicals in Cancer and Immune Deficiency Disorders, J. Groopman, C. Evans, and D. Golde, *Editors*

Volume 101
Molecular Biology of Plant-Pathogen Interactions, Brian Staskawicz, Paul Ahlquist, and Olen Yoder, *Editors*

Volume 102
Growth Factors and Their Receptors: Genetic Control and Rational Application, Russell Ross, Anthony W. Burgess, and Tony Hunter, *Editors*

Volume 103
Growth Inhibitory and Cytotoxic Polypeptides, Harold L. Moses, Peter Lengyel, and Charles D. Stiles, *Editors*

Volume 104
Cellular Proteases and Control Mechanisms, Tony Hugli, *Editor*

Volume 105
Molecular and Cellular Mechanisms of Human Hypersensitivity and Autoimmunity, Edward J. Goetzl, *Editor*

Volume 106
Cell Activation and Signal Initiation: Receptor and Phospholipase Control of Inositol Phosphate, PAF, and Eicosanoid, Production, Edward A. Dennis and Tony Hunter, *Editors*

Volume 107
Tissue Engineering, Richard Skalak and C. Fred Fox, *Editors*

Please contact the publisher for information about previous titles in this series.

UCLA Symposia Board

Growth Factors and Their Receptors

Genetic Control and Rational Application

Growth Factors and Their Receptors

Genetic Control and Rational Application

Proceedings of an Abbott-Cetus-Genentech-
Smith, Kline, & French-UCLA Symposium
Held in Keystone, Colorado
January 24-30, 1988

Editors

Russell Ross
Department of Pathology
University of Washington School of Medicine
Seattle, Washington

Anthony W. Burgess
Melbourne Tumor Biology Branch
Ludwig Institute for Cancer Research
Royal Melbourne Hospital
Victoria, Australia

Tony Hunter
Molecular Biology and Virology Laboratory
Salk Institute
San Diego, California

Alan R. Liss, Inc. • New York

Address all Inquiries to the Publisher
Alan R. Liss, Inc., 41 East 11th Street, New York, NY 10003

Library of Congress Cataloging-in-Publication Data

Abbott-Cetus-Genentech-Smith, Kline and French-UCLA Symposium (1988: Keystone, Colo.)
Growth factors and their receptors: genetic control and rational application: proceedings of an Abbott-Cetus-Genentech-Smith, Kline and French-UCLA Symposium held in Keystone, Colorado, January 24–30, 1988 / editors, Russell Ross, Tony W. Burgess, Tony Hunter.
p. cm.—(UCLA symposia on molecular and cellular biology; new ser., v. 102)
Includes bibliographies and index.
ISBN 0-8451-4701-3
1. Growth promoting substances—Congresses. 2. Growth promoting substances—Receptors—Congresses. 3. Gene expression—Congresses. 4. Genetic regulation—Congresses. I. Ross, Russell. II. Burgess, Anthony, 1946- . III. Hunter, Tony. 1943— . IV. Abbott Laboratories. V. Title. VI. Series.
[DNLM: 1. Cell Transformation, Neoplastic—congresses. 2. Growth Substances—congresses. 3. Receptors, Endogenous Substances—congresses. W3 U17N new ser. v. 102 / QZ 202 A125 1988g]
QP801.G74A23 1988
616.99'2071—dc 19
DNLM/DLC
for Library of Congress 89–2469
CIP

Pages 1–217 of this volume are reprinted from the Journal of Cellular Biochemistry, Volumes 38 and 39. The Journal is the only appropriate literature citation for the articles printed on these pages. The page numbers in the table of contents, contributors list, and index of this volume correspond to the page numbers at the foot of these pages.

The table of contents does not necessarily follow the pattern of the plenary sessions. Instead, it reflects the thrust of the meeting as it evolved from the combination of plenary sessions, poster sessions, and workshops, culminating in the final collection of invited papers, submitted papers, and workshop summaries. The order in which articles appear in this volume does not follow the order of citation in the table of contents. Many of the articles in this volume were published in the Journal of Cellular Biochemistry, and they are reprinted here. These articles appear in the order in which they were accepted for publication and then published in the Journal. They are followed by papers which were submitted solely for publication in the proceedings.

Contents

Contributors

Robert Abraham, Department of Immunology, Mayo Clinic, Rochester, MN 55905 **[229]**

Claude Asselin, Biochemistry Department, SUNY at Stony Brook, Stony Brook, NY 11794-5215 **[239]**

D. Barritault, Université Paris XII, Laboratoire de Biotechnologie des Cellules Eucaryotes, 94010 Créteil, Paris, France **[95, 163]**

Laura Beguinot, Laboratory of Molecular Biology, National Cancer Institute, National Institutes of Health, Bethesda, MD 20892; present address: University Institute of Microbiology, 1353 Copenhagen K, Denmark **[131]**

P.M.P. van Bergen en Henegouwen, Department of Molecular Cell Biology, University of Utrecht, 3584 CH Utrecht, The Netherlands **[207]**

J. Boonstra, Department of Molecular Cell Biology, University of Utrecht, 3584 CH Utrecht, The Netherlands **[207]**

R. Bordoni, V.A. Medical Center and Department of Medicine, Emory University School of Medicine, Atlanta, GA 30033 **[173]**

Daniel F. Bowen-Pope, Department of Pathology, University of Washington, Seattle, WA 98195 **[297]**

Alton L. Boynton, Cancer Research Center of Hawaii, Basic Science Program, University of Hawaii, Honolulu, HI 96813 **[21]**

Hal E. Broxmeyer, Departments of Medicine (Hematology/Oncology), Microbiology and Immunology, and The Walther Oncology Center, Indiana University School of Medicine, Indianapolis, IN 46223 **[11]**

B. Busch, Department of Biochemistry and Molecular Biology, Merck Research Laboratories, Rahway, NJ 07065 **[253]**

J. Capparella, Department of Biochemistry and Molecular Biology, Merck Research Laboratories, Rahway, NJ 07065 **[253]**

D. Caruelle, Université Paris XII, Laboratoire de Biotechnologie des Cellules Eucaryotes, 94010 Créteil Cedex, Paris, France **[95]**

J.P. Caruelle, Université Paris XII, Laboratoire de Biotechnologie des Cellules Eucaryotes, 94010 Créteil Cedex, Paris, France **[95]**

David T. Chang, Johns Hopkins Oncology Center, Baltimore, MD 21205 **[117]**

Loredana Chiadó-Piat, Department of Biomedical Sciences and Oncology, University of Torino Medical School, Torino, Italy **[49]**

Mary Elizabeth Cicione, Department of Medicine, Roger Williams General Hospital, Brown University, Providence, RI 02908 **[107]**

P.G. Comber, Departments of Pathology and Laboratory Medicine, School of Medicine, University of Pennsylvania, Philadelphia, PA 19104-6082 **[39]**

Paolo M. Comoglio, Department of Biomedical Sciences and Oncology, University of Torino Medical School, Torino, Italy **[49]**

G. Conn, Department of Biochemistry and Molecular Biology, Merck Research Laboratories, Rahway, NJ 07065 **[253]**

Scott Cooper, Department of Medicine (Hematology/Oncology), Indiana University School of Medicine, Indianapolis, IN 46223 **[11]**

G. Coscas, Université Paris XII, Laboratoire de Biotechnologie des Cellules Eucaryotes, 94010 Créteil Cedex, Paris, France **[95]**

J. Courty, Laboratoire de Biotechnologie des Cellules Eucaryotes, Université Paris XII, 94010 Créteil, Paris, France **[163]**

R. Andrew Cuthbertson, PO Royal Melbourne Hospital and Howard Florey Institute for Experimental Physiology and Medicine, University of Melbourne, Victoria 3050, Australia **[219]**

M.C. Dauchel, Laboratoire de Biotechnologie des Cellules Eucaryotes, Université Paris XII, 94010 Créteil, Paris, France **[163]**

L.H.K. Defize, Hubrecht Laboratory, Netherlands Institute for Developmental Biology, 3584 CT Utrecht, The Netherlands **[207]**

J. de Kroon, Department of Molecular Cell Biology, University of Utrecht, 3584 CH Utrecht, The Netherlands **[207]**

Maria Flavia Di Renzo, Department of Biomedical Sciences and Oncology, University of Torino Medical School, Torino, Italy **[49]**

J. DiSalvo, Department of Biochemistry and Molecular Biology, Merck Research Laboratories, Rahway, NJ 07065 **[253]**

Ashley R. Dunn, Molecular Biology Laboratory, Melbourne Tumor Biology Branch, Ludwig Institute for Cancer Research, PO Royal Melbourne Hospital, Victoria 3050, Australia **[219]**

Larry R. Ellingsworth, Collagen Corporation, Palo Alto, CA 94303 **[153]**

Paul Fanning, Children's Hospital, Harvard Medical School, Boston, MA 02115 **[83]**

A.H. Fischer, Departments of Pathology and Laboratory Medicine, School of Medicine, University of Pennsylvania, Philadelphia, PA 19104-6082 **[39]**

John W. Forstrom, ZymoGenetics, Inc., Seattle, WA 98105 **[297]**

A. Raymond Frackelton, Jr., Department of Medicine, Roger Williams General Hospital, Brown University, Providence, RI 02908 **[107,229]**

A. Gaudric, Université Paris XII, Laboratoire de Biotechnologie des Cellules Eucaryotes, 94010 Créteil Cedex, Paris, France **[95]**

Elena V. Gerbaudo, Department of Biomedical Sciences and Oncology, University of Torino Medical School, Torino, Italy **[49]**

Heinz A. Gerber, Department of Pathology, University of Bern, 3010 Bern, Switzerland **[145]**

Brenda Gerwin, Laboratory of Human Carcinogenesis, National Cancer Institute, Bethesda, MD 20892 **[269]**

G. Gimenez-Gallego, Department of Biochemistry and Molecular Biology, Merck Research Laboratories, Rahway, NJ 07065 **[253]**

Silvia Giordano, Department of Biomedical Sciences and Oncology, University of Torino Medical School, Torino, Italy **[49]**

Bernd Groner, Ludwig Institute for Cancer Research, Bern Branch, Inselspital, 3010 Bern, Switzerland **[145]**

B. Groux-Muscatelli, Université Paris XII, Laboratoire de Biotechnologie des Cellules Eucaryotes, 94010 Créteil Cedex, Paris, France **[95]**

Giao Hangoc, Department of Medicine (Hematology/Oncology), Indiana University School of Medicine, Indianapolis, IN 46223 **[11]**

Curtis C. Harris, Laboratory of Human Carcinogenesis, National Cancer Institute, Bethesda, MD 20892 **[269]**

Charles E. Hart, ZymoGenetics, Inc., Seattle, WA 98105 **[297]**

Stephen D. Hauschka, Department of Biochemistry, University of Washington, Seattle, WA 98195 **[195]**

Timothy D. Hill, Cancer Research Center of Hawaii, Basic Science Program, University of Hawaii, Honolulu, HI 96813 **[21]**

Richard D. Huhn, Department of Medicine, Roger Williams General Hospital, Brown University, Providence, RI 02908; present address: Department of Biochemistry, St. Jude Children's Research Hospital, Memphis, TN 38101 **[107]**

Nancy E. Hynes, Ludwig Institute for Cancer Research, Bern Branch, Inselspital, 3010 Bern, Switzerland **[145]**

James N. Ihle, Department of Molecular Mechanisms of T Cell Leukemogenesis, NCI-Frederick Cancer Research Facility, Frederick, MD 21701; present address: Department of Biochemistry, St. Jude Children's Research Hospital, Memphis, TN 38101 **[229]**

Robert J. Isfort, Department of Molecular Mechanisms of T Cell Leukemogenesis, NCI-Frederick Cancer Research Facility, Frederick, MD 21701; present address: Department of Biochemistry, St. Jude Children's Research Hospital, Memphis, TN 38101 **[229]**

C. Ronald Kahn, Research Division, Joslin Diabetes Center, Department of Medicine, Brigham and Women's Hospital and Harvard Medical School, Boston, MA 02215 **[181]**

Yang Ke, Laboratory of Human Carcinogenesis, National Cancer Institute, Bethesda, MD 20892 **[269]**

Jonathan R. Keller, Biological Carcinogenesis Development Program, Program Resources, Inc., NCI-Frederick Cancer Research Facility, Frederick, MD 21701 **[153]**

James D. Kelly, ZymoGenetics, Inc., Seattle, WA 98105 and the Department of Pathology, University of Washington, Seattle, WA 98195 **[297]**

L. Kelly, Department of Biochemistry and Molecular Biology, Merck Research Laboratories, Rahway, NJ 07065 **[253]**

Henrik Kindmark, Cancer Research Center of Hawaii, Basic Science Program, University of Hawaii, Honolulu, HI 96813; present address: Department of Medical Cell Biology, University of Uppsala Biomedicum, Uppsala, Sweden **[21]**

Michael Klagsbrun, Children's Hospital, Harvard Medical School, Boston, MA 02115 **[83]**

Richard A. Lang, Molecular Biology Laboratory, Melbourne Tumor Biology Branch, Ludwig Institute for Cancer Research, PO Royal Melbourne Hospital, Victoria 3050, Australia **[219]**

John F. Lechner, Laboratory of Human Carcinogenesis, National Cancer Institute, Bethesda, MD 20892 **[269]**

David B. Lewis, Department of Pediatrics (Infectious Diseases), University of Washington, Seattle, WA 98195 **[1]**

D. Linemeyer, Department of Biochemistry and Molecular Biology, Merck Research Laboratories, Rahway, NJ 07065 **[253]**

Douglas R. Lowy, Laboratory of Cellular Oncology, National Cancer Institute, National Institutes of Health, Bethesda, MD 20892 **[131]**

Kenneth B. Marcu, Biochemistry Department, SUNY at Stony Brook, Stony Brook, NY 11794-5215 **[239]**

David Mark, Department of Molecular Biology, Cetus Corporation, Emeryville, CA 94608 **[71]**

Jamey D. Marth, Howard Hughes Medical Institute and the Department of Medicine (Medical Genetics), University of Washington, Seattle, WA 98195 **[1]**

Tohru Masui, Laboratory of Human Carcinogenesis, National Cancer Institute, Bethesda, MD 20892 **[269]**

W. Stratford May, The Johns Hopkins Oncology Center, Baltimore, MD 21231 **[229]**

Kevin H. Mayo, Department of Chemistry, Temple University, Philadelphia, PA 19122 **[283]**

Kimberly Meade-Cobun, Johns Hopkins Oncology Center, Baltimore, MD 21205 **[117]**

T. Mellin, Department of Biochemistry and Molecular Biology, Merck Research Laboratories, Rahway, NJ 07065 **[253]**

J. Menke, Department of Biochemistry and Molecular Biology, Merck Research Laboratories, Rahway, NJ 07065 **[253]**

A. Mereau, Laboratoire de Biotechnologie des Cellules Eucaryotes, Université Paris XII, 94010 Créteil, Paris, France **[163]**

Donald Metcalf, Cancer Research Unit, Walter and Eliza Hall Institute for Medical Research, PO Royal Melbourne Hospital, Victoria 3050, Australia **[219]**

Masao Miyashita, Laboratory of Human Carcinogenesis, National Cancer Institute, Bethesda, MD 20892 **[269]**

Mark J. Murray, ZymoGenetics, Inc., Seattle, WA 98105 **[297]**

Radha P. Narsimhan, Department of Biomedical Sciences and Oncology, University of Torino Medical School, Torino, Italy **[49]**

David G. Nathan, Department of Pediatrics, Harvard Medical School, and Department of Medicine and Division of Hematology and Pediatric Oncology, Children's Hospital and Dana Farber Cancer Institute, Boston, MA 02115 **[259]**

Alain Nepveu, Biochemistry Department, SUNY at Stony Brook, Stony Brook, NY 11794-5215; present address: Department of Molecular Biology, Ludwig Institute, Montreal Branch, Montreal H3A 1A1, Canada **[239]**

Bradley B. Olwin, Department of Biochemistry, University of Wisconsin, Madison, WI 53706 **[195]**

S. Ortega, Department of Biochemistry and Molecular Biology, Merck Research Laboratories, Rahway, NJ 07065 **[253]**

Ira Pastan, Laboratory of Molecular Biology, National Cancer Institute, National Institutes of Health, Bethesda, MD 20892 **[131]**

Richard Peet, the Department of Medicine (Medical Genetics), University of Washington, Seattle, WA 98195 **[1]**

Roger M. Perlmutter, Howard Hughes Medical Institute, and the Department of Medicine (Medical Genetics), University of Washington, Seattle, WA 98195 **[1]**

Andrea Pfeifer, Laboratory of Human Carcinogenesis, National Cancer Institute, Bethesda, MD 20892 **[269]**

M.B. Prystowsky, Departments of Pathology and Laboratory Medicine, School of Medicine, University of Pennsylvania, Philadelphia, PA 19104-6082 **[39]**

Peter Ralph, Department of Cell Biology, Cetus Corporation, Emeryville, CA 94608 **[11]**

Daniel A. Rappolee, Laboratory of Radiobiology and Environmental Health, University of California, San Francisco, CA 94143 **[71]**

Roger Reddel, Laboratory of Human Carcinogenesis, National Cancer Institute, Bethesda, MD 20892 **[269]**

Carl W. Rettenmier, Department of Tumor Cell Biology, St. Jude Children's Research Hospital, Memphis, TN 38101 **[29]**

A. Richmond, V.A. Medical Center and Department of Medicine, Emory University School of Medicine, Atlanta, GA 30033 **[173]**

Snezna Rogelj, The Whitehead Institute for Biomedical Research, Cambridge, MA 02142 **[83]**

Russell Ross, Department of Pathology, University of Washington, Seattle, WA 98195 **[297]**

Martine F. Roussel, Department of Tumor Cell Biology, St. Jude Children's Research Hospital, Memphis, TN 38101 **[29]**

Enrique Rozengurt, Imperial Cancer Research Fund, London WC2A 3PX, England **[57]**

Francis W. Ruscetti, Laboratory of Molecular Immunoregulation, Biological Response Modifiers Program, NCI-Frederick Cancer Research Facility, Frederick, MD 21701 **[153]**

D.E. Sabath, Departments of Pathology and Laboratory Medicine, School of Medicine, University of Pennsylvania, Philadelphia, PA 19104-6082 **[39]**

Suzanne Saurer, Ludwig Institute for Cancer Research, Bern Branch, Inselspital, 3010 Bern, Switzerland **[145]**

M.-T. Schaeffer, Department of Biochemistry and Molecular Biology, Merck Research Laboratories, Rahway, NJ 07065 **[253]**

C. Sestier, Université Paris XII, Laboratoire de Biotechnologie des Cellules Eucaryotes, 94010 Créteil Cedex, Paris, France **[95]**

Charles J. Sherr, Department of Tumor Cell Biology, St. Jude Children's Research Hospital, Memphis, TN 38101 **[29]**

P.M. Shipman, Departments of Pathology and Laboratory Medicine, School of Medicine, University of Pennsylvania, Philadelphia, PA 19104-6082 **[39]**

Colin A. Sieff, Department of Pediatrics, Harvard Medical School, and Department of Medicine and Division of Hematology and Pediatric Oncology, Children's Hospital and Dana Farber Cancer Institute, Boston, MA 02115 **[259]**

Garwin K. Sing, Laboratory of Molecular Immunoregulation, Biological Response Modifiers Program, NCI-Frederick Cancer Research Facility, Frederick, MD 21701 **[153]**

James Sinnett-Smith, Imperial Cancer Research Fund, London WC2A 3PX, England **[57]**

Robert A. Smith, ZymoGenetics, Inc., Seattle, WA 98105 **[297]**

D. Soderman, Department of Biochemistry and Molecular Biology, Merck Research Laboratories, Rahway, NJ 07065 **[253]**

Don A. Stevens, The Johns Hopkins Oncology Center, Baltimore, MD 21231 **[229]**

K. Sullivan, The W.M. Keck Foundation, Autoimmune Disease Center, Department of Basic and Clinical Research, Scripps Clinic and Research Foundation, La Jolla, CA 92037 **[39]**

Luca Tamagnone, Department of Biomedical Sciences and Oncology, University of Torino Medical School, Torino, Italy **[49]**

E.M. Tan, The W.M. Keck Foundation, Autoimmune Disease Center, Department of Basic and Clinical Research, Scripps Clinic and Research Foundation, La Jolla, CA 92037 **[39]**

G. Thomas, V.A. Medical Center and Department of Medicine, Emory University School of Medicine, Atlanta, GA 30033 **[173]**

K.A. Thomas, Department of Biochemistry and Molecular Biology, Merck Research Laboratories, Rahway, NJ 07065 **[253]**

Robert W. Tucker, Johns Hopkins Oncology Center, Baltimore, MD 21205 **[117]**

H. van Damme, Department of Molecular Cell Biology, University of Utrecht, 3584 CH Utrecht, The Netherlands **[207]**

William C. Vass, Laboratory of Cellular Oncology, National Cancer Institute, National Institutes of Health, Bethesda, MD 20892 **[131]**

Thierry J. Velu, Laboratory of Cellular Oncology, National Cancer Institute, National Institutes of Health, Bethesda, MD 20892 **[131]**

A.J. Verkleij, Department of Molecular Cell Biology, University of Utrecht, 3584 CH Utrecht, The Netherlands **[207]**

Alice Wang, Department of Molecular Biology, Cetus Corporation, Emeryville, CA 94608 **[71]**

Robert A. Weinberg, The Whitehead Institute for Biomedical Research, Cambridge, MA 02142 **[83]**

Zena Werb, Laboratory of Radiobiology and Environmental Health, University of California, San Francisco, CA 94143 **[71]**

Morris F. White, Research Division, Joslin Diabetes Center, Department of Medicine, Brigham and Women's Hospital and Harvard Medical School, Boston, MA 02215 **[181]**

Douglas E. Williams, Department of Medicine (Hematology/Oncology), Indiana University School of Medicine, Indianapolis, IN 46223 **[11]**

Christopher B. Wilson, Department of Pediatrics (Infectious Diseases), University of Washington, Seattle, WA 98195 **[1]**

Vincent L. Wilson, Laboratory of Human Carcinogenesis, National Cancer Institute, Bethesda, MD 20892 **[269]**

Jian-Qing Yang, Biochemistry Department, SUNY at Stony Brook, Stony Brook, NY 11794-5215; present address: Department of Molecular Biology, Imclone Systems Inc., New York, NY 10014 **[239]**

Ian Zachary, Imperial Cancer Research Fund, London WC2A 3PX, England **[57]**

Ke Zhang, Laboratory of Cellular Oncology, National Cancer Institute, National Institutes of Health, Bethesda, MD 20892 **[131]**

Steven F. Ziegler, Howard Hughes Medical Institute and the Department of Biochemistry, University of Washington, Seattle, WA 98195 **[1]**

Preface

The UCLA Symposium on *Growth Factors and Their Receptors: Genetic Control and Rational Application* brought together a diverse array of scientists who presented many exciting new findings on growth factors and their receptors. The outcome was a gamut of chemical, biological, and clinical studies that melded together to create a stimulating meeting on the mechanisms of tissue and cellular growth control. The conference began with a keynote address that set the stage for the week with a thoughtful and provocative discussion of the complexity of the signals controlling cell proliferation and differentiation, and the multistep nature of carcinogenesis. With regard to the latter issue, oncogenes such as *ras* cause the initial loss of growth control associated with carcinogenesis but rarely induce a complete oncogenic phenotype. This was illustrated by the properties of v-*ras*-transformed keratinocytes, which, although tumorigenic by themselves, do not form tumors when inoculated together with dermal fibroblasts. However, if the v-*ras*-transformed keratinocytes acquire another oncogenic perturbation (e.g., a v-*myc* gene), they form tumors even when mixed with the "suppressor" dermal fibroblasts.

As the signals for controlling the growth and differentiation of cells are being identified, it is becoming clear that tissue homeostasis is an extremely complex process. The four presentations on the structure of the receptor for platelet-derived growth factor (PDGF) illustrated some of the difficulties we are facing. Since the initial description of the structure for the PDGF A-B heterodimer, which is composed of one A and one B chain, two homodimer forms of PDGF (A-A and B-B) have been identified. Recombinant forms of all three PDGF dimers have been prepared, and from receptor binding studies with these forms it appears that there are two PDGF receptors which have different affinities for each dimer. The two types of PDGF receptor may be displayed on the same cell (e.g., osteosarcoma cells) or separately, as in some uterine cells.

The critical structural determinants for several growth factor receptor interactions were presented in both poster and symposium sessions. Using chimeric human and chicken epidermal growth factor (EGF) receptors, it is possible to define a region between the cysteine-rich domains as the ligand-binding site on the EGF receptor. Interestingly, human transforming growth factor α (hTGFα) binds tightly to the chicken EGF receptor, whereas hEGF binds only weakly, suggesting that the so-called EGF receptor may really be the TGFα receptor. Studies analyzing the effects of site-directed mutation on the functions of growth factor receptors with protein-tyrosine kinase activity were reported. There is general agreement that protein-tyrosine kinase activity is intimately involved in the process of ligand-induced downregulation of receptors and is absolutely required for mitogenic activity. However, a mutant of the PDGF receptor from which the long insert in the middle of the protein kinase domain has been eliminated is not able to sustain a mitogenic response, even though it retains PDGF-stimulated protein-tyrosine kinase activity and can induce many of the characteristic early growth factor mediated events such as phosphatidylinositol turnover. This implies that protein-tyrosine kinase activity is necessary but not sufficient for mitogenic signalling by the PDGF receptor.

Another important theme was that many growth factor molecules appear to be designed for multiple interactions. For example, the activity of acidic fibroblast growth factor (FGF) is stimulated more than a hundredfold by heparin sulphate to which it binds. This heparin dependence can be overcome by specific structural mutations in acidic FGF. It is clear that the interaction between growth factors and the extracellular matrix elements must be considered if we are to understand the action of growth factors in tissues.

The presentations on the cellular responses to growth factors emphasized the complexity of cytoplasmic biochemistry and emphasized the need for caution in attempts to probe the relevant events triggered (or sustained) by growth factors. There were several examples of indirect stimulation of proliferation where a "mitogen" such as IL-1 is capable of inducing the production of an array of growth factors (G-CSF, GM-CSF, and IL-6). These secondary signals my be responsible for the proliferative and differentiative responses to the primary agent (e.g., IL-1) on bone marrow progenitor cells. Similarly, one of the PDGF-inducible gene products (KC) is identical to a potent melanoma growth-stimulating activity (MGSA) capable of inducing autocrine proliferation of melanoma cells. The expression of the same protein has also been associated with the transformed phenotype of some cells. Indeed, although many of the new gene products detected after growth factor stimulation will be needed for the induction of DNA synthesis and/or division, it appears that a significant number will be connected with intercellular signalling.

The candidate molecules on the critical pathways for mitogenic signalling continue to multiply. However, the substrates for growth factor receptor and intracellular protein-tyrosine kinases still proved a strong focus of attention for scientists investigating the growth factor stimulated cascades leading to mitosis. Protein kinase C, phospholipase C, and the adenylate cyclase enzyme systems all act in close juxtaposition to the cell membrane. The interaction of the type I phosphatidylinositol kinase with the polyoma virus middle T antigen/pp60$^{c\text{-}src}$ complex, pp60$^{v\text{-}src}$, and the PDGF receptor revealed a new phosphatidylinositol product (phosphatidylinositol-1, 3-diphosphate) that could provide a novel mitogenic-signalling pathway. Bombesin and PDGF are capable of both stimulating protein kinase C and activating adenylate cyclase. To what extent either or both of these events are involved in mitogenesis is still not understood. The activation of secondary protein kinases by the primary signalling protein kinases was well documented, and the idea that there are protein kinase cascades involved in signal amplification and arborization is rapidly being converted from fantasy into fact.

Growth factors induce rapid changes in the transcription rates for a number of genes associated with proliferation and differentiation. For instance, transcription of the c-*fos* and c-*myc* genes is induced by mitogen treatment of many cell types. Several new members of this "immediate early" family of genes were described. These genes fall into two classes. As discussed above, some of these genes encode secreted proteins that may be growth factors. In the other category are genes encoding nuclear proteins, such as c-*myc* and c-*fos*, which appear to be general transcriptional activators. Two new members of this family, *jun*B and a protein with three predicted "zinc fingers," seem likely to be sequence-specific transcription factors. Presumably these nuclear proteins are themselves involved in regulating expression of a secondary set of genes needed for progression through G_1 into S phase. Evidence was presented that this is the case for the c-*fos* protein.

The role of growth factors in tumorigenesis and/or normal cell production is not yet defined in detail. However, it is now known that most of the growth factors characterized by their ability to stimulate cells in culture also have potent effects on cell differentiation and proliferation in vivo. The hemopoietic growth factors (HGF) are powerful inducers of the production of a range of white blood cells. The chronic overproduction of one of the HGFs is associated with tissue macrophage infiltration and damage (e.g., in GM-CSF transgenic mice). Promising results of clinical trials with several HGFs were reported, and realization of the therapeutic potential of HGFs may be near at hand. Fibroblast growth factor, EGF, and PDGF appear to be excellent candidate molecules for accelerating the repair of both full-thickness wounds and the religation of nerve axons. Activated macrophages produce a range of growth factors as well as potent biological response modifiers including the HGFs, the interleukins, TGFα, PDGF, and tumor necrosis factor (TNF). Overexpression of these growth factors by endothelial cells and macrophages may be responsible for both rheumatoid arthritis and atherosclerosis.

The enthusiasm and energy of the five hundred conference participants and the opportunity to meet with colleagues investigating a range of peptides and proteins initiating cell proliferation or accelerating differentiation made for an exciting week for all of us. The joint sessions with the concurrent meeting on growth inhibitors provided a reminder that growth control not only requires stimulatory molecules but also inhibitory molecules. As the scientific organizers, we found the conference most enjoyable and we would like to thank all involved—session conveners, presenters, and organizers—for making our task so rewarding.

Tony Burgess
Tony Hunter
Russell Ross

Journal of Cellular Biochemistry 38:117–126 (1988)
Growth Factors and Their Receptors: Genetic Control and Rational Application 1–10

Structure and Expression of *lck* Transcripts in Human Lymphoid Cells

Roger M. Perlmutter, Jamey D. Marth, David B. Lewis, Richard Peet, Steven F. Ziegler, and Christopher B. Wilson

Howard Hughes Medical Institute (R.M.P., J.D.M., R.P., S.F.Z.) and the Departments of Medicine (Medical Genetics) (R.M.P., J.D.M., R.P.), Biochemistry (S.F.Z.), and Pediatrics (Infectious Diseases) (D.B.L., C.B.W.), University of Washington, Seattle, Washington 98195

The murine *lck* gene encodes a membrane-associated protein tyrosine kinase that has been implicated in lymphocyte oncogenesis. Here we report the structure of normal human *lck* transcripts and the pattern of expression of these transcripts in developing thymus and in peripheral T cell subsets. The human *lck* gene encodes a 509 amino acid polypeptide that is closely related to the murine *lck*-encoded protein throughout its length. Analysis of the deduced amino acid sequence of human $p56^{lck}$ demonstrates that an amino-terminal domain, widely divergent among the seven known *src* family members, has been conserved between murine and human $p56^{lck}$, and thus probably includes sequences crucial to the lymphocyte-specific function of this molecule. Human *lck* transcripts were detected in $CD4^+$ and $CD8^+$ T cells, in partially purified B cells, and in Epstein-Barr virus-immortalized B cell lines, but not in monocytes, granulocytes, or in nonhematopoietic cell types. Human *lck* transcripts are readily detectable in fetal thymocytes at 70 days of gestation, but not at 57 days of gestation, indicating that *lck* expression appears coordinately with the appearance of lymphoid cells in the developing thymus. Thus *lck* gene expression is a marker for cells of the lymphocyte lineage in man. We conclude that the *lck* gene probably participates in a signal transduction pathway uniquely present in lymphoid cells.

Key words: lymphocyte signaling, *src* gene family, protein tyrosine kinases

Protein tyrosine kinases may be grouped into two distinct classes, both of which are implicated in the control of cell growth. The first class includes a series of transmembrane receptors that transduce signals supplied by growth factors, e.g., epidermal growth factor or platelet-derived growth factor [1–5]. A second class of protein tyrosine kinases includes molecules that while typically membrane associated appear to reside entirely intracellularly, lacking transmembrane or ligand-binding domains [6]. Many members of this second class are encoded by cellular homologues of retroviral oncogenes (e.g., c-*src*, c-*yes*, and c-*fgr*) and thus clearly may under certain circumstances provide a growth-promoting stimulus. The mechanism whereby tyrosine phosphorylation regu-

Received February 1, 1988; accepted April 5, 1988.

lates cellular proliferation is enigmatic, since relevant substrates remain for the most part undefined [7].

We have focused attention on a member of the *src*-like gene family, *lck*. Although structurally closely related to *src, lck* is specifically expressed in lymphocytes among normal cell types [8–10]. Moreover, the level of *lck* transcripts in T lymphocytes responds dramatically to combinations of signals that stimulate lymphokine secretion [9]. The murine *lck* gene encodes a 56-kd polypeptide that is membrane associated, presumably by virtue of covalently linked myristate [11]. Rearrangement and overexpression of the *lck* gene is a characteristic of some murine lymphomas, suggesting that $p56^{lck}$ may assist in controlling lymphocyte proliferation [8,10,12–14]. Since considerable information exists regarding the factors required for T cell growth [15,16], there is reason to hope that analysis of the regulation of the *lck* gene will provide insights into its possible functions. Here we describe the structure of normal human *lck* transcripts and the pattern of expression of these transcripts in normal blood cells and in developing thymus.

MATERIALS AND METHODS

cDNA Libraries and Library Screening

A mitogen-stimulated human peripheral blood leukocyte cDNA library was the kind gift of Drs. P. Concannon and L. Hood [17]. Dr. D. Littman kindly provided two additional human cDNA libraries, one generated using RNA from peripheral blood leukocytes [18] and a second generated using RNA from the PEER cell line [19]. All three libraries were propagated and screened as previously described using the 1.75-kb *Eco*RI fragment of the NT18 plasmid that contains the entire murine *lck* coding region sequence [8].

DNA Sequencing

Recombinant phage containing human *lck* sequences were plaque purified and phage DNA isolated using a plate-lysate technique [8]. The *Eco*RI inserts were subcloned into pUC18 and pUC19 plasmid vectors and into M13mp18 and M13mp19 phage vectors for sequence analysis using the dideoxynucleotide chain termination method [20,21]. All sequences were determined on both strands and on overlapping fragments. The majority of the sequence was obtained using *lck*-specific oligonucleotides, synthesized using an Applied Biosystems DNA synthesizer, as primers [22].

Isolation and Fractionation of Human Peripheral Blood Cells

Human mononuclear cells, T lymphocytes, granulocytes, and monocytes were purified from adult peripheral blood as previously described [9,28]. Isolation of T cell subsets was achieved by treatment of appropriate fractions with monoclonal antibodies (mAbs) and complement as follows: for $CD4^+$ cells, mAbs FC-2 (CD16) and 10.1 (anti-CD8), and for $CD8^+$ cells, mAbs FC-2 and 66.1 (anti-CD4). Viable cells were recovered by Ficoll-Hypaque centrifugation. By indirect immunofluorescence, $CD4^+$, and $CD8^+$ T cell subfractions were greater than 90% reactive with their appropriate mAbs. These mAbs were kindly provided by Genetic Systems Inc., Seattle, WA, with the exception of FC-2, which was the generous gift of Dr. E. Clark, University of Washington. B lymphocytes were partially purified (approximately 70%) using B lymphokwik (One Lambda, Los Angeles, CA) according to the manufacturer's instructions.

RNA Isolation and RNA Blotting

Total RNA from cell lines, purified cells, or from thymic tissue was obtained and analyzed as previously described [8].

RESULTS

Using the murine *lck* gene as a probe, we previously identified *lck* transcripts in human thymus RNA and localized the human *lck* gene to chromosome 1p32–35 at a site of frequent chromosomal abnormalities in human lymphomas and neuroblastomas [23]. To characterize normal human *lck* transcripts, we used the murine *lck* probe to screen cDNA libraries generated using RNA from human peripheral blood lymphocytes and a human T cell line [17–19].

Cloning of the Human *lck* Gene

Figure 1A presents restriction maps for three human *lck* cDNA clones. Clone HK28 was derived from a partially processed transcript and contains a 173-bp intron sequence that was not present in any of 22 other clones spanning this region. The entire sequence of the human *lck* cDNA, derived from normal peripheral blood lymphocytes, is compared with that of a nearly full-length murine *lck* cDNA in Figure 1B, along with the deduced amino acid sequences of each. Trevillyan et al. [24] have previously reported the 3′ 850 bp of this sequence, which they obtained by screening a cDNA library made using RNA from the Jurkat T cell line. Similarly, Koga et al. [25] have reported the sequence of what is clearly a human *lck* cDNA derived from the Jurkat T cell line. The HK28 human *lck* cDNA clone includes 56 bp of presumed 5′ untranslated region, an open reading frame of 509 codons, and a 365 bp 3′ untranslated region including a single prototypical recognition site for polyadenylation (underlined). Overall, the human and murine *lck* coding sequences are 91% identical at the nucleotide level and differ by only 17 of 509 amino acids (Figs. 1B, 2). Our sequence differs significantly (26 amino acid substitutions) from that reported by Koga et al. [25]. In particular, their sequence lacks a conserved tyrosine (codon 505), which is the site of in vivo phosphorylation in murine $p56^{lck}$ [26]. We have previously shown that mutations at this site activate the *lck* kinase and unmask its transforming potential [26], thus the differences between the Koga et al. sequence and our normal T cell sequence may be biologically important. At the same time, comparison of the Koga et al. sequence with that of Trevillyan et al. [24], both of which were derived from cDNAs from Jurkat cells, suggests that these differences represent sequencing errors [27]. The incomplete sequence of Trevillyan et al. [24] differs from our sequence by a single base deletion in the 3′ untranslated region but is identical within the 133 codons that they report (the 5′-most G of their sequence is presumably artifactual). Similarly, the incomplete sequence of Veillette et al. [27] is identical except for a typographical error that substitutes His (CAT) for Leu (CTC) at codon 472.

Evolutionary Conservation of the *lck* Gene

Human *lck* is more closely related to murine *lck* than it is to any of the other six *src*-family kinases. This is shown graphically in Figure 2 where the deduced amino acid sequences of the human *lck*- and *hck*-encoded proteins are aligned. Seven of the 17

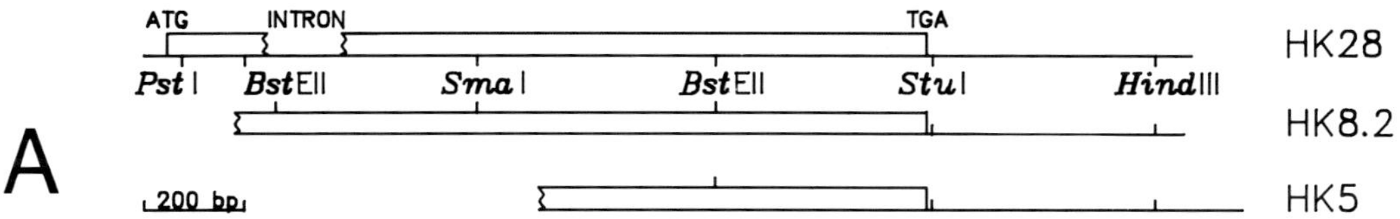

Fig. 1. Structure of human *lck* transcripts deduced from cDNA clones. **A**: Restriction maps of three overlapping human *lck* cDNA clones. The translation, initiation, and termination codons are indicated and the coding region is boxed. Clone HK28, from a normal peripheral blood lymphocyte cDNA library [17], represents a partially processed transcript and contains a 173-bp intron sequence at the indicated position. **B**: The complete nucleotide sequence encoded by the human *lck* gene (with the intron sequence of HK28 deleted) is compared with the full-length sequence encoded by the murine *lck* gene [8,10]. The coding region is presented as triplets with the deduced amino acid sequence for the human *lck*-encoded protein indicated in single-letter code. The consensus poly-A addition site is underlined.

amino acid substitutions separating the murine and human *lck* sequences are concentrated in the first 60 amino acids. Nevertheless, although sequence conservation in the amino-terminal domain may be slightly more relaxed than elsewhere in the molecule, the close relationship between murine and human *lck* sequences in this region is consistent with the view that this "domain" is specialized for unique, and in many cases lineage-specific, functions [28,29].

Expression of *lck* in Human Lymphocytes

Evidence for the conservation of *lck* function in man and mouse was sought through analysis of *lck* transcripts in human cells. Using a probe encompassing the entire *lck* coding region, cross-hybridizing transcripts were detected in human peripheral blood mononuclear cells and in fractionated T lymphocytes, but not in granulocytes or monocytes (Fig. 3). In a separate experiment, appropriately sized *lck* transcripts were also detected in partially purified peripheral blood B lymphocytes (Fig. 3).

Since our best B lymphocyte preparations were surely contaminated with other cell types, and since Koga et al. [25] failed to identify *lck* transcripts in three human B cell lines, we sought to confirm the presence of *lck* transcripts in B lymphocytes. To this end, we examined Epstein-Barr virus-immortalized human B lymphocytes for *lck* expression. To verify that the *lck* transcripts identified were in each case encoded by the *lck* gene and not by a closely related but distinct genetic element, we used a unique 288-bp *Stu*I/*Hind*III probe derived from the 3′ untranslated region of the human *lck* cDNA (Fig. 1A). Since 3′ untranslated regions are, in most cases, under little selection pressure, it is unlikely that this sequence would be shared by other protein tyrosine kinase genes. Indeed, the murine and human *lck* transcripts are only 45% homologous within this region (Fig. 1B) and hence do not cross-hybridize using our experimental conditions. Figure 4 demonstrates that the 3′ untranslated region *lck* probe identifies transcripts in both $CD4^+$ and $CD8^+$ T lymphocytes and at a lower level in partially purified peripheral blood B lymphocytes as well. These results are consistent with previous studies of the expression of the *lck* gene in murine lymphoid cell lines [8]. Importantly, transcripts were also present in two Epstein-Barr virus (EBV)-transformed polyclonal B cell lines (e.g., line PG) but not in a human neuroblastoma cell line (SKN), a human melanoma cell line, or human fetal adrenal cells (Fig. 4). Veillette et al. [27] also identified *lck* transcripts in EBV-transformed peripheral blood cells and in 4/5 Burkitt

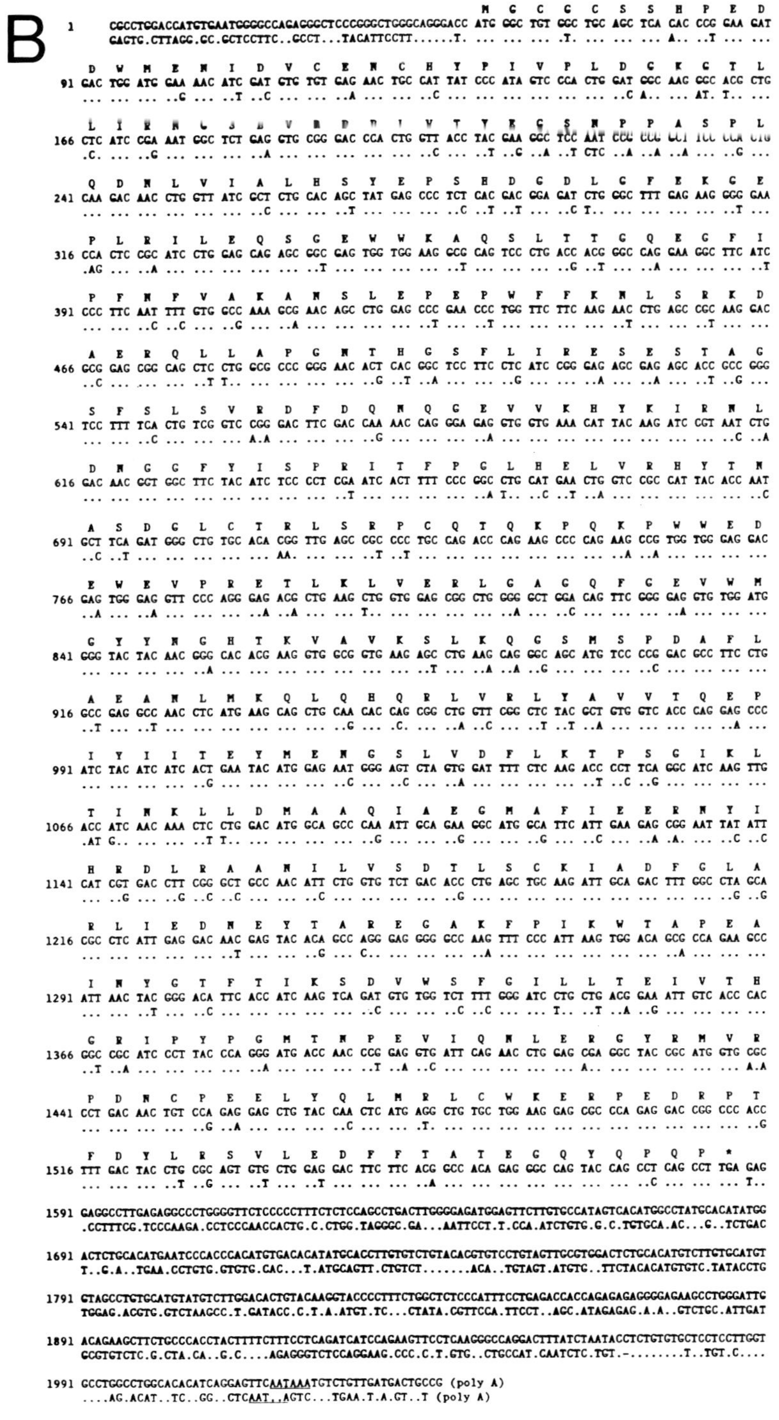

Figure 1B

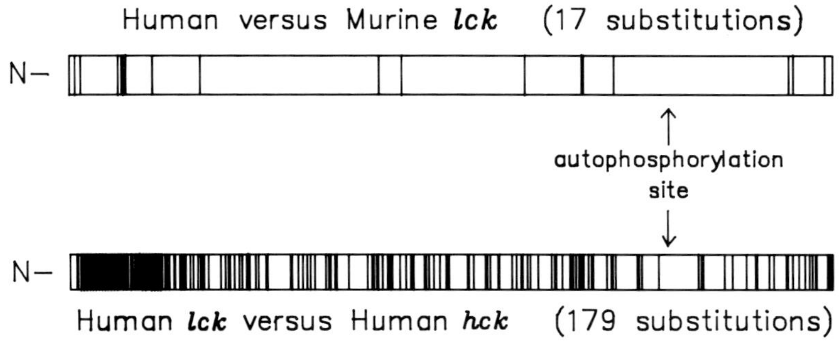

Fig. 2. Conservation of the *lck* amino-terminal region in man and mouse. Shown are two histograms derived by comparing the human (this paper) and murine [8,10] *lck* sequences (**top**) or the human *lck* sequence with its closest known *src*-family counterparty, the human *hck* [27,31] sequence (**bottom**). In each case, a vertical line has been drawn at each position where the two proteins compared differ in sequence. The site of in vitro autophosphorylation in p56*lck* is indicated.

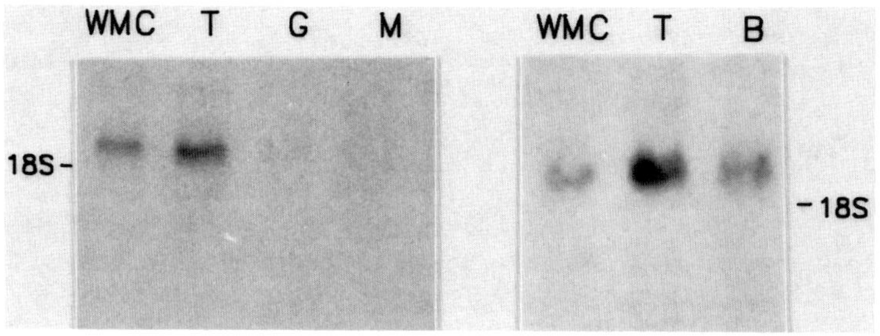

Fig. 3. *lck* expression in human peripheral blood leukocytes. Total RNA was isolated from whole mononuclear cells (**WMC**) and granulocyte (**G**), monocyte (**M**), T lymphocyte (**T**), and B lymphoctye (**B**) fractions and 10 μg of each RNA sample was subjected to RNA blot analysis using the 2.1-kb HK28 cDNA insert as a probe. The left and right panels represent two independent experiments. The position of 18S rRNA markers is indicated.

lymphoma cell lines. We conclude that the *lck* gene product is probably expressed in most (if not all) lymphocyte classes.

Expression of *lck* Is a Marker for the Appearance of Lymphoid Elements in the Developing Human Thymus

We have previously demonstrated that the murine *lck* gene is expressed at approximately fivefold higher levels in thymus as compared with spleen [8]. In man, adult spleen contains only about 1 pg *lck* mRNA per microgram total RNA as compared to 12 pg *lck* mRNA per μg total RNA in thymus [9]. To ascertain when in development *lck* expression is first observed, we probed RNA isolated from human fetal thymus tissue obtained at 57, 70, 100, and 110 days of gestation and from 2.5-month-, 2.5-year-, and 8.5-year-old individuals using the human *lck* cDNA. Figure 5 demonstrates that *lck* transcripts are identifiable at 70 but not at 57 days of gestation. Lymphoid elements first appear in the human thymus at 8 to 9 weeks of gestation [30]. Thus the development of *lck* expression correlates precisely with the arrival of bone marrow-derived immigrants in the developing thymus. In separate experiments, *lck* transcripts were detectable in

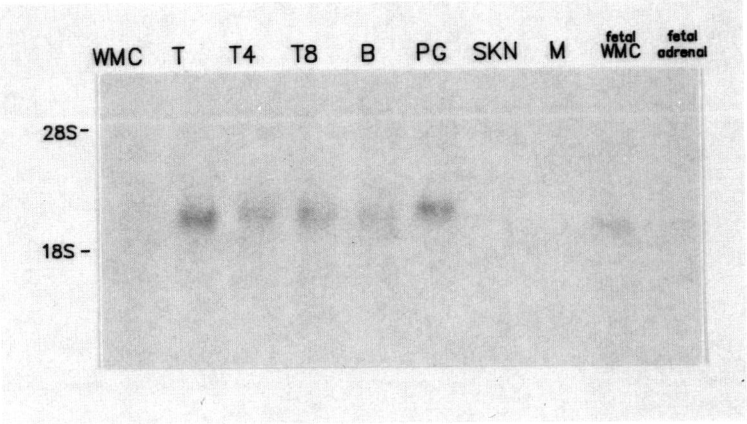

Fig. 4. Expression of *lck* mRNA in human T cell subsets and in cell lines. Total RNA (10 μg) was isolated from whole mononuclear cells (**WMC**), fractionated T lymphocytes (**T**), $CD4^+$ lymphocytes (**T4**), $CD8^+$ lymphocytes (**T8**), B lymphocytes (**B**), an Epstein-Barr virus-transformed cell line (**PG**), the human neuroblastoma cell line SKN-BE2 (**SKN**), a human melanoma cell line (**M**), and from fetal whole mononuclear cells and adrenal cells, and was subjected to RNA blot analysis using a probe derived from the 3′ untranslated region of the human *lck* cDNA by digestion with *Stu*I and *Hind*III (see Fig. 1A). The position of 18S and 28S rRNA markers is indicated.

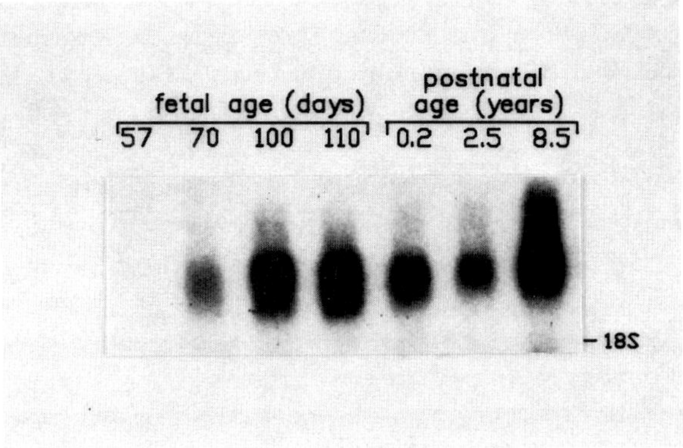

Fig. 5. Ontogeny of *lck* expression in human thymus. Total RNA was isolated from human fetal and postnatal thymic specimens and *lck* expression examined by RNA blot analysis of 10 μg of total RNA using the full-length HK28 cDNA as a probe. The approximate developmental age of each specimen is indicated at the top of the figure and the position of 18S rRNA markers is shown. The presence of RNA in each lane was verified by reprobing the blot with a labeled human elongation factor 1α cDNA (data not shown).

mature $CD4^+$ and $CD8^+$ thymocytes, in $CD4^+/CD8^+$ (double-positive) cells, and in thymocytes lacking both markers (data not shown).

DISCUSSION

Seven *src*-like protein tyrosine kinases have been described: c-*fgr* [31,32], *hck* [28,33], *lck* [8,120], *lyn* [34], c-*src* [35], *fyn* [36,37], and c-*yes* [38]. All are similar in size and contain a characteristic protein tyrosine kinase "domain" defined by sterotyped sequence motifs believed to participate in catalytic function [2]. Three of the *src*-like kinases were originally identified as products of retroviral oncogenes, directly implicating these molecules in the control of replication. In addition, the *lck* gene is rearranged and overexpressed in some murine lymphomas [8,10] and both *lck* and *hck* are located at sites of frequent chromosomal aberrations in human malignant disease [23,33]. Nevertheless, high-level expression of *src*-like kinases is often observed in cells with little replicative potential. Thus c-*src* itself is expressed in neurons and platelets [39,40] and *hck* is maximally expressed in granulocytes [28]. These observations suggest that *src*-family members must either, under some circumstances, actively suppress proliferation, or must be primarily active in metabolic processes unrelated to cell growth.

Restricted Expression of *lck*

Additional perspective on the function of the *src*-like protein tyrosine kinases can be gained by comparing the expression patterns of each. For example, the human *hck* gene is expressed in myeloid cells [28], and *hck* expression appears to increase when immature myeloid cells become committed to the myeloid lineage ([33] and S.F. Ziegler and R.M. Perlmuter, unpublished observations). By analyzing murine tissues and transformed cell lines, we previously proposed that the murine *lck* gene is expressed primarily in lymphocytes [8]. Here we have used an *lck*-specific probe to document that the human *lck* gene is expressed predominantly in T lymphocytes of both the $CD4^+$ and the $CD8^+$ subclasses. Thus both the structure and the expression pattern of the *lck* gene have been conserved since the time that mouse and man diverged. Veillette et al. [27] have also demonstrated that the human *lck* gene is specifically expressed in lymphoid cells, although many colon carcinomas and some lung tumors also contain appreciable levels of *lck* transcripts.

Structural Variation Among *src*-Like Protein Tyrosine Kinases

Differences among *src*-family members are concentrated in the amino-terminal 70 residues of each, suggesting that the unique function(s) of *lck* as compared with *hck* is defined by this amino-terminal region. Both $p60^{src}$ and $p56^{lck}$ are myristylated at their amino-terminal glycine residues [11,41], and this myristylation is probably involved in the association of these proteins with cell membranes [42,43]. Since all *src*-like kinases share a common Met-Gly-Ser/Cys motif at their amino termini, it is likely that all are myristylated and that all will prove to be similarly membrane associated. In this sense, the variable amino-terminal structures of each *src*-family member may be viewed as ideally positioned to interact with cytoplasmic portions of transmembrane receptors, permitting the carboxy-terminal kinase domain to serve as the catalytic effector in a transmembrane signal transduction mechanism. For the *lck* gene, the putative receptor must be assumed to be present on lymphoid cells in general. Conservation of this amino-terminal region in both the human and murine *lck* sequences (Fig. 2) is consistent with this view. Thus it is provocative that treatment of resting human T lymphocytes with activating regimens of mitogen and phorbol ester induces rapid reductions in the level of *lck* transcripts in these cells as well as an altered pattern of expression of $p56^{lck}$ protein [9].

lck Expression as a Marker for the Lymphoid Lineage

By analyzing the pattern of appearance of *lck* transcripts in developing human fetal thymus, we demonstrated that *lck* transcripts are first detectable at the time when the thymic rudiment becomes colonized with hematopoietic elements (Fig. 5). We cannot, however, determine whether lymphoid stem cells, either fetal or adult, already contain *lck* transcripts. In general, *lck* mRNA is barely detectable in murine bone marrow (data not shown). Since lymphoid stem cells capable of reconstituting irradiated hosts are present at frequencies of 10^{-3} or less in bone marrow populations [44], we cannot determine whether these precursors express *lck* transcripts before arrival in the thymus. Nevertheless, the production of *lck* mRNA appears to occur uniquely in lymphocytes, among normal cells, and may well be a marker for commitment within the lymphoid lineage.

Complexity of Protein Tyrosine Kinase-Mediated Regulation

In addition to *lck* transcripts, lymphocytes frequently contain mRNA encoding other related *src*-like protein tyrosine kinases, for example c-*fgr* [45] and *fyn* [36,37]. While this array of potential growth-regulating elements may be viewed as complementary to the large repertoire of signals to which lymphocytes respond, it is clear that considerable effort will be required to identify specific phosphorylation pathways within this complex web of kinases and their substrates. Under these circumstances, the unique lineage specificity of the *lck* gene in man and mouse is especially interesting since it suggests that the molecules with which the *lck* gene product interacts may be similarly specialized.

ACKNOWLEDGMENTS

We thank Patrick Chou and Yim-Foon Lee for preparing oligonucleotides, Pat Concannon, Lee Hood, and Dan Littman for providing human lymphocyte cDNA libraries, Ed Baetge for the SKN neuroblastoma cell line, Betsy Mellins for melanoma and EBV-transformed lymphoid cell lines, and our colleagues for helpful discussions. This work was supported in part by grants from the American Cancer Society and the National Institutes of Health.

REFERENCES

1. Downward J, Yarden Y, Mayes E, Scrace G, Totty N, Stockwell P, Ullrich A, Schlessinger J, Waterfield MD: Nature 307:521, 1984.
2. Hunter T, Cooper JA: Annu Rev Biochem 54:897, 1985.
3. Sherr OJ, Rettenmeir CW, Sacca R, Roussel MF, Look AT, Stanley ER: Cell 41:665, 1985.
4. Ullrich A, Bell JR, Chen EY, Herrera R, Petruzelli LM, Dull TJ, Gray A, Coussens L, Lia Y-C, Tsubokawa M, Mason A, Seeburg PH, Grunfeld C, Rosen OM, Ramachandran J: Nature 313:756, 1985.
5. Yarden Y, Escobedo JA, Kuang WJ, Yang-Feng TL, Daniel TO, Tremble PM, Chen EY, Ando ME, Harkins RN, Francke U, Fried VA, Ullrich A, Williams LT: Nature 323:226, 1986.
6. Cooper JA: In Kemp B, Alewood PF (eds): ''Peptides and Protein Phosphorylation.'' Boca Raton, FL: CRC Press, 1988 (in press).
7. Cooper JA: BioEssays 4:9, 1986.
8. Marth JD, Peet R, Krebs EG, Perlmutter RM: Cell 43:393, 1985.
9. Marth JD, Lewis DB, Wilson CB, Gearn ME, Krebs EG, Perlmuter RM: EMBO J 6:2727, 1987.
10. Voronova AF, Sefton BM: Nature 313:682, 1986.
11. Marchildon GA, Casnellie JE, Walsh KA, Krebs EG: Proc Natl Acad Sci USA 81:7679, 1984.
12. Voronova AF, Buss JE, Patschinsky T, Hunter T, Sefton BM: Mol Cell Biol 4:2705, 1984.
13. Casnellie JE, Harrison ML, Hellstrom KE, Krebs EG: J Biol Chem 258:10,738, 1983.
14. Gacon G, Gisselbrecht S, Pau JP, Boissel JP, Tolle J, Fischer S: EMBO J 1:1579, 1982.

15. MacDonald HR, Nabholz M: Annu Rev Cell Biol 2:231, 1986.
16. Weiss A, Imboden J, Hardy K, Manger B, Terhorst C, Stobo J: Annu Rev Immunol 4:593, 1986.
17. Concannon PL, Pickering LA, Kung P, Hood L: Proc Natl Acad Sci USA 83:6598, 1986.
18. Littman DR, Thomas Y, Maddon PJ, Chess L, Axel R: Cell 40:237, 1985.
19. Littman DR, Newton M, Crommie D, Ang S-L, Seidman JG, Gettner SN, Weiss A: Nature 326:85, 1987.
20. Sanger F, Nicklen S, Coulsen AR: Proc Natl Acad Sci USA 74:5463, 1977.
21. Yanisch-Perron C, Vieira J, Messing J: Gene 33:103, 1985.
22. Strauss E, Kobori J, Siu G, Hood L: Anal Biochem 154:353, 1986.
23. Marth JD, Disteche C, Pravtcheva D, Ruddle F, Krebs EG, Perlmutter RM: Proc Natl Acad Sci USA 83:7400, 1986.
24. Trevillyan JM, Norstrom A, Linna TJ: Biochim Biophys Acta 845:1, 1985.
25. Koga Y, Caccia C, Toyonaga B, Spolski R, Yanagi Y, Yoshikai Y, Mak TW: Eur J Immunol 16:1643, 1986.
26. Marth JD, Cooper JA, King CS, Ziegler SF, Tinker DA, Overell RW, Krebs EG, Perlmutter RM: Mol Cell Biol 8:540, 1988.
27. Veillette A, Foss FM, Sausville EA, Bolen JB, Rosen N: Oncogene Res 1:357, 1987.
28. Ziegler SF, Marth JD, Lewis DB, Perlmutter RM: Mol Cell Biol 7:2276, 1987.
29. Tanaka A, Gibbs CP, Arthur RR, Anderson SK, Kung H-J, Fujita, D: Mol Cell Biol 7:1978, 1987.
30. Lobach DF, Haynes BF: J Clin Immunol 7:81, 1987.
31. Nishizawa M, Semba K, Yoshida MC, Yamamoto T, Sasaki M, Toyoshima K: Mol Cell Biol 6:511, 1986.
32. Katamine S, Notario V, Rao CD, Miki T, Cheah MSC, Tronick SR, Robbins KC: Mol Cell Biol 8:259, 1988.
33. Quintrell N, Lebo R, Varmus H, Bishop JM, Petenati MJ, LeBeau MM, Diaz MO, Rowley JD: Mol Cell Biol 7:2267, 1987.
34. Yamanashi Y, Fukushige S-L, Semba K, Sukegawa J, Miyajima N, Matsubara K-L, Yamamoto T, Toyoshima K: Mol Cell Biol 7:237, 1987.
35. Takeya T, Hanafusa H: Cell 32:881, 1983.
36. Semba K, Nishizawa M, Miyajima N, Yoshida MC, Sukegawa J, Yamanashi Y, Sasaki M, Yamamoto T, Toyoshima K: Proc Natl Acad Sci USA 83:5459, 1986.
37. Kawakami T, Pennington CY, Robbins KC: Mol Cell Biol 6:4195, 1986.
38. Sukegawa J, Semba K, Yamanashi Y, Nishizawa M, Miyajima N, Yamamoto T, Toyoshima K: Mol Cell Biol 7:41, 1987.
39. Brugge JS, Cotton PC, Queral AE, Baret JN, Nonner D, Keane RW: Nature 316:554, 1985.
40. Golden A, Nemeth SP, Brugge JS: Proc Natl Acad Sci USA 83:852, 1986.
41. Schultz AM, Henderson LE, Oroszlan S, Garber EA, Hanafusa H: Science 227:427, 1985.
42. Cross RF, Garber EA, Pellman D, Hanafusa H: Mol Cell Biol 4:1834, 1984.
43. Buss, JE, Kamps MP, Fould K, Sefton BM: J Virol 58:468, 1986.
44. Muller-Sieburg CE, Whitlock CA, Weissman IL: Cell 44:653, 1986.
45. Cheah MS, Ley TJ, Tronick SR, Robbins KC: Nature 319:238, 1986.

Journal of Cellular Biochemistry 38:127–136 (1988)
Growth Factors and Their Receptors: Genetic Control and Rational Application 11–20

Recombinant Human Granulocyte-Colony Stimulating Factor and Recombinant Human Macrophage-Colony Stimulating Factor Synergize In Vivo to Enhance Proliferation of Granulocyte-Macrophage, Erythroid, and Multipotential Progenitor Cells in Mice

Hal E. Broxmeyer, Douglas E. Williams, Scott Cooper, Giao Hangoc, and Peter Ralph

Departments of Medicine (Hematology/Oncology) (H.E.B., D.E.W., S.C., G.H.), Microbiology and Immunology (H.E.B.), and The Walther Oncology Center (H.E.B.), Indiana University School of Medicine, Indianapolis, Indiana 46223; Department of Cell Biology, Cetus Corporation, Emeryville, California 94608 (P.R.)

Combinations of low dosages of purified recombinant human (rh) macrophage-colony stimulating factor (M-CSF; also termed CSF-1) and rh granulocyte-colony stimulating factor (G-CSF) were compared alone and in combination for their influence on the cycling rates and numbers of bone marrow and splenic granulocyte-macrophage, erythroid, and multipotential progenitor cells in vivo in mice pretreated with iron-saturated human lactoferrin (LF). LF was used to enhance detection of the stimulating effects of exogenously added CSFs. Concentrations of each CSF that were not active in vivo when given alone were active when given together, with the other CSF. The concentrations of rhM-CSF and rhG-CSF needed to increase progenitor cell cycling in the marrow and spleen were reduced by factors of 40–200 when these CSFs were administered in combination with low dosages of the other CSF. At the concentrations of rhM-CSF and rhG-CSF tested, synergism was not noted on absolute numbers of progenitor cells or total nucleated cell counts per organ or circulating in the blood. These findings may

Abbreviations: BFU-E, erythroid progenitor; CFU-GEMM, multipotential progenitor; CFU-GM, GM progenitor; CSF, colony stimulating factor; CSF-1, macrophage CSF; G, granulocyte; GM, granulocyte-macrophage; h, human; IL, interleukin; LF, lactoferrin; M, macrophage; n, natural; PWMSCM, pokeweed mitogen mouse spleen cell-conditioned medium; r, recombinant.

Received January 20, 1988; accepted May 2, 1988.

have potential relevance when considered in a clinical setting where the CSFs might be used in combination with other biotherapy and/or chemotherapy.

Key words: **colony stimulating factor, macrophage-colony stimulating factor, granulocyte-colony stimulating factor, in vivo action, synergism, hematopoietic progenitor cell, multipotential hematopoietic progenitor cell, granulocyte-macrophage progenitor cell, erythroid progenitor cell, lactoferrin**

The genes for the hematopoietic colony stimulating factors (CSFs)—interleukin-3 (IL-3), granulocyte-macrophage (GM)-CSF, granulocyte (G)-CSF, and macrophage (M)-CSF (also termed CSF-1)—have been cloned and expressed and the CSFs purified [reviewed in 1–3]. Their actions in vitro have been well studied [1–3], and there are now numerous reports documenting the preclinical actions of the CSFs. IL-3 [4–14] and M-CSF [7,8,10–12,15–18] have thus far been studied in mice, while GM-CSF [7,10,12,19–22] and G-CSF [23–29] have been evaluated in mice and monkeys. Clinical trials with recombinant human (rh)GM-CSF [30–32] and rhG-CSF [33] are ongoing and it is only a short time before rhIL-3 and rhM-CSF will be evaluated in a clinical setting.

Cell regulation in vitro encompasses a complex set of biomolecule-cell interactions [1,2] and molecules can act together in an additive or synergistic manner to enhance hematopoietic progenitor cell proliferation [1,2]. While it is not always clear whether synergism noted with combinations of molecules is manifested by a direct action of the multiple molecules on the progenitor cell itself, there is some evidence that this does happen [34]. Synergism between molecules is an important concept since it allows for amplification of an effector function. We have recently reported that combinations of r murine (m) GM-CSF and rmIL-3, or rm GM-CSF and natural (n) mM-CSF, or rmIL-3 and nmM-CSF could synergize when administered to mice pretreated with human lactoferrin (LF) to enhance the cycling status and absolute number of bone marrow granulocyte-macrophage (CFU-GM), erythroid (BFU-E), and multipotential (CFU-GEMM) progenitor cells [12]. Additionally rm IL-3 and nmM-CSF synergized in vivo to enhance the cycling rates of bone marrow high proliferative potential colony forming cells [11]. LF was used to dampen steady-state myelopoiesis in the mice [8,35], thus making the detection of the actions of low dosages of CSFs in the mice more apparent [7,10].

Experiments were set up to evaluate the possibility that rhG-CSF and rh M-CSF might synergize in vivo when administered to mice for the following reasons. Clinical trials have begun with rhG-CSF [33], M-CSF is detectable in human serum [36–38], and both rhG-CSF [39] and rhM-CSF [40] can stimulate colony formation from murine CFU-GM in vitro. Mice were pretreated with LF to reduce endogenous levels of CSF [35] prior to addition of rhG-CSF and rh M-CSF, alone or in combination. Effects were assessed on the cycling rates and numbers of CFU-GM, BFU-E, and CFU-GEMM in the marrow and spleens of the mice.

MATERIALS AND METHODS

Mice

(C57BL/6 × DBA/2)F_1 (BDF_1) mice, 6–8 weeks old, were purchased from Cumberland View Farms (Clinton, TN).

Molecules

Recombinant hG-CSF [95% pure, specific activity > 5 × 10^7 units/mg) [41] and rh CSF-1 from a truncated long clone (lot DCP-006, 95% pure, specific activity >5 × 10^7 units/mg) [42] from Cetus Corporation (Emeryville, CA), were diluted in sterile pyrogen-free saline prior to use. Purified human milk LF, purchased from Sigma Chemical Co. (St. Louis, MO), was fully iron saturated [43] and depleted of endotoxin by removing the material that gelled in the presence of *Limulus* lysate (Sigma) [7]. The concentration of LF was measured by an immunoradiometric assay for LF [43]. Injections of factors to mice were given i.v. in volumes of 0.2 ml. LF was given 3 h prior to administration of the CSFs, and when both CSFs were given to mice, each CSF was injected separately with in a few minutes of the other CSF. By using the *Limulus* lysate test for endotoxin, which has a sensitivity range down to 0.5 ng/ml, no endotoxin (<0.1 ng per injection) was detected in the preparations of rhG-CSF, rhM-CSF, or LF.

Analysis of Hematopoietic Progenitor Cells In Vitro

The assays for CFU-GM, BFU-E, and CFU-GEMM were performed as previously described [7,12]. Granulocyte-macrophage colony formation (>50 cells cells/aggregate) was stimulated using 10% (vol/vol) pokeweed mitogen mouse spleen cell-conditioned medium (PWMSCM) as a stimulus. Erythroid and multipotential colony formation was stimulated with 1 unit erythropoietin (Hyclone, Logan, UT), 0.1 mM hemin (Eastman Kodak, Rochester, NY), and 1% PWMSCM. Bone marrow and spleen cells were plated at respective concentrations of 7.5 × 10^4 and 1.0 × 10^6 cells/ml. Cells were incubated in a humidified environment at lowered (5%) oxygen tension using an Oxyreducer (Reming Bioinstruments, Redfield, NY). Colonies were scored after 6–7 days of incubation.

Cycling Status of Hematopoietic Progenitor Cells

The proportion of progenitors in DNA synthesis (S-phase) was estimated as reported previously [7,12]. The high specific activity (20 Ci/mmol) tritiated thymidine (50 μCi/ml) (New England Nuclear, Boston, MA) kill technique is based upon calculation of the reduction in the number of colonies after pulse exposure of cells for 20 min to tritiated thymidine compared to control (McCoy's medium, Gibco, Grand Island, NY). A negative number, as seen in Table II, means that more colonies formed after cells were exposed to tritiated thymidine than after cells were exposed to McCoy's medium and that essentially no cells were in S-phase at that specific time of treatment. The percentage of cells in S-phase was based on control colony counts for CFU-GM BFU-E and CFU-GEMM that were, respectively, in the range 22–138, 9–45, 8–28 for marrow and 30–73, 10–148, and 8–60 for spleen.

Statistical Analysis

Three plates were scored for each CFU-GM sample and two plates were scored for each BFU-E/CFU-GEMM sample. The results are expressed as the mean ± 1 SEM, and these are derived from the averages of the colony counts from each of the individual mice within a group. The probability of significant differences between groups was determined with the use of Student's t-test.

RESULTS

Effects of CSFs on Hematopoietic Progenitor Cell Cycling Rates

Recombinant hM-CSF and rhG-CSF were assessed, alone or in combination, for their effects on the cycling status of bone marrow (Table I) and splenic (Table II) CFU-GM, BFU-E and CFU-GEMM in vivo. Mice were first pretreated i.v. with 100 μg human LF in order to reduce endogenous levels of CSFs and to dampen myelopoiesis [8,35]. Three hours later control mice were treated i.v. with sterile pyrogen-free saline and test mice were given rhM-CSF and/or rhG-CSF. Mice were sacrificed 22 hours later and femoral marrow and spleen from these mice evaluated in vitro. Less than or equal to 11% of marrow and splenic hematopoietic progenitor cells were in S-phase in mice receiving saline. The cycling rates of the hematopoietic progenitors were significantly enhanced ($P < 0.005$) by either 5,000–20,000 units rhM-CSF or 2,000–10,000 units rhG-CSF (Tables I, II). Lower concentrations of either CSF were ineffective in enhancing progenitor cell proliferation ($P > 0.05$). In contrast, adminsitration of both CSFs together, at concentrations as low as 50 units each, was sufficient to enhance significantly ($P < 0.005$) the proliferative rates of the three progenitor cell compartments in the marrow (Table I) and in the spleen (Table II).

Effects of CSFs on Absolute Numbers of Hematopoietic Progenitor Cells and on Peripheral Blood Counts

Concentrations of 20,000 units rhM-CSF alone significantly ($P < 0.005$) enhanced absolute numbers of CFU-GM, BFU-E, and CFU-GEMM per femur, while 10,000 units rhG-CSF enhanced absolute numbers of marrow CFU-GEMM (Table III). Lower concentrations of either CSF alone, or concentrations of rhM-CSF plus rhG-CSF as high as 500 units each, had no significant effect on marrow progenitor cell numbers (Table III). Concentrations as high as 20,000 units rhM-CSF or 10,000 units rhG-CSF or the combination of 500 units rhM-CSF plus 500 units rhG-CSF had no influence ($P > 0.05$) on numbers of splenic progenitor cells (Table IV). No effect was noted on numbers of nucleated cells within the femoral marrow or spleen at any of the concentrations of CSFs assessed (Tables III, IV) and peripheral blood counts and differentials taken at 22, 48, or 72 h were not significantly different from control in any of the test mice (data not shown).

DISCUSSION

LF pretreated mice were used to enhance the capacity to detect the proliferative effects in vivo of low dosages of CSFs [7,8,35]. The reversible myelosuppressive effects of LF in vivo probably relate to the ability of LF to decrease the release of CSFs from cells of the mononuclear phagocytic lineage [44] and/or to decrease the release of a monokine [45–47], most probably IL-1 [48–51], which then triggers other cell types to release CSFs. In the LF-pretreated mice, rhM-CSF can synergize with rhG-CSF to enhance the percentage of marrow and splenic CFU-GM, BFU-E, and CFU-GEMM in S-phase of the cell cycle; 40–200-fold less of each CSF is needed in comparison to concentrations of each which induce the same conditions when each of the CSFs is administered only by itself (Table V). This study extends our previous reports [11,12] in which combinations of murine preparations of IL-3, GM-CSF, and M-CSF (CSF-1) manifest similar synergistic effects on marrow progenitors in the

TABLE I. Influence of Purified Recombinant Human M-CSF and Recombinant Human G-CSF, Alone or in Combination, on the Cycling Status of Bone Marrow Hematopoietic Progenitor Cells In Vivo After Mice Have Been Pretreated With Purified Human LF†

Test material given in vivo (units)	No. of mice	% Of progenitor cells in S-phase		
		CFU-GM	BFU-E	CFU-GEMM
Saline	19	11 ± 3	8 ± 5	6 ± 5
M-CSF (20,000)	6	55 ± 7*	61 ± 3*	67 ± 5*
M-CSF (10,000)	12	50 ± 4*	31 ± 8*	34 ± 7*
M-CSF (5,000)	12	34 ± 5*	12 ± 4	6 ± 4
M-CSF (2,000)	12	13 ± 5	3 ± 3	3 ± 3
M-CSF (1,000)	15	10 ± 5	13 ± 5	3 ± 4
G-CSF (10,000)	11	47 ± 5*	54 ± 6*	61 ± 7*
G-CSF (5,000)	12	28 ± 5*	43 ± 9*	41 ± 10*
G-CSF (2,000)	12	15 ± 3	14 ± 7	30 ± 9*
G-CSF (1,000)	15	11 ± 4	12 ± 6	16 ± 6
M-CSF (500) + G-CSF (500)	9	56 ± 3*	69 ± 4*	64 ± 5*
M-CSF (100) + G-CSF (100)	9	49 ± 5*	68 ± 5*	58 ± 4*
M-CSF (50) + G-CSF (50)	9	44 ± 3*	25 ± 8*	22 ± 7*
M-CSF (10) + G-CSF (10)	9	17 ± 5	12 ± 8	4 ± 7

†Mice were pretreated with 100 μg endotoxin-depleted iron-saturated human LF i.v. Pretreatment with LF placed the progenitors into a slow, or noncycling, state. Three hours later they were treated i.v. with sterile pyrogen-free saline or various concentrations of M-CSF or G-CSF, alone or in combination. Mice were sacrificed 22 h later.
*Significant change from saline group, P at least <0.005.

TABLE II. Influence of Purified Recombinant Human M-CSF and Recombinant G-CSF, Alone or in Combination, on the Cycling Status of Splenic Hematopoietic Progenitor Cells In Vivo After Mice Have Been Pretreated With Purified Human LF†

Test material given in vivo (units)	No. of mice	% Of progenitor cells in S-phase		
		CFU-GM	BFU-E	CFU-GEMM
Saline	13	6 ± 3	−4 ± 2	−4 ± 2
M-CSF (10,000)	6	30 ± 5*	51 ± 3*	54 ± 5*
M-CSF (5,000)	6	18 ± 6*	6 ± 3	0.3 ± 4
M-CSF (2,000)	6	9 ± 4	−3 ± 3	−3 ± 3
M-CSF (1,000)	9	12 ± 4	3 ± 3	−3 ± 4
G-CSF (10,000)	6	43 ± 4*	53 ± 4*	53 ± 3*
G-CSF (5,000)	6	14 ± 6	−4 ± 5	−10 ± 4
G-CSF (2,000)	6	7 ± 4	−6 ± 3	−0.3 ± 3
G-CSF (1,000)	9	13 ± 5	−7 ± 3	−0.3 ± 4
M-CSF (500) + G-CSF (500)	6	46 ± 1*	56 ± 3*	54 ± 3*
M-CSF (100) + G-CSF (100)	6	58 ± 4*	48 ± 1*	49 ± 3*
M-CSF (50) + G-CSF (50)	6	22 ± 7*	29 ± 5*	11 ± 7
M-CSF (10) + G-CSF (10)	6	11 ± 2	1 ± 3	−0.2 ± 4

†Mice were treated as described in the legend to Table I. A negative sign next to the No. means that there are few or no cells in S-phase.
*Significant change from saline group, P at least <0.005.

TABLE III. Influence of Purified Recombinant Human M-CSF and Recombinant Human G-CSF, Alone or in Combination, on Numbers of Bone Marrow Nucleated Cells and Hematopoietic Progenitors In Vivo After Mice Have Been Pretreated With Purified Human LF†

Test material given in vivo (units)	Nucleated cells (No. × 10^{-6} per femur)	Colonies, no. × 10^{-3} per femur		
		CFU-GM	BFU-E	CFU-GEMM
Saline	18.7 ± 0.7	14.4 ± 2.6	5.0 ± 0.5	2.5 ± 0.3
M-CSF (20,000)	21.1 ± 1.5	38.4 ± 11.6*	9.8 ± 1.6*	5.7 ± 0.7*
M-CSF (10,000)	17.9 ± 1.0	19.1 ± 4.9	4.7 ± 0.5	2.7 ± 0.3
M-CSF (5,000)	17.3 ± 0.8	15.2 ± 3.5	4.0 ± 0.4	2.3 ± 0.2
M-CSF (2,000)	18.6 ± 1.2	16.2 ± 3.8	4.6 ± 0.4	2.3 ± 0.2
M-CSF (1,000)	19.0 ± 0.8	14.8 ± 2.8	4.4 ± 0.4	2.0 ± 0.2
G-CSF (10,000)	17.8 ± 0.8	19.3 ± 5.4	5.8 ± 1.0	4.0 ± 0.8*
G-CSF (5,000)	18.6 ± 1.7	17.5 ± 4.5	5.4 ± 0.7	3.4 ± 0.5
G-CSF (2,000)	19.1 ± 1.0	16.2 ± 3.6	4.8 ± 0.4	2.8 ± 0.3
G-CSF (1,000)	18.8 ± 1.0	16.1 ± 3.1	5.1 ± 0.5	2.7 ± 0.2
M-CSF (500) + G-CSF (500)	18.0 ± 1.0	12.0 ± 1.6	6.7 ± 1.0	2.7 ± 0.5
M-CSF (100) + G-CSF (100)	19.0 ± 1.0	10.4 ± 2.1	5.1 ± 0.7	2.1 ± 0.2
M-CSF (50) + G-CSF (50)	17.0 ± 1.0	11.5 ± 0.5	4.2 ± 0.4	2.3 ± 0.2
M-CSF (10) + G-CSF (10)	20.0 ± 1.0	10.6 ± 1.7	4.7 ± 0.2	1.9 ± 0.4

†Protocol and No. of mice assessed are the same as those described in Table I.
*Significant change from saline group, P at least <0.005.

TABLE IV. Influence of Purified Recombinant Human M-CSF and Recombinant Human G-CSF, Alone or in Combination, on Numbers of Splenic Nucleated Cells and Hematopoietic Progenitors In Vivo After Mice Have Been Pretreated With Purified Human LF†

Test material given in vivo (units)	Nucleated cells (No. × 10^{-6} per spleen)	Colonies, no. × 10^{-3} per spleen		
		CFU-GM	BFU-E	CFU-GEMM
Saline	92.0 ± 9	5.1 ± 0.5	6.1 ± 1.4	1.6 ± 0.3
M-CSF (10,000)	79.0 ± 5	3.9 ± 0.4	4.3 ± 0.5	0.9 ± 0.4
M-CSF (5,000)	80.0 ± 4	4.1 ± 0.4	3.4 ± 1.4	0.9 ± 0.4
M-CSF (2,000)	81.0 ± 6	3.9 ± 0.6	4.6 ± 0.4	1.3 ± 0.1
M-CSF (1,000)	105.0 ± 10	6.3 ± 0.6	8.8 ± 2.2	2.0 ± 0.2
G-CSF (10,000)	102.0 ± 12	5.6 ± 0.9	6.1 ± 1.0	1.0 ± 0.2
G-CSF (5,000)	89.0 ± 8	4.8 ± 0.6	5.0 ± 0.5	0.9 ± 0.3
G-CSF (2,000)	88.0 ± 4	3.8 ± 0.7	4.7 ± 0.3	0.8 ± 0.5
G-CSF (1,000)	108.0 ± 9	5.8 ± 0.5	8.0 ± 1.8	2.2 ± 0.4
M-CSF (500) + G-CSF (500)	94.0 ± 8	6.2 ± 1.1	7.7 ± 1.8	2.3 ± 0.5
M-CSF (100) + G-CSF (100)	84.0 ± 7	5.8 ± 0.9	8.6 ± 2.3	2.2 ± 0.2
M-CSF (50) + G-CSF (50)	94.0 ± 11	5.2 ± 0.4	6.6 ± 1.6	1.4 ± 0.1
M-CSF (10) + G-CSF (10)	113.0 ± 7	5.9 ± 0.4	7.3 ± 2.9	2.2 ± 0.5

†Protocol is same as that described in legend to Table I and Nos. of mice assessed are the same as in Table II.

same model system, by evaluating recombinant preparations of human CSFs, assessing rhG-CSF, and determining effects in the spleen, as well as in the marrow.

The concept of synergism, roughly defined as the total effect's being greater than the sum of the two effects taken independently, is an intriguing one, especially in terms of the actions of biomolecules in vivo. Synergism adds further complexity to known biomolecule-cellular interactions uncovered through studies in vitro [1,2,34]. Synergistic interactions can reflect cascading effects mediated indirectly through

TABLE V. Lowest Concentrations of CSFs Having Activity for Marrow and Splenic Hematopoietic Progenitor Cell Cycling When M-CSF or G-CSF Are Administered, Alone or in Combination, to Mice Pretreated With LF

	CFU-GM			BFU-E			CFU-GEMM		
	CSF alone[a]	CSF + other CSF[b]	Fold change[c]	CSF alone[a]	CSF + other CSF[b]	Fold change[c]	CSF alone[a]	CSF + other CSF[b]	Fold change[c]
Bone marrow									
M-CSF	5,000	50	100	10,000	50	200	10,000	50	200
G-CSF	5,000	50	100	5,000	50	100	2,000	50	40
Spleen									
M-CSF	5,000	50	100	10,000	50	200	10,000	100	100
G-CSF	10,000	50	200	10,000	50	200	10,000	100	100

[a]Lowest concentration of CSFs having activity when administered alone.
[b]Lowest concentration of CSF having no activity alone, but having activity when administered with a dosage of the other CSF, that also has no activity when used alone.
[c]Fold change refers to differences between lowered CSF concentrations that are active when CSFs are given together compared to when CSFs are administered to mice in the absence of the other form of CSF.

accessory cells, effects mediated directly at both accessory and progenitor cell levels, or effects manifested entirely at the progenitor cell level. Our past [11,12] and present studies done in vivo don't allow us to distinguish which of the above possibilities apply, but we consider it likely that many of the effects noted, especially those of rhM-CSF and rhG-CSF on BFU-E and CFU-GEMM, are accessory cell mediated.

The results described here are also consistent with our previous findings that less CSF, alone or in combination with other CSFs, is required to enhance the cycling rates of hematopoietic progenitor cells than is required to increase the absolute numbers of these progenitors, and increases in numbers of progenitors per organ do not necessarily translate into increased total nucleated blood cell counts in the marrow, spleen or blood [7,12]. It is clear that multiple or continuous infusions of much higher concentrations of CSFs can increase circulating levels of blood leukocytes [20–23,52] and that single injections of CSFs, while having a proliferative effect on the progenitor cells, result in what appears to be ineffective myelopoiesis. Nevertheless, low dosages of CSFs may have potential clinical relevance. Clinical trials with rhGM-CSF [30–32] and rhG-CSF [33] have demonstrated efficacy in increasing circulating leukocyte levels, and we have noted that administration of rhGM-CSF to patients with the myelodysplastic syndrome and with cancer [30,31] place their slowly or noncycling marrow CFU-GM into a more rapid, but reversible, proliferative phase [53]. Marrow CFU-GM from the patients administered rhGM-CSF were increased in sensitvity to the suppressive effects in vitro of rh acidic isoferritin in two of eight cases, and to lower dosages of tumor necrosis factor-alpha in all cases evaluated [53]. Thus, it is possible that enhancement of the percentage of progenitors in S-phase of the cell cycle, without increases in progenitor or nucleated cell numbers, might be able to be utilized in a clinical setting in combination with other biotherapy and/or with chemotherapy.

ACKNOWLEDGMENTS

These studies were supported by Public Health Service Grants CA 36464 and CA 36740 from the National Cancer Institute and by a grant from the Cetus Corporation (to H.E.B.). D.E.W. was supported by NIH training program IT32 AM 07519 during the course of these studies. We thank Stephanie Moore for typing the manuscript, Dr. Robert J. Drummond for rhG-CSF, and Dr. Cynthia Cowgill and the Department of Process and Product Development, Cetus Corporation, for rhM-CSF (CSF-1).

REFERENCES

1. Broxmeyer HE: Int J Cell Cloning 4:378–405, 1986.
2. Broxmeyer HE, Williams DE: CRC Crit Rev Oncol Hematol, in press, 1988.
3. Metcalf D: Science 229:16–22, 1985.
4. Kindler V, Thorens B, deKossodo S, Allet B, Eliason JF, Thatcher D, Farber N, Vassalli P: Proc Natl Acad Sci USA 83:1001–1005, 1986.
5. Lord BI, Malineux G, Testa NG, Keely M, Spooncer E, Dexter TM: Lymphokine Res 5:97–104, 1986.
6. Metcalf D, Begley CG, Johnson GR, Nicola NA, Lopez AF, Williamson DJ: Blood 68:46–57, 1986.
7. Broxmeyer HE, Williams DE, Cooper S, Shadduck RK, Gillis S, Waheed A, Urdal DL, Bicknell DC: J Clin Invest 79:721–730, 1986.

8. Broxmeyer HE, Williams DE, Hangoc G, Cooper S, Gentile P, Shen RN, Ralph P, Gillis S, Bicknell DC: Blood Cells 13:31–48, 1987.
9. Broxmeyer HE, Williams DE, Cooper S: Leuk Res 201–205, 1987.
10. Broxmeyer HE, Williams DE, Boswell HS, Cooper S, Shadduck RK, Gillis S, Waheed A, Urdal DL: Immunobiology 172:168–174, 1986.
11. Williams DE, Hangoc G, Cooper S, Boswell HS, Shadduck RK, Gillis S, Waheed A, Urdal D, Broxmeyer HE: Blood 70:401–403, 1987.
12. Broxmeyer HE, Williams DE, Hangoc G, Cooper S, Gillis S, Shadduck RK, Bicknell DC. Proc Natl Acad Sci USA 84:3871–3875, 1987.
13. Metcalf D, Begley CG, Nicola NA, Johnson GR: Exp Hematol 15:288–295, 1987.
14. Hangoc G, Lu L, Oliff A, Gillis S, Hu W, Bicknell DC, Williams D, Broxmeyer HE: Leukemia 1:762–764, 1987.
15. Broxmeyer HE, Williams DE, Cooper S, Waheed A, Shadduck RK: Blood 69:913–918, 1987.
16. Shadduck RK, Waheed A, Boegel F, Pigoli G, Porcellini A, Rizzoli V: Blood Cells 13:49–63, 1987.
17. Yanoi N, Yamada M, Watanabe Y, Saito M, Kuboyama M, Motoyoshi K, Takaku F, Funakoshi S, Watanabe M: Exp Hematol 11:1027–1036, 1983.
18. Motoyoshi K, Takaku F, Maekawa T, Mirua Y, Kimura K, Furusawa S, Hattori M, Nomura T, Mizoguchi H, Ogawa M, Kinugasa K, Tominaga T, Shimoyama M, Deura K, Ohta K, Taguchi T, Masaoka T, Kimura I: Exp Hematol 14:1069–1075, 1986.
19. Metcalf D, Begley CG, Williamson DJ, Nice EC, deLamarter J, Mermod J, Thachter D, Schmidz A: Exp Hematol 15:1–9, 1987.
20. Donahue RE, Wang E, Stone DK, Kamen R, Wong CG, Seghal PK, Nathan DC, Clark SC: Nature 321:872–875, 1986.
21. Mayer P, Lam C, Obenaus H, Liehl E, Besemer J: Blood 70:206–213, 1987.
22. Monroy RL, Skelly RR, MacVittie TJ, Davis TA, Sauber JJ, Clark SC, Donahue RE: Blood 70:1696–1999, 1987.
23. Welte K, Bonilla MA, Gillio AP, Boone TC, Potter GK, Gabrilove JL, Moore MAS, O'Reilly RJ, Souza LM: J Exp Med 165:941–948, 1987.
24. Tsuchuja M, Nomura H, Asano S, Kaziro Y, Nagata S: EMBO J 6:611–616, 1987.
25. Shimamura M, Kobayashi Y, Yuo A, Urabe A, Okabe T, Komatsu Y, ltoh S, Takaku F: Blood 69:353–355, 1987.
26. Cohen AM, Zsebo KM, lnove H, Hines D, Boone TC, Chazin VR, Tsai L, Ritch T, Souza LM: Proc Natl Acad Sci USA 84:2484–2488, 1987.
27. Tsuchiya M, Nomura H, Asano S, Kaziro Y, Nagata S: EMBO J 6:611–616, 1987.
28. Broxmeyer HE, Williams DE, Cooper S, Hangoc G, Ralph P: Blood 70(Suppl 1):168a, 1987.
29. Welte K, Bonilla MA, Gabrilove JL, Gillio AP, Potter GK, Moore MAS, O'Reilly RJ, Boone TC, Souza LM: Blood Cells 13:17–30, 1987.
30. Groopman JE, Mitsyasu RT, DeLeo MJ, Oette DH, Golde DW: N Engl J Med 317:593–598, 1987.
31. Vadhan-Raj S, Keating M, LeMaistre A, Hittelman WH, McCredie JM, Trujillo JM, Broxmeyer HE, Henney C, Gutterman JU: N Engl J Med 317:1546–1552, 1987.
32. Vadhan-Raj S, Buescher S, LeMaistre A, Keating M, Walters R, Kantarjian H, Hittelman WN, Broxmeyer HE, Gutterman JU: Blood, 72:134–141, 1988.
33. Gabrilove J, Jakubowski A, Fain K, Scher H, Grous J, Sternberg C, Yagoda A, Clarkson B, Moore MAS, Bonilla MA, Oettgen HF, Alton K, Downing M, Welte K, Souza LM: Blood 70(Suppl 1):135a, 1987.
34. Williams DE, Straneva JE, Cooper S, Shadduck RK, Waheed A, Gillis S, Urdal D, Broxmeyer HE: Exp Hematol 15:1007–1012, 1987.
35. Broxmeyer HE, Smithyman A, Eger RR, Meyers PA, DeSousa M: J Exp Med 148:1052–1067, 1978.
36. Das SK, Stanley ER, Guilbert LJ, Forman LW: Blood 58:630–641, 1981.
37. Gilbert HS, Praloran V, Stanley ER: Blood 70(Suppl 1):135a, 1987.
38. Shadle E, Ralph P, Broxmeyer HE: Unpublished observation.
39. Nagata S, Tsuchiya M, Asano S, Kaziro Y, Yamazaki T, Yamamoto O, Hirata Y, Kubota N, Oheda M, Nomura H, Ono M: Nature 319:415–418, 1986.
40. Ralph P, Ladner MB, Wang AM, Kawasaki ES, McConlogue L, Weaver JF, Weiss SA, Shadle P, Koths K, Warren MK, Stanley ER, Broxmeyer HE: In Webb DR, Pierce CW, Cohen S (eds): "Molecular Basis of Lymphokine Action." Clifton NJ: Humana Press Inc., 1987, pp 295–311.

41. Devlin PE, Drummond RJ, Toy P, Mark DF, Watt K, Devlin JJ: Gene, 65:13–22, 1988.
42. Ladner MB, Martin GA, Noble JA, Nikoloff DM, Tal R, Kawasaki ES, White TJ: EMBO J 6:2693–2698, 1987.
43. Broxmeyer HE, Bicknell DC, Gillis S, Harris EL, Pelus LM, Sledge GW Jr: Blood Cells 11:429–446, 1986.
44. Broxmeyer HE, Platzer E: J Immunol 133:306–314, 1984.
45. Bagby GC, Regas VD, Bennett RM, Vandenbark AA, Garewal HS: J Clin Invest 68:56–63, 1981.
46. Bagby GC, McCall E, Layman DL: J Clin Invest 71:340–344, 1983.
47. Bagby GC, McCall E, Bergstrom KA, Burger D: Blood 62:663–668, 1983.
48. Zucali JR, Dinarello CA, Oblon DJ, Gross MA, Anderson L., Weiner RS: J Clin Invest 77:1857–1863, 1986.
49. Zucali JR, Broxmeyer HE, Dinarello CA, Gross MA, Weiner RS: Blood 69:33–37, 1987.
50. Bagby GC, Dinarello CA, Wallace P, Wagner C, Hefeneider S, McCall E: J Clin Invest 78:1316–1323, 1986.
51. Zucali JR, Levy DA, Broxmeyer HE, Morse CA: Blood 70(Suppl 1):191a, 1987.
52. Broxmeyer HE, Geissler K, Cooper S, Williams DE: Biotechnology Therapeutics, in press, 1988.
53. Broxmeyer HE, Cooper S, Williams DE, Hangoc G, Gutterman JU, Vadhan-Raj S: Exp. Hematol. 16:594–602, 1988.

Journal of Cellular Biochemistry 38:137–144 (1988)
Growth Factors and Their Receptors: Genetic Control and Rational Application 21–28

Epidermal Growth Factor-Stimulated DNA Synthesis Requires an Influx of Extracellular Calcium

Timothy D. Hill, Henrik Kindmark, and Alton L. Boynton

Cancer Research Center of Hawaii, Basic Science Program, University of Hawaii, Honolulu, Hawaii 96813

The dependency of normal cell proliferation on adequate extracellular Ca^{2+} levels was further investigated by determining the role of Ca^{2+} influx in epidermal growth factor (EGF)-induced rat liver epithelial (T51B) cell DNA synthesis. Fura-2-loaded T51B cells responded with an increase in $[Ca^{2+}]_i$ to EGF (5–50 ng/ml) that was blocked by low (25 μM) extracellular Ca^{2+} or by pretreatment with 50 μM La^{3+} to inhibit plasma membrane Ca^{2+} flux. Confluent T51B cells treated for 24 h with EGF (0.1–50 ng/ml) dose-dependently incorporated [^{3}H]-thymidine into cell nuclei. Low extracellular Ca^{2+} or addition of La^{3+} prevented the EGF-stimulated rise in labeled nuclei, indicating that a movement of Ca^{2+} into the cell was required for DNA synthesis. This was supported by our findings that bradykinin, which induced a rise in $[Ca^{2+}]_i$ by opening plasma membrane Ca^{2+} channels in T51B cells (but not A23187, thrombin or ATP, which raise $[Ca^{2+}]_i$ primary through mobilization of intracellular Ca^{2+} stores), potentiated DNA synthesis stimulated by submaximal doses of EGF. Potentiation of the action of EGF by the tumor promoter 12-0-tetradecanoyl-phorbol-13-acetate (TPA), indicates that activation of protein kinase C and an influx of Ca^{2+} share a common mechanism for initiating DNA synthesis.

Key words: EGF, DNA synthesis, Bradykinin, TPA, rat liver epithelial cells, calcium

Extracellular Ca^{2+} has been demonstrated to be required for cell proliferation in a wide variety of non-neoplastic cells of mesenchymal and epithelial origin both in vivo and in vitro [1–5]. Studies of rat liver epithelial (T51B) cell division have revealed two specific extracellular Ca^{2+}-dependent stages within the cell-division

Henrik Kindmark is on leave from the Department of Medical Cell Biology, University of Uppsala, Biomedicum, Uppsala, Sweden.

Address all correspondence to A.L. Boynton, Cancer Research Center of Hawaii, Basic Science Program, University of Hawaii, Honolulu, Hawaii 96813.

Received February 1, 1988; accepted May 4, 1988.

cycle [6–8]. The first is a brief 30 min period necessary for cells to transit the G_0 to G_1 boundary, and the second occurs during the 2 h immediately preceding the G_1 to G_S transition. In contrast, neoplastic cells require 10- to 50-fold less extracellular Ca^{2+} for proliferative activity and for these specific Ca^{2+}-dependent cell-cycle transitions than do their non-neoplastic counterparts [9,10].

The biochemical basis for the Ca^{2+} sensitivity in normal cells has been the focus of several investigations, which demonstrated that calmodulin, Ca^{2+}/calmodulin-dependent protein kinase, cyclic AMP, cyclic AMP-dependent protein kinase, and Ca^{2+} are involved in the progression of cells through the G_1 phase of their cell-cycle [for review, see 11]. These observations suggest that a surge of Ca^{2+} into the cytosolic compartment of cells is a necessary event for initiation of G_0 to G_1, and G_1 to S cell-cycle transitions. However, the requirement for a rise in free Ca^{2+}, or the source of Ca^{2+}, whether extracellular or intracellular, has not been established.

Ca^{2+} influx across the plasma membrane in response to mitogens binding to surface receptors occurs in a variety of cell types, yet the importance of membrane Ca^{2+} fluxes in cell proliferation is unknown [12]. This study examines the role of plasma membrane Ca^{2+} influx in T51B cell DNA synthesis induced by epidermal growth factor (EGF)—an agent shown to cause phosphotidylinositol turnover and a rise in intracellular free Ca^{2+} ($[Ca^{2+}]_i$) [13,23].

MATERIALS AND METHODS

Materials

EGF, bradykinin, thrombin, ATP, A23187, and $LaCl_3$ were purchased from Sigma (St. Louis, MO). TPA (12-0-tetradecanoyl-phorbol-13-acetate) was obtained from LC Services Corp (Woburn, MA). Fura-2am (acetoxymethyl ester) was obtained from Molecular Probes (Eugene, OR). [^{3}H] thymidine was acquired from ICN (Cambridge, MA).

Cell Culture

T51B cells were plated at a density of 7,000 cells/cm^2 on either 24 $\times$ 30-mm rectangular Thermanox coverslips (DNA synthesis) or 12-mm round glass coverslips (Ca^{2+} measurement) placed in multiwell sterile culture plates (Miles Lab., Naperville, IL) and grown to confluency (3 $\times$ 10^4 cells/cm^2) in Eagle's Basal Medium (BME; Gibco, Grand Island, NY) supplemented with 10% bovine calf serum (BCS; Colorado Serum Co., Denver, CO) and 25 μg/ml gentamicin (37°C; 95% air/5% CO_2, water saturated).

Measurement of Cytoplasmic Free Ca^{2+} Concentration

Monolayers of confluent T51B cells on glass coverslips (12 mm) were loaded with 5 μM fura-2am for 45 min in 90% BME:10% BCS, washed 5 min in HEPES buffer (25 mM HEPES, 125 mM NaCl, 6.0 mM KCl, 1.0 mM MgCl, and 0.55 mM glucose, pH 7.4), and placed in plastic cuvettes (1-cm light path) containing either low Ca^{2+} (0.025 mM) HEPES buffer or high Ca^{2+} (1.8mM) HEPES buffer Ca^{2+} (37°C). Changes in fluorescence were monitored at 340-nm excitation and 500-nm emission using a fluorescence spectrophotometer (Perkin Elmer). Calculations of $[Ca^{2+}]_i$ were made according to the formua $[Ca^{2+}]=K_d\,(F-F_{min})/(F_{max}-F)$; where

F_{min} is determined by the formula $F_{min} = (F_{Mn}{}^{2+}) + 0.243 (F_{max} - F_{Mn}{}^{2+})$ as previously reported [14].

Measurement of DNA Synthesis

DNA synthesis was assessed according to the method of Boynton and Whitfield by determing the proportion of cell nuclei labeled with [^{3}H]thymidine (2.5 μCi/ml) following a 24 h incubation [15].

RESULTS

The ability of EGF to increase T51B whole cell cytosolic free Ca^{2+} was dependent on a high extracellular [Ca^{2+}] (Fig. 1). Maximum changes in intracellular Ca^{2+} levels were obtained with doses of EGF ranging between 25 and 50 ng/ml. Lowering the buffer [Ca^{2+}] from 1.8 mM to 0.025 mM abolished the rise in $[Ca^{2+}]_i$ induced by 50 ng/ml EGF (Fig. 1B). Plasma membrane Ca^{2+} channel blockage by addition of 50 μM La^{3+} prior to EGF eliminated the cell Ca^{2+} response (Fig. 1C). La^{3+} alone was unable to penetrate the plasma membrane, as evidenced by a lack of change in the fura-2 fluorescence signal upon treatment with La^{3+} alone (data not shown). Ca^{2+} mobilization from intracellular storage sites induced by EGF was not detected in T51B cells, although this response has been reported for other cell types [16,17].

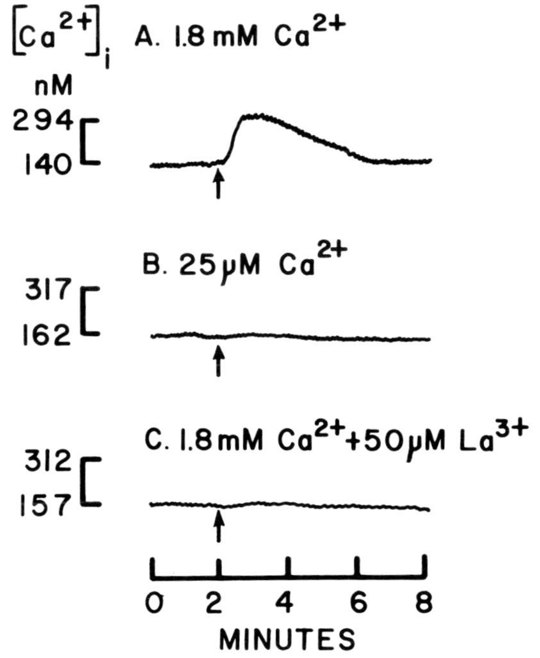

Fig. 1. Effect of low extracellular Ca^{2+}, or La^{3+} addition on the EGF-induced rise in free intracellular Ca^{2+}. T51B cells grown to a confluent monolayer on glass coverlsips and loaded with 5 μM fura-2 for 45 min (37°C) were placed in buffer that contained either **A)** 1.8 mM Ca^{2+}; **B)** 0.025 mM Ca^{2+}; or **C)** 1.8 mM Ca^{2+} and 50 μM La^{3+} (37°C). Two minutes later (arrow) 50 ng/ml EGF was added and the change in fluoresence recorded. Low extracellular Ca^{2+}, or La^{3+} addition, eliminated the Ca^{2+} response of EGF. The above tracings are single experiments which were repeated 20 times.

Addition of the surface binding agents thrombin, bradykinin, and ATP to fura-2-loaded cells in the presence of high extracellular Ca^{2+} (1.8 mM; Fig. 2A–C), or low extracellular Ca^{2+} (0.025 mM; Fig. 2D–F) resulted in the identification of the source of Ca^{2+} mobilized by these agents. Thrombin produced a rise in $[Ca^{2+}]_i$ through the liberation of intracellular Ca^{2+} stores alone, which was unaffected by the extracellular $[Ca^{2+}]$ (Fig. 2A,D). In contrast, the effect of bradykinin on $[Ca^{2+}]_i$ is completely dependent on a high level of extracellular Ca^{2+} (Fig. 2B,E). T51B cell response to ATP is biphasic, having a rapid, short-lasting spike of intracellularly released Ca^{2+}, followed with a prolonged tailing that is sustained from an influx of extracellular Ca^{2+} (Fig. 2C,F). In low Ca^{2+} buffer, the ATP-induced Ca^{2+} response pattern was condensed and lacked a broad tail indicating that the mobilized Ca^{2+} was derived exclusively from sites within the cell.

Fresh bovine calf serum (BCS) stimulated T51B cell DNA synthesis dose-dependently, causing a maximum proportion of labeled cells at a dose of 20% BCS/80% BME (Fig. 3A). EGF induced DNA synthesis in the absence of added fresh BCS (Fig. 3B) and caused a further twofold increase in the proportion of labeled nuclei when administered to cells in 1% BCS/90% BME medium (Fig. 3C). Low extracellular Ca^{2+} (0.025mM) or addition of La^{3+} (50 μM) inhibited the EGF-induced mitogenic response of T51B cells (Fig. 3D).

To investigate further the requirement of Ca^{2+} flux in mediating EGF mitogenic activity, various Ca^{2+} mobilizing agents and the tumor promoter TPA were examined alone or with a submaximal mitogenic dose of EGF to determine if Ca^{2+} fluxes could enhance the DNA synthetic response. Shown in Figure 4A,B bradykinin and TPA potentiated the effects of EGF 2- to 3-fold in media containing 1% BCS; with a similar, yet deminished effect on EGF in 100% BME media (Fig. 4C,D). Lowering the extracellular $[Ca^{2+}]$ completely abolished the potentiating action of these agents. The submaximal dose of EGF (0.1 ng/ml) used in these studies was unable to generate

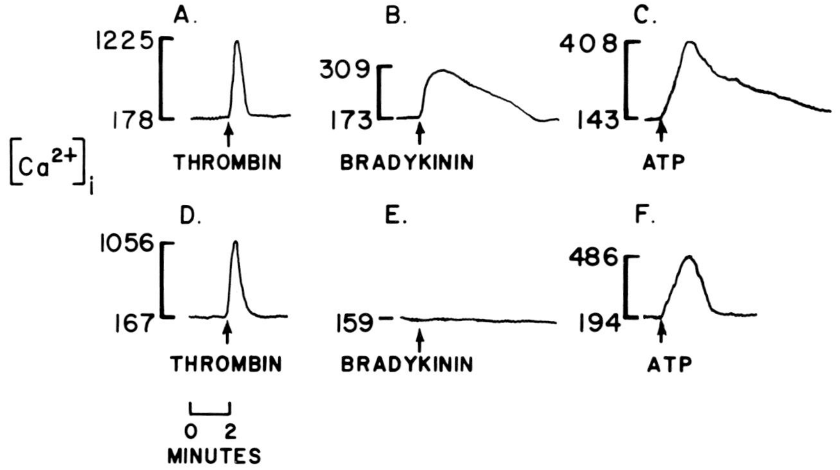

Fig. 2. Effect of thrombin, bradykinin, and ATP on T51B cell intracellular Ca^{2+} levels. Fura-2-loaded cells were placed in buffer containing either 1.8 mM Ca^{2+} (**A–C**) or 0.025 mM Ca^{2+} (**D–F**) and exposed to thrombin (0.05 U/ml), bradykinin (50 μM), and ATP (100 μM). The action of bradykinin and the prolonged elevated Ca^{2+} component of the ATP response were blocked by low extracellular Ca^{2+} levels, while the response to thrombin remained unaffected. These are representative tracings from experiments performed 20 times.

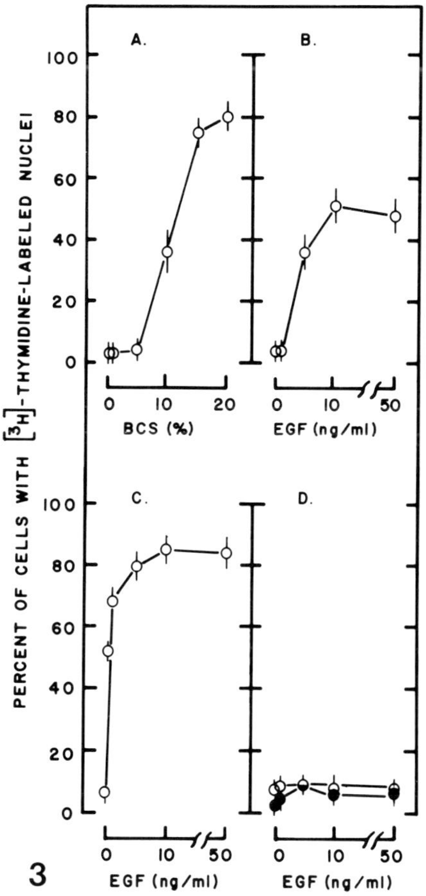

3

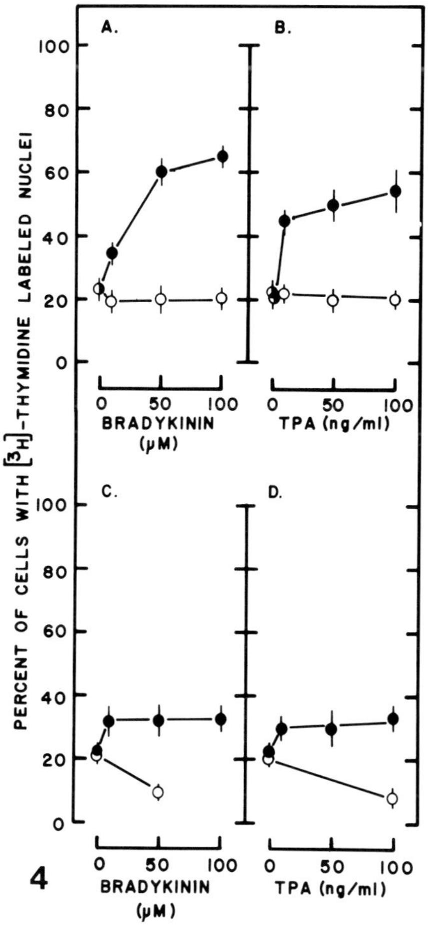

4

Fig. 3. Influence of low extracellular Ca^{2+}, or La^{3+} addition on EGF-stimulated DNA synthesis. Confluent and proliferatively quiescent T51B cells were treated at the start of the experiment as follows: **A:** Fresh bovine calf serum in BME (1.8 mM Ca^{2+}). **B:** EGF in fresh BME (1.8 mM Ca^{2+}). **C:** EGF in fresh 1% BCS/99% BME (1.8 mM Ca^{2+}) **D:** EGF in fresh 1% BCS/99% BME (0.025 mM Ca^{2+}) (open circles); or 1% BCS/99% BME (1.8 mM Ca^{2+} and 50 μM La^{3+}) (closed circles). Cells were incubated 24 h with [^{3}H] thymidine and fixed; and labeled nuclei were visualized by autoradiography. Effects of EGF were inhibited by lowering the media Ca^{2+} content, and by La^{3+} blockage of plasma membrane Ca^{2+} channels. Values are mean ± SEM; n = 4.

Fig. 4. Influence of bradykinin and TPA on EGF-induced DNA synthesis. At the start of the experiment confluent and quescent T51B cells were treated as follows. **A:** Bradykinin + 0.1 ng/ml EGF in fresh 1% BCS/99% BME. **B:** TPA + 0.1 ng/ml EGF in fresh 1% BCS/99% BME. **C:** Bradykinin + 3 ng/ml EGF in 100% BME. **D:** TPA + 3 ng/ml EGF in 100% BME. The above treatments contained either 1.8 mM Ca^{2+} (closed circles) or 0.025 mM Ca^{2+} (open circles). Cells were incubated 24 h in the presence of [^{3}H] thymidine, and the fraction of labeled nuclei determine by autoradiography. In the presence of serum, the proportion of label nuclei induced by EGF was increased 2-3-fold by both bradykinin and TPA. The effect was reduced in the absence of serum. Values are mean ± SEM; n = 3.

a Ca^{2+} influx and had no effect on the Ca^{2+} influx induced by bradykinin (Table I). T51B cell $[Ca^{2+}]_i$ was unaffected by TPA treatment (Table I).

The potentiating action of bradykinin and TPA was not an additive effect since neither agent is mitogenic for T51B cells (Table II). Agents that raise the $[Ca^{2+}]_i$ through complete or partial depletion of intracellular Ca^{2+} stores (thrombin, ATP, A23187) were ineffective as mitogens or in potentiating EGF-induced DNA synthesis (Table II). In addition, combinations of bradykinin and thrombin (Ca^{2+} influx and intracellular Ca^{2+} release), bradykinin and TPA (Ca^{2+} influx and protein kinase C activation), and TPA and A23187 (Ca^{2+} influx, intracellular Ca^{2+} release, and protein kinase C activation) were all incapable of stimulating DNA synthesis (Table II).

DISCUSSION

The results presented here demonstrate that EGF-induced DNA synthesis requires an influx of extracellular Ca^{2+}. Stimulation of T51B cells with EGF causes an initial rapid and transient influx of Ca^{2+}, and culminates in DNA replication and cell division several hours later. Both early and late events in cell proliferation were inhibited by placing cells in low Ca^{2+} media, or by treatment with La^{3+} to displace

TABLE I. Effect of EGF, Bradykinin, and TPA on $[Ca^{2+}]_i$

Treatment	Change in $[Ca^{2+}]_i$ (nM)
50 ng/ml EGF	169
5 ng/ml EGF	87
0.1 ng/ml EGF	0
50 μM Bradykinin	217
0.1 ng/ml EGF + 50 μM Bradykinin	224
100 ng/ml TPA	0
0.1 ng/ml EGF + 100 ng/ml TPA	0

Fura-2-loaded cells grown on glass coverslips were placed in 1.8 mM Ca^{2+} buffer and measured by spectrofluorometry.

TABLE II. Influence of Ca^{2+}-Mobilizing Agents on EGF-Induced DNA Synthesis

	Ca^{2+} flux[a]		% [^{3}H] thymidine-labeled nuclei	
	Extra	Intra	−0.1 ng/ml EGF	+0.1 ng/ml EGF
Control (EGF)	+	−	3.6 ± 1.4[b]	25.1 ± 3.8
Bradykinin (50 μM)	+	−	4.3 ± 0.8	63.8 ± 1.1
Thrombin (0.05 U/ml)	−	+	3.1 ± 1.3	24.4 ± 2.7
ATP (100 μM)	+	+	4.2 ± 1.5	9.3 ± 0.6
A23187 (1 μM)	+	+	1.8 ± 0.9	3.2 ± 1.1
TPA (100 ng/ml)	−	−	3.4 ± 1.2	56.1 ± 3.6
Bradykinin (50 μM) + Thrombin (0.05) U/ml)	+	+	3.6 ± 1.7	—
Bradykinin (50 μM) + TPA (100 ng/ml)	+	−	5.8 ± 0.9	—
TPA (100 ng/ml) + A23187 (1 μM)	+	+	2.2 ± 1.5	—

[a]Source of Ca^{2+}: Extra, extracellular Ca^{2+}; Intra, intracellular Ca^{2+}.
[b]Values are mean ± SEM.

cell-surface Ca^{2+} and block transmembrane Ca^{2+} fluxes [18]. Previous reports of DNA synthesis induced by serum [19] and insulinlike growth factor II [20] have shown a similar requirement for a flux of Ca^{2+}, which was sensitive to both La^{3+} and the depletion of extracellular Ca^{2+}. This evidence suggests that cellular reactions leading to DNA synthesis are initiated by a surge of Ca^{2+} (down its concentration gradient) across the plasma membrane, which temporarily raises intracellular free Ca^{2+} levels. This concept is further supported by the observation that bradykinin, an agent that induces a rise in $[Ca^{2+}]_i$ only through the opening of plasma membrane Ca^{2+} channels and has no mitogenic activity with T51B cells, was able to potentiate the effects of EGF on DNA synthesis. Moreover, dose-dependency of EGF stimulation of T51B cells may be related to the extent of Ca^{2+} influx produced at each dose, rather than the number of EGF receptors occupied on the cell plasma membrane surface.

Intracellular Ca^{2+}-mobilizing agents, thrombin and ATP, are known to cause phosphotidylinositol turnover and formation of inositol trisphosphate ($Ins(1,4,5)P_3$) [21,22]. Since these agents are not mitogenic with T51B cells and showed no potentiation of EGF-induced DNA synthesis, a rise in $[Ca^{2+}]_i$ supplied by emptying internal Ca^{2+} stores appears to be insufficient for initiating cellular processes involved in cell proliferation. The Ca^{2+} ionophore A23187, and ATP were inhibitory toward the action of EGF, indicating that high $[Ca^{2+}]_i$ may activate opposing regulatory mechanisms. In support of this hypothesis is the observation that EGF induces $Ins(1,4,5)P_3$ formation and a high transient increase in $[Ca^{2+}]_i$ in A431 cells, but inhibits DNA synthesis [23].

TPA alone was not mitogenic but in combination with EGF enhanced the fraction of T51B cells that had initiated DNA synthesis to a similar degree as bradykinin, despite the absence of a detectable increase in $[Ca^{2+}]_i$. One possible explanation for this discrepancy between bradykinin and TPA is that the sensitivity of the spectrofluorometric method is incapable of detecting a Ca^{2+}flux induced by TPA. Video-imaging microscopy has recently been utilized to demonstrate a $[Ca^{2+}]_i$ increase in response to TPA in Balb/c 3T3 cells [24]. However, activation of protein kinase C in the absence of Ca^{2+} surge could also explain the potentiating effects of TPA [25]. Both EGF binding to surface receptors and TPA activation of protein kinase C in A431 cells were demonstrated to cause phosphorylation of the same threonine residue in EGF receptors, suggesting that EGF indirectly influences protein kinase C activation [26]. In another study, TPA initiated DNA synthesis within an hour in Ca^{2+}-deprived T51B cells arrested in G_1, demonstrating the ability of active protein kinase C to participate in the processes which are necessary to elicit a DNA synthetic response [27].

Recent reports of the influence of TPA and EGF on arachidonic acid release and prostaglandin production in porcine thyroid cells indicate a possible common pathway for these agents in mediating cell proliferative activity [28,29]. Prostaglandins are known to be mitogenic for certain cell types [30]. In addition, bradykinin stimulates arachidonic acid release and prostaglandin synthesis in a variety of tissues [31], which could be the basis for its potentiating action. Further work to establish the mechanism of action of protein kinase C and Ca^{2+} influx in potentiating the proliferative action of EGF is clearly indicated.

ACKNOWLEDGMENTS

This work was supported by NCI grant CA 39745 and CA 42942. The excellent technical assistance of Nathan Yanagi and Sherilyn Boynton is greatly appreciated.

REFERENCES

1. Boynton AL, Whitfield JF: Proc Natl Acad Sci USA 73:1651-1654, 1976.
2. Boynton AL, Whitfield JF, Isaacs RJ, Morton HJ: In Vitro 10:12–17, 1974.
3. Boynton AL, Swierenga SHH, Whitfield JF: In Stich H, Sun R (eds): "Short-Term Tests for Chemical Carcinogens." New York, NY: Springer Verlag, 1981, p 362.
4. Hennings H, Michael D, Chen C, Steinert P, Holbrook K, Yuspa S: Cell 19:245–254, 1980.
5. Lechner JF, Kaighn ME: Exp Cell Res 121:432–435, 1981.
6. Boynton AL, Kleine LP, Whitfield JF, Bossi D: Exp Cell Res 160:197–205, 1985.
7. Hazelton B, Mitchell, B, Tupper J: J Cell Biol 83:487–498, 1979.
8. Hesketh R, Smith GA, Metcalfe JC: In Boynton AL, Leffert HL (eds): "Control of Animal Cell Proliferation." Orlando, FL: Academic Press, 1987, p 395.
9. Swierenga SHH, Whitfield JF, Boynton AL, MacManus JP, Rixon RH, Sikorska M, Tsang BK, Walker PR: Ann NY Acad Sci 349:294–311, 1980.
10. Boynton AL, Whitfield JF, Issacs RJ, Tremblay RG: Cancer Res 37:2657–2661, 1977.
11. Boynton AL, Zwiller J, Hill TD, Nilsson T, Arkhammer P, Berggren PO: In Fiskum G (ed): "Cell Calcium Metabolism." New York, NY: Plenum Press, 1988.
12. Putney JW: Physiol (Lond) 268:139–149, 1977.
13. Pike LJ, Eakes AT: J Biol Chem 262:1644–1651, 1987.
14. Hill TD, Berggren PO, Boynton AL: Biochem Biophys Res Commun 149:897–901, 1987.
15. Boynton AL, Whitfield JF: In Vitro 12:479–484, 1976.
16. Moolenaar WH, Aerts RJ, Tertoolen LGJ, de Laat SW: J Biol Chem 261:279–284, 1986.
17. Pandiella A, Malgoroli A, Meldolesi J, Vincentini LM: Exp Cell Res 170:175–185, 1987.
18. Langer GA, Frank JS; J Cell Biol 54:441–447, 1972.
19. Jones A, Boynton AL, MacManus JP, Whitfield JF: Exp Cell Res 138:87–93, 1982.
20. Nishimoto I, Hata Y, Ogata E, Kojima I: J Biol Chem 262:12120–12126, 1987.
21. Cunningham DD, Raben DM: J Cell Physiol 125:582–590, 1985.
22. Charest R, Blackmore PF, Exton JH: J Biol Chem 260:15789–15794, 1985.
23. Hepler JR, Nakahata N, Lovenberg TW, DiGuiseppi J, Herman B, Earp HS, Harden Tk: J Biol Chem 262:2951–2956, 1987.
24. Stephens M, Zagari M: J Cell Biol 105:22a, 1987.
25. Mauerus PW, Neufield EJ, Wilson DB: Cell 37:701–703, 1984.
26. King CS, Copper JA: J Biol Chem 261:10073–10078, 1986.
27. Boynton AL, Whitfield JF: Cancer Res 40:4541–4545, 1980.
28. Kasai K, Hiraiwa M, Emoto T, Akimoto K, Takaoka T, Shimoda SI: Life Sci 41:241–247, 1987.
29. Takasu N, Sato S, Yamada T, Shimizu Y: Biochem Biophys Res Commun 143:880–884, 1987.
30. Jimenez de Asua L, Otto AM, Drummond AH, Macphee CH, Smith C: Adv Prostaglandin Thromboxine Leukotriene Res 15:391–393, 1985.
31. Regoli D, Barabe J: Pharmacol Rev 32:1–46, 1980.

Journal of Cellular Biochemistry 38:179–187 (1988)
Growth Factors and Their Receptors: Genetic
Control and Rational Application 29–37

Colony-Stimulating Factor-1 Receptor (c-*fms*)

Charles J. Sherr, Martine F. Roussel, and Carl W. Rettenmier

Department of Tumor Cell Biology, St. Jude Children's Research Hospital, Memphis, Tennessee 38101

The macrophage colony-stimulating factor, CSF-1 (M-CSF), is a homodimeric glycoprotein required for the lineage-specific growth of cells of the mononuclear phagocyte series. Apart from its role in stimulating the proliferation of bone marrow-derived precursors of monocytes and macrophages, CSF-1 acts as a survival factor and primes mature macrophages to carry out differentiated functions. Each of the actions of CSF-1 are mediated through its binding to a single class of high-affinity receptors expressed on monocytes, macrophages, and their committed progenitors. The CSF-1 receptor (CSF-1R) is encoded by the c-*fms* proto-oncogene, and is one of a family of growth factor receptors that exhibits an intrinsic tyrosine-specific protein kinase activity. Transduction of c-*fms* sequences as a viral oncogene (v-*fms*) in the McDonough (SM) and HZ-5 strains of feline sarcoma virus has resulted in alterations in receptor coding sequences that affect its activity as a tyrosine kinase and provide persistent signals for cell growth in the absence of its ligand. The genetic alterations in the c-*fms* gene that unmask its latent transforming potential abrogate its lineage-specific activity and enable v-*fms* to transform a variety of cells that do not normally express CSF-1 receptors.

Key words: **c-*fms* proto-oncogene, v-*fms* oncogene, macrophage colony-stimulating factor, (CSF-1, M-CSF), cell transformation, tyrosine kinases, leukemogenesis**

Oncogenes of RNA tumor viruses arose through recombination between retroviruses and normal cellular proto-oncogene sequences that are generally assumed to play a vital role in regulating cell growth and differentiation [1]. Transduction of different oncogene sequences as cDNA copies in the genomes of retroviruses confers their ability to transform cells morphologically in tissue culture and to produce tumors in animals after a short latency period. Investigations of the genome structures of acutely transforming viruses and comparison of their viral oncogene sequences to their proto-oncogene counterparts has revealed that recombined oncogenes have frequently undergone structural alterations that contribute to their transforming potential. Moreover, the expression of oncogenes under the control of retroviral promoter/enhancer elements distorts their tissue-specific transcription and enables these genes to be expressed in cells that may not normally transcribe their respective proto-oncogene RNAs. Thus, the types of tumors induced by these viruses appear to reflect the susceptibility of host cells to viral infection

Received January 19, 1988; accepted April 13, 1988.

and the tissue-specificity of viral transcriptional control signals, as well as the ancestral origins of the oncogenes themselves.

The v-*fms* oncogene of the Susan McDonough (SM) and HZ-5 strains of feline sarcoma virus (FeSV) was derived from c-*fms* proto-oncogene sequences of the domestic cat [2–4] which are now recognized to encode the receptor for the macrophage colony-stimulating factor, CSF-1 [5]. Although the CSF-1 receptor (CSF-1R) is normally restricted in its expression to hematopoietic cells of the mononuclear phagocyte series [6,7] and to placental trophoblasts during fetal development [8–10], the v-*fms* gene can transform fibroblasts, epithelial cells, and hematopoietic target cells that do not normally express the c-*fms* gene. Nucleotide sequencing analyses of the v-*fms* and c-*fms* genes, combined with their genetic manipulation and assays of their transforming potential in the context of retroviral vectors, have begun to pinpoint structural alterations in CSF-1R that alter its biological activity and unmask its latent transforming potential.

BIOSYNTHESIS AND BIOCHEMICAL PROPERTIES OF THE CSF-1 RECEPTOR

The receptor for CSF-1 is a member of a family of growth factor receptor genes —including the receptors for epidermal growth factor (EGF), platelet-derived growth factor (PDGF), insulin, and insulin growth factor-1 (IGF-1) [11–15]—which encode polypeptides with tyrosine-specific protein kinase activity. The CSF-1 receptor is synthesized on membrane-bound polyribosomes as an integral transmembrane glycoprotein, oriented with its aminoterminal ligand-binding domain (ca. 485 amino acids) in the cisternae of the rough endoplasmic reticulum (ER) and its carboxylterminal tyrosine kinase domain (ca. 435 amino acids) in the cytoplasm. The polypeptide acquires mannose-rich, asparagine(N)-linked oligosaccharide chains during its synthesis and is transported through the Golgi complex to the cell surface, undergoing concomitant modification of the oligosaccharides to complex carbohydrate chains. Post-translational processing of N-linked oligosaccharides is accompanied by an increase in the apparent molecular weight of CSF-1R on polyacrylamide gels containing sodium dodecyl sulfate, so that the intracellular and mature cell surface forms of the receptor can be readily distinguished, both by their sugar content and molecular mass [5,10,16,17].

Fusion of secretory vesicles with the plasma membrane orients the receptor with its glycosylated ligand binding domain outside the cell and its tyrosine kinase domain at the inner surface of the membrane. Thus, antibodies directed to epitopes in the aminoterminal half of the polypeptide can react with receptor molecules on the surfaces of viable cells and can be used to purify receptor-bearing populations by fluorescence-activated cell sorting procedures [5,17]. In the absence of extracellular CSF-1, the receptors turn over with a half-life of 3–4 h, but binding of saturating quantities of CSF-1 to the receptor results in internalization of receptor-ligand complexes, loss of detectable CSF-1 binding sites at the cell surface, and rapid receptor degradation (downmodulation) [17,19–22]. In the absence of ligand, this is followed by a refractory period of several hours during which the receptor is reexpressed at the cell surface following de novo synthesis.

In purified membrane preparations incubated in the presence of [γ^{32}P]-ATP and manganese, binding of CSF-1 to its receptor activates the receptor kinase and leads to autophosphorylation of CSF-1R on tyrosine [5,21–24]. Similarly, exposure of murine macrophages to saturating quantities of CSF-1 results in phosphorylation of the receptor on tyrosine within 1 min and virtually complete receptor downmodulation within 15 min of stimulation [21,22]. Only the mature cell surface form of the receptor is phosphorylated on tyrosine in response to the growth factor [22]. Although CSF-1 binding must induce a conformational change in the cell surface receptor that activates its intrinsic tyrosine kinase activity, the physical nature of the activating event remains unclear. For the EGF receptor, several lines of evidence suggest that the formation of receptor dimers may alter the affinity of the receptor for its ligand and induce the receptor kinase activity [25–30]. To date, CSF-1R has not been subjected to the same experimental scrutiny.

CSF-1R is also phosphorylated on serine residues, and phorbol esters that activate protein kinase C also induce rapid receptor degradation [20,31]. In macrophage cell lines chronically exposed to phorbol esters such as TPA, the activity of protein kinase C is itself down-regulated, and CSF-1 receptors eventually reappear at the cell surface. These receptors can still be down-modulated in response to CSF-1 (unpublished observations), suggesting that the biochemical mechanisms underlying ligand- and phorbol ester-induced receptor degradation differ from one another.

By Western blotting analysis using antibodies to phosphotyrosine, CSF-1 binding has been shown to induce phosphorylation of a series of cellular substrates [22] whose identity and physiological role, if any, in mediating signal transduction are unknown. CSF-1 induces immediate changes in macrophage membrane structure, including formation of filopodia and vesicles and enhanced phagocytic activity [32], and it increases the transcription of several cellular genes, including the c-*fos* and c-*myc* proto-oncogenes, within minutes to hours after stimulation [33,34]. In susceptible cells, these early events are followed by decreases in the rates of protein turnover within 2 h [35] and eventual mitogenesis [36]. In addition to its role in stimulating cell growth, CSF-1 augments the production of other macrophage cytokines, such as interleukin-1 and tumor necrosis factor; potentiates the release of plasminogen activator, thromboplastin, prostaglandins, and biocidal oxygen metabolites; stimulates killing of microorganisms; and promotes antibody-dependent lysis of tumor cells [37–41]. Thus, activation of the receptor kinase appears to initiate a cascade of intracellular events that ultimately affect the proliferation and functional activities of receptor-bearing cells.

Introduction of retrovirus vectors containing the human c-*fms* gene into mouse NIH/3T3 fibroblasts enables the cells to form colonies in semisolid medium in response to exogenous human recombinant CSF-1 [42]. Similarly, introduction of the murine c-*fms* gene into mouse fibroblasts confers their ability to express specific high-affinity binding sites for the murine growth factor [43]. Cotransfection of NIH/3T3 cells with vectors encoding the human c-*fms* and CSF-1 genes also induces cell transformation and tumorigenicity in nude mice by an autocrine mechanism [42]. In the latter cells, CSF-1R is persistently down-modulated, and neutralizing antibodies to CSF-1 do not efficiently revert the transformed phenotype [17]. This suggests that an intracellular interaction between CSF-1 and its receptor in autocrine transformed cells may potentially occur within the secretory compartment, thereby allowing ligand-receptor complexes to be rapidly directed to lysosomes. Together, these data provide formal genetic proof that c-*fms* encodes CSF-1R and indicate that the receptor gene specifies all the information necessary to initiate a biologic response, even in cells of heterologous species and in differentiated cell types that do not normally express the receptor [42].

STRUCTURAL AND FUNCTIONAL DIFFERENCES BETWEEN *c-fms* AND *v-fms*

The v-*fms* gene of the SM and HZ5 strains of FeSV was recombined into the open reading frame of the retroviral *gag* gene, so that the full genome length mRNAs of both viruses encode *gag-fms* fusion proteins of ca. 180 kDa (gP180$^{gag\text{-}fms}$) [3,4,44]. Recombination between FeLV and c-*fms* occurred in a region of the proto-oncogene encoding the 5′ untranslated sequences of c-*fms* mRNA, prematurely opening the c-*fms* reading frame and preserving the intact ligand-binding domain of CSF-1R, including its signal peptide sequence [45,46]. During its synthesis, the viral polyprotein is cleaved by signal peptidase near the *gag-fms* junction [46]. This process generates an immature v-*fms*-coded glycoprotein (gp120$^{v\text{-}fms}$) that undergoes modification of its N-linked oligosaccharides in a manner analogous to CSF-1R to yield the mature cell surface form (gp140$^{v\text{-}fms}$) [47,48]. Unlike normal CSF-1R synthesis, processing of SM-FeSV gp120$^{v\text{-}fms}$ to gp140$^{v\text{-}fms}$ occurs very inefficiently, suggesting that the immature precursor has undergone some alteration that retards its intracellular transport from the endoplasmic reticulum [47]. Deletion of *gag* sequences from SM-FeSV does not affect the synthesis or transport of gp120$^{v\text{-}fms}$, because translation can begin at the normal c-*fms* initiator codon [46]. As expected, cells infected with SM-FeSV express CSF-1 binding sites, and chemical crosslinking established that CSF-1 was specifically bound to gp140$^{v\text{-}fms}$ molecules at the cell surface [23].

Expression of the v-*fms* gene product at the cell surface is required for cell transformation [49–52]. Unlike the normal CSF-1 receptor, gp140$^{v\text{-}fms}$ expressed in transformed cells is constitutively phosphorylated on tyrosine [22,53], and the viral receptor analog does not undergo rapid down-modulation in response to CSF-1 or phorbol esters [20]. Moreover, in membrane preparations, the v-*fms* gene product can be phosphorylated on tyrosine in the absence of its ligand, and addition of murine CSF-1 does not appear to up-regulate the receptor kinase activity [23]. However, stimulation of SM-FeSV-transformed cells with saturating concentrations of human CSF-1 can further up-regulate the receptor kinase activity [22], suggesting that its constitutive activity as an enzyme does not preclude a functional interaction with the homologous feline growth factor. The persistence of the v-*fms*-coded glycoprotein at the cell surface and its activity as a constitutive tyrosine kinase in the absence of its ligand contribute unregulated signals for cell growth. These properties must be due to particular genetic alterations in the v-*fms* gene that render its product transforming.

Nucleotide sequence comparison and conceptual translation of the feline v-*fms* [44] and human c-*fms* genes [45] reveals scattered amino acid substitutions, many of which appear to be due to interspecies divergence. However, the carboxylterminal 40 amino acids of CSF-1R were replaced by 11 unrelated residues in the SM-FeSV v-*fms* product, resulting in the loss of a single tyrosine residue (tyr^{969}) four amino acids from the normal receptor C-terminus. A frameshift in this region of the v-*fms* gene of HZ5-FeSV results in a structurally different C-terminal modification that also deletes tyr^{969} [4]. Neither the full length c-*fms* gene nor C-terminally truncated or mutated analogs lacking tyr^{969} are active in transformation [42], whereas the v-*fms* gene transforms many different types of cells in the absence of ligand [2,3,20,54,55]. Replacement of the C-terminal region of the v-*fms* gene product with sequences from the normal receptor attenuates, but does not eliminate, the gene's transforming activity [42,56]. When tyr^{969} in these chimeric constructs was mutated to a phenylalanine residue, full transforming efficiency was restored, suggesting that tyr^{969} negatively regulates the receptor tyrosine kinase activity, possibly by acting as a phosphorylation site [42]. However, the mechanism of negative regulation remains unclear since CSF-1R does not contain detectable phosphotyrosine until it is stimulated by ligand [22]. In a cotransfection assay in which human c-*fms* was introduced into NIH/3T3 cells together with the CSF-1 gene, a mutant c-*fms* allele containing phe^{969} proved more efficient than the wild-type gene in inducing autocrine transformation [42,57]. Thus, C-terminal truncation of the v-*fms* gene product accentuates its transforming activity but is not sufficient in itself to activate the gene's transforming potential. Other mutation(s) elsewhere in c-*fms* must therefore be necessary to render it transforming and ligand-independent in its activity.

We recently found that a chimeric gene specifying an aminoterminal portion of the extracellular domain of the human c-*fms* gene joined to the remainder of the feline v-*fms* oncogene encodes a functional CSF-1 receptor [69]. When expressed in NIH/3T3 cells, the chimeric gene product was efficiently transported to the cell surface, was phosphorylated on tyrosine only in response to human recombinant CSF-1, underwent ligand-induced downmodulation, and stimulated colony formation of the cells in semi-solid medium containing CSF-1. Since the chimeric gene product retained the complete v-*fms*-coded kinase domain, including the C-terminal truncation that eliminated tyr^{969} from the c-*fms* product, these data further emphasize the fact that the C-terminal alteration is not sufficient to activate the gene's oncogenic potential. Nor do other genetic alterations in the kinase domain of v-*fms* inherently abrogate its ability to undergo ligand-induced downmodulation. The putative "activating" mutations in v-*fms* might well be localized within the extracellular domain, inducing a receptor conformation that is enzymatically active. However, we cannot exclude the possibility that replacement of extracellular v-*fms* sequences with a functional human CSF-1 ligand-binding domain suppresses an activating mutation elsewhere in the v-*fms*-coded molecule. Clearly, phosphorylation of the receptor on tyrosine per se does not provide the signal for receptor degradation, since the v-*fms* gene product contains phosphotyrosine even in the absence of ligand. Moreover, the v-*fms* gene could contain several independent mutations, one of which affects kinase activity and another of which affects processing of $gp120^{v\text{-}fms}$ to $gp140^{v\text{-}fms}$ and down-modulation of the mature form of the glycoprotein. Further studies will be necessary to pinpoint more precisely putative mutation(s) that activate the latent transforming potential of c-*fms*.

TRANSFORMING POTENTIAL OF THE v-*fms* GENE

SM- and HZ5-FeSV were isolated from multicentric fibrosarcomas of domestic cats [4,58], and cell-free filtrates of the SM-FeSV-containing tumor cells induced fibroblast transformation [59] and fibrosarcomas when inoculated into kittens [58]. Although SM-FeSV transforms cultured fibroblast cell lines from cats, rats, and mice in vitro, it was not originally demonstrated to induce hematopoietic malignancies in vivo. Introduction of helper-free SM-FeSV into a CSF-1-dependent, SV40-immortalized murine macrophage cell line abrogated their factor dependence and rendered the cells tumorigenic in nude mice [20]. Animals inoculated with these transformed macrophages developed high-grade histiocytic sarcomas metastatic to liver, lung, brain, pancreas, and bone, and cell lines reestablished from the tumors continued to express v-*fms*-coded kinase activity in culture. Because these cells did not express CSF-1 mRNA, their transformation was induced by a nonautocrine mechanism. Conceptually analogous results were obtained after introducing a murine retroviral vector containing the v-*fms* gene into interleukin-3 (IL-3)-dependent mouse FDC-P1 cells [54]. Again, cells expressing high levels of the v-*fms*-coded kinase became factor independent for growth in culture and tumorigenic in animals. In both cases, expression of the v-*fms* gene did not transmodulate the synthesis, expression, or affinity of normal CSF-1 or IL-3 receptors coexpressed in the target cells. This suggests that the v-*fms*-coded kinase provides persistent signals for cell proliferation that short circuit the proximal signal transduction pathways engaged by the normal receptors. In addition, the fact that v-*fms* can transform both fibroblasts and immature myeloid cell lines indicated that its activity was not restricted to normal CSF-1 receptor-bearing cells.

To determine if v-*fms* would induce hematopoietic malignancies in an in vivo setting, normal murine bone marrow cells infected with helper-free SM-FeSV were used to reconstitute lethally irradiated mice [55]. Southern blotting analysis demonstrated integrated SM-FeSV proviruses in spleen cells from 50% of the engrafted recipients one month after transplantation, and 30–50% of the splenocytes from individual recipients were found to be derived by clonal expansion of single SM-FeSV-infected progenitor cells. Reinoculation of provirus-positive spleen cells into secondary lethally irradiated mice induced clonal erythroleukemias and B cell lymphomas in some of the animals. By contrast, other secondary recipients receiving similar grafts remained disease-free and were repopulated by uninfected stem cells. The provirus-positive cells present in the spleens of primary recipients therefore did not have an obligate proliferative advantage when transferred to naive animals, suggesting that the formation of clonal malignancies in a proportion of the secondary recipients must have involved further genetic events in vivo. As expected, once clonal malignancies were established, the tumor cells could be transplanted at 100% efficiency into tertiary recipients. Thus, expression of the v-*fms* gene in the spleen cells of primary recipients initiated their expansion as premalignant populations that ultimately gave rise to tumor cells through a multistep mechanism. The fact that v-*fms* induced tumors of multiple hematopoietic lineages involving cells that do not express CSF-1 receptors again suggests that the v-*fms*-coded kinase has lost the target cell specificity of the normal receptor and acts more promiscuously.

AUTOCRINE TRANSFORMATION BY CSF-1 AND c-*fms*

Cointroduction of the human c-*fms* and CSF-1 genes into NIH/3T3 fibroblasts transforms these cells by an autocrine mechanism and renders them tumorigenic in nude mice [42, and see above]. In recent experiments, we used retroviral vectors to introduce the human CSF-1 gene into a murine SV40-immortalized macrophage cell line that depends on CSF-1 for its proliferation and survival in vitro. Although murine CSF-1 does not stimulate human CSF-1 receptors, human CSF-1 is equally active on both murine and human cells. Whereas the infected cells were able to proliferate in the absence of exogenous CSF-1, they were not tumorigenic when inoculated into nude mice [24]. Control experiments confirmed that murine CSF-1 receptors were synthesized at high levels in these cells and underwent persistent down-modulation in response to the endogenously produced growth factor. Since fibroblasts could be fully transformed by an autocrine mechanism [17,42], and because the SV40-immortalized macrophage cell line was not resistant to transformation by v-*fms* [20, and see above], it remains unclear why the latter cells were not tumorigenic after introduction of the CSF-1 gene. Indeed, others have demonstrated that a mouse macrophage tumor induced by a c-*myc*-containing retrovirus expressed CSF-1 after a DNA rearrangement at the CSF-1 locus. Because the proliferation of these tumor cells was blocked by specific antiserum to the growth factor, CSF-1 gene rearrangement appeared to be a secondary event in tumor cell development [60].

Human monocytes can synthesize CSF-1 after treatment with other cytokines, including γ-interferon and granulocyte-macrophage (GM)-CSF [61,62]. Thus, CSF-1 may normally function as an autocrine or paracrine growth factor to potentiate the activity of monocytes during an inflammatory response. The number of CSF-1 receptors per cell increases significantly as cells of the mononuclear phagocyte lineage differentiate, and mature monocytes and macrophages express the highest numbers of receptors per cell [19]. Whereas the mature cells require CSF-1 primarily as a survival and priming factor, earlier progenitors exhibit a proliferative response. If the pleiotropic effects of CSF-1 are mediated through its interaction with a single class of high-affinity receptors, it may prove that different physiologic substrates for the CSF-1R kinase are expressed as precursors of mononuclear phagocytes differentiate. Monocytes and macrophages might be relatively refractory to the proliferative effects of CSF-1 and, therefore, might not be readily converted to a tumorigenic phenotype in response to autocrine signals. Alternatively, these cells might synthesize negative regulatory cytokines, such as transforming growth factor β [63], which limit their tumorigenic potential in vivo.

POSSIBLE ROLES OF *fms* GENES IN HUMAN DISEASE

At least three different mechanisms involving *fms* expression could contribute to tumor formation: First, in cells in which c-*fms* is normally expressed, genetic alterations in the proto-oncogene analogous to those observed in v-*fms* might lead to persistent ligand-independent signals for cell proliferation. Second, unscheduled expression of the c-*fms* gene could render cells inappropriately responsive to CSF-1, leading to their clonal expansion, and increasing their probability of undergoing further genetic changes that contribute to malignancy. Third, inappropriate expression of CSF-1 in cells of the mononuclear phagocyte lineage might stimulate their growth through an autocrine mechanism.

In mice, the Friend murine leukemia virus (F-MuLV) can produce clonal myelogenous leukemias after insertional mutagenesis at the c-*fms* locus [64]. Integration of F-MuLV upstream of the CSF-1R coding region (and probably downstream of the normal c-*fms* promoter) induces the transcription of c-*fms* mRNA in immature myeloid cells, leading to the synthesis of unaltered receptor molecules. It has been postulated that these cells are rendered CSF-1 responsive and may be inappropriately expanded in the bone marrow microenvironment, ultimately becoming leukemic as part of a multistep process [65,66]. Although immature myeloid cells containing clonally integrated F-MuLV proviruses at the c-*fms* locus are initially factor-dependent and nontransplantable, factor-independent leukemic cell lines can eventually be established that cause myelogenous leukemias when passaged in vivo. At least some of the leukemic cell lines have been found to synthesize CSF-1 [65,66], whereas others have lost the unrearranged c-*fms* allele [64]. This suggests that secondary events occurring subsequent to F-MuLV insertion might involve either the CSF-1 or c-*fms* genes themselves. In certain circumstances, rearrangement at the CSF-1 locus in mononuclear phagocytes can serve as a second event that contributes to a tumorigenic, factor-independent phenotype [60, and see above].

We have recently used antibodies to human CSF-lR to screen cases of acute myelogenous leukemia (AML) for c-*fms* expression [18]. Approximately 40% of human AMLs express apparently normal CSF-1 receptors that undergo downmodulation in response to CSF-1 or phorbol esters. Although these receptors should be expressed as a differentiation-specific phenotypic marker of malignant cells of the mononuclear phagocyte series, there was no strict correlation between receptor expression and other morphological or immunophenotypic parameters characteristic of monocytic differentiation.

We would therefore predict that the expression of c-*fms* in those AMLs that lack monocytic characteristics might either be due to rearrangements upstream of CSF-1R coding sequences or, alternatively, to the aberrant expression of *trans*-acting factors that regulate tissue-specific expression of the c-*fms* promoter. Although the human c-*fms* gene has been molecularly cloned [67,68], and its nucleotide sequence has been recently determined (A. Hampe, personal communication), the promoter has not been identified and may map at a considerable distance from sequences encoding CSF-1R. The detailed molecular characterization of this region will be necessary in order to elucidate the precise mechanisms by which CSF-1 receptor expression is regulated during normal myelopoiesis and in myeloid malignancies.

ACKNOWLEDGMENTS

We thank Drs. Esther F. Wheeler, Jean Michel Heard, James R. Downing, Richard Ashmun, and A. Thomas Look for contributing to these experiments. This work was supported by U.S. Public Health Service grants CA38187 and CA20180 from the National Cancer Institute, NIH, and by the American Lebanese Syrian Associated Charities of St. Jude Children's Research Hospital.

REFERENCES

1. Bishop JM: Cell 42:23–28, 1985.
2. Frankel AE, Neubauer RL, Fishinger PJ: J Virol 18:481–490, 1976.
3. Donner L, Fedele LA, Garon CF, Anderson SJ, Sherr CJ: J Virol 41:489–500, 1982.
4. Besmer P, Lader E, George PC, Bergold PJ, Qiu F-H, Zuckerman EE, Hardy WD: J Virol 60:194–203, 1986.
5. Sherr CJ, Rettenmier CW, Sacca R, Roussel MF, Look AT, Stanley ER: Cell 41:665–676, 1985.
6. Guilbert LJ, Stanley ER: J Cell Biol 85:153–159, 1980.
7. Byrne PV, Guilbert LJ, Stanley ER: J Cell Biol 91:848–853, 1981.
8. Müller R, Slamon DJ, Adamson ED, Tremblay JM, Müller D, Cline MJ, Verma IM: Mol Cell Biol 3:1062–1069, 1983.
9. Müller R, Tremblay JM, Adamson ED, Verma IM: Nature 304:454–456, 1983.
10. Rettenmier CW, Sacca R, Furman WL, Roussel MF, Holt JT, Nienhuis AW, Stanley ER, Sherr CJ: J Clin Invest 77:1740–1746, 1986.
11. Ushiro H, Cohen S: J Biol Chem 255:8363–8365, 1980.
12. Ek B, Westermark B, Wasteson A, Heldin C-H: Nature 295:419, 420, 1982.
13. Nishimura J, Huang JS, Deuel TF: Proc Natl Acad Sci USA 79:4303–4307, 1982.
14. Kasuga M, Zick Y, Blithe DL, Crettaz, M, Kahn, CR: Nature 298:667–669, 1982.
15. Jacobs S, Kull FC, Jr, Earp HS, Svoboda ME, van Wyk JJ, Cuatracasas P: Biol Chem 258:9581–9584, 1983.
16. Rettenmier CW, Chen JH, Roussel MF, Sherr CJ: Science 228:320–322, 1985.
17. Rettenmier CW, Roussel MF, Ashmun RA, Ralph P, Price K, Sherr CJ: Mol Cell Biol 7:2378–2387, 1987.
18. Ashmun RA, Look AT, Furman WL, Rettenmier CW, Seremetis S, Sherr CJ: Blood 70 (Suppl 1): 274a, 1987.
19. Guilbert LJ, Stanley ER: J Biol Chem 261:4024–4032, 1986.
20. Wheeler EF, Rettenmier CW, Look AT, Sherr CJ: Nature 324:377–380, 1986.
21. Yeung YG, Jubinsky PT, Sengupta A, Yeung DCY, Stanley ER: Proc Natl Acad Sci USA 84:1268–1271, 1987.
22. Downing JR, Rettenmier CW, Sherr CJ: Mol Cell Biol 8:1795–1799, 1988.
23. Sacca R, Stanley ER, Sherr CJ, Rettenmier CW: Proc Natl Acad Sci USA 83:3331–3335, 1986.
24. Roussel MF, Rettenmier CW, Sherr CJ: Blood 71:1218–1225, 1988.
25. Schechter Y, Hernaez L, Schlessinger J, Cuatrecasas P: Nature 278:835–838, 1979.

26. Schreiber AB, Libermann TA, Lax I, Yarden Y, Schlessinger J: J Biol Chem 258:846–853, 1983.
27. Fanger OB, Austin KS, Earp HS, Cidlowski JA: Biochemistry 25:6414–6420, 1986.
28. Schlessinger J: J Cell Biol 103:2067–2072, 1986.
29. Yarden Y, Schlessinger J: Biochemistry 26:1434–1442, 1987.
30. Boni-Schnetzler M, Pilch PF: Proc Natl Acad Sci USA 84:7832–7836, 1987.
31. Chen BD, Lin HS, Hsu S: J Cell Physiol 116:207–212, 1983.
32. Tushinski RJ, Oliver IT, Guilbert LJ, Tynan PW, Warner JR, Stanley ER: Cell 28:71–81, 1982.
33. Bravo R, Neuberg M, Burckhardt J, Almendral J, Wallich R, Müller R: Cell 48:251–260, 1987.
34. Orlofsky A, Stanley ER: EMBO J 6:2947–2952, 1987.
35. Tushinski RJ, Stanley ER: J Cell Physiol 116:67–75, 1983.
36. Tushinski RJ, Stanley ER: J Cell Physiol 122:221–228, 1985.
37. Warren MK, Ralph P: J Immunol 137:2281–2285, 1986.
38. Ralph P, Warren MK, Ladner MD, Kawasaki ES, Boosman A, White TJ: Cold Spring Harbor Symp Quant Biol 51:679–683, 1986.
39. Lyberg T, Stanley ER, Prydz H: J Cell Physiol 132:367–370 1987.
40. Lee MT, Warren MK: J Immunol 138:3019–3022, 1987.
41. Karbassi A, Becker JM, Foster JS, Moore RN: J Immunol 139:417–421, 1987.
42. Roussel MF, Dull TJ, Rettenmier CW, Ralph P, Ullrich A, Sherr CJ: Nature 325:549–552, 1987.
43. Rothwell VM, Rohrschneider LR: Oncogene Res 1:311–324, 1987.
44. Hampe A, Gobet M, Sherr CJ, Galibert F: Proc Natl Acad Sci USA 81:85–89, 1984.
45. Coussens L, Van Beveren C, Smith D, Chen E, Mitchell RL, Isacke CM, Verma IM, Ullrich A: Nature 320:277–280, 1986.
46. Wheeler EF, Roussel MF, Hampe A, Walker MH, Fried VA, Look AT, Rettenmier CW, Sherr CJ: J Virol 59:224–233, 1986.
47. Anderson SJ, Gonda MA, Rettenmier CW, Sherr CJ: J Virol 51:730–741, 1984.
48. Rettenmier CW, Roussel MF, Quinn CO, Kitchingman GR, Look AT, Sherr CJ: Cell 40:971–981, 1985.
49. Roussel MF, Rettenmier CW, Look AT, Sherr CJ: Mol Cell Biol 4:1999–2009, 1984.
50. Nichols EJ, Manger R, Hakomori S, Herscovics A, Rohrschneider LR: Mol Cell Biol 5:3467–3475, 1985.
51. Hadwiger A, Niemann H, Käbisch A, Bauer H, Tamura T: EMBO J 5:689–694, 1986.
52. Lyman SD, Rohrschneider LR: Mol Cell Biol 7:3287–3296, 1987.
53. Tamura T, Simon E, Niemann H, Snoek GT, Bauer H: Mol Cell Biol 6, 4745–4748, 1986.
54. Wheeler EF, Askew D, May S, Ihle JN, Sherr CJ: Mol Cell Biol 7:1673–1680, 1987.
55. Heard JM, Roussel MF, Rettenmier CW, Sherr CJ: Cell 51:663–673, 1987.
56. Browning PJ, Bunn HF, Cline A, Shuman M, Nienhuis AW: Proc Natl Acad Sci USA 83, 7800–7804, 1986.
57. Heard JM, Roussel MF, Rettenmier CW, Sherr CJ: Oncogene Res 1:423–440 1987.
58. McDonough SK, Larsen S, Brodey RS, Stock ND, Hardy Jr, WD: Cancer Res 31:953–956, 1971.
59. Sarma PS, Sharar AL, McDonough S: Proc Soc Exp Biol Med 140:1365–1368, 1972.
60. Baumbach WR, Stanley ER, Cole MD: Mol Cell Biol 7:664–671, 1987.
61. Rambaldi A, Young DC, Griffin JD: Blood 69:1409–1413, 1987.
62. Horiguchi J, Warren MK, Ralph P, Kufe D: Biochem Biophys Res Commun 141:924–930, 1986.
63. Assoian RK, Fleurdelys BE, Stevenson HC, Miller PJ, Madtes DK, Raines EW, Ross R, Sporn MB: Proc Natl Acad Sci USA 84:6020–6024, 1987.
64. Gisselbrecht S, Fichelson S, Sola B, Bordereaux D, Hampe A, André C, Galibert F, Tambourin P: Nature 329:259–261, 1987.
65. Heard JM, Fichelson S, Sola B, Martial MA, Varet B, Levy JP: Mol Cell Biol 4:216–220, 1984.
66. Heard JM, Sola B, Martial MA, Fichelson S, Gisselbrecht S: Blood 68:193–199, 1986.
67. Roussel MF, Sherr CJ, Barker PE, Ruddle FH: J Virol 48:770–773, 1983.
68. Heisterkamp N, Groffen J, Stephenson JR: Virology 126:248–258, 1983.
69. Roussel MF, Downing JR, Ashmun RA, Rettenmier CW, Sherr CJ: Proc Natl Acad Sci USA (in press).

Journal of Cellular Biochemistry 38:189–198 (1988)
Growth Factors and Their Receptors: Genetic
Control and Rational Application 39–48

Cyclin mRNA and Protein Expression in Recombinant Interleukin 2-Stimulated Cloned Murine T Lymphocytes

P.M. Shipman, D.E. Sabath, A.H. Fischer, P.G. Comber, K. Sullivan, E.M. Tan, and M.B. Prystowsky

Departments of Pathology and Laboratory Medicine School of Medicine, University of Pennsylvania, Philadelphia, Pennsylvania 19104-6082 (P.M.S., D.E.S., A.H.F., P.G.C., M.B.P.); The W.M. Keck Foundation, Autoimmune Disease Center, Department of Basic and Clinical Research, Scripps Clinic and Research Foundation, La Jolla, California 92037 (K.S., E.M.T.)

Expression of cyclin, a non-histone nuclear protein, during recombinant interleukin 2 (rIL2)-driven cell-cycle progression of cloned T lymphocytes has been assessed. We found that expression of cyclin protein, as detected by immunofluorescence, is tightly associated with proliferation, and not merely S-phase, of L2 cells stimulated with rIL2. Cyclin immunofluorescence was detected in all cell-cycle phases ($G_1/S/G_2/M$, as detected by flow cytometry) of proliferating L2 cells. Accumulation of cyclin mRNA levels was induced as early as 1 h after stimulation, was maximal at 25–49 h, and remained elevated throughout stimulation, as detected by Northern blot analysis. A cDNA- encoding murine cyclin was cloned from a cDNA library prepared from IL2-stimulated cloned T cells. The sequence of the 5′ end of the murine cyclin cDNA was determined and found to be 88% and 82% similar to the sequences of cDNA clones encoding rat and human cyclin, respectively. The present studies demonstrate that cyclin protein and mRNA accumulation are highly regulated during IL2-induced proliferation of a cloned T cell. These data provide a framework for addressing the molecular mechanisms regulating cyclin gene expression during cellular proliferation.

Key words: cyclin, proliferating cell nuclear antigen, cloned T lymphocytes, interleukin 2

Cellular proliferation is the consequence of a complex array of biochemical and molecular events generally initiated in response to membrane-associated signals. The nature of many of these intracellular events has been described. For example, gene transcription [1] and translation [2] are required for cell-cycle progression. Further, a variety of "early" events have been associated with cellular proliferation, including changes in intracellular ion concentrations [3], phospholipid metabolism [4], and protein phosphorylation [5]. However, the mechanisms by which transcriptional and

Received February 1, 1988; accepted May 19, 1988.

postranscriptional events are regulated and related to these "early" events remain enigmatic. A general approach to the study of regulation of cellular proliferation has been the examination of genes whose expression is required for entry into the cell cycle. A more specific approach is focused on analyzing molecules which may control DNA synthesis [6]. The nuclear protein cyclin (proliferating cell nuclear antigen, PCNA) has been studied using both approaches. Initial studies showed that cyclin expression is highly regulated in proliferating cells [7,8]. Recent evidence indicates that cyclin functions as part of a multi-enzyme complex responsible for the replication of DNA [9,10].

T lymphocytes are induced to proliferate in a two-signal manner. The first signal for T-cell proliferation is the interaction of antigen (in the context of self major histocompatibility molecules) with its receptor which results in lymphokine production (e.g., interleukin 2 [IL2]) and IL2 receptor expression. The second signal is provided by the interaction of IL2 and the IL2 receptor which drives G_1 progression. Unstimulated cloned T lymphocytes exist in a G_1 resting stage and can be induced to progress through cell division upon stimulation with IL2 [11], thus providing a model system for the evaluation of biochemical and molecular regulation of T-lymphocyte proliferation. Our laboratory has previously demonstrated that biosynthesis of cyclin, as assessed by biosynthetic labeling and SDS-PAGE, increases 1,000-fold during the mid-G_1 phase of the cell cycle in an IL2-stimulated cloned T lymphocyte, L2 [12]. Immunofluorescence studies have demonstrated that cyclin accumulates in the nucleus of L2 cells during IL2-induced proliferation [13]. Thus, the present investigations were designed to characterize further modifications in expression of cyclin protein as well as mRNA as a function of progression through the cell cycle.

MATERIALS AND METHODS

Cell Line

Derivation and maintenance of the cloned murine T-helper cell line, L2, has been previously described [14]. Cells were used 6–8 days following exposure to antigen and IL2.

Human IL2

Purified human recombinant IL2 (rIL2), a product of recombinant DNA technology in *Escherichia coli* [15], was a gift of Cetus (Emeryville, CA). This substance is 98% pure as assessed by SDS-PAGE analysis and contains .01 ng endotoxin per 3.6×10^6 units of IL2 according to the Limulus assay. Specific activity was assessed by the manufacturer, using a standard bioassay for IL2.

Immunofluorescence

Cyclin antiserum obtained from a patient with systemic lupus erythematosus was used as the primary antibody for indirect immunofluorescence, according to the method of Clevenger et al. [16]. L2 cells ($1–3 \times 10^6$) were washed in PBS then fixed for 10 min with 10 ml of 1.0% paraformaldehyde (Polysciences, Warrington, PA) in PBS. The cells were centrifuged (200*g*, 10 min), resuspended in 0.1% Triton X-100 (Sigma, St. Louis, MO) in PBS for 3 min, and recentrifuged. Cells were incubated in a 1:400 dilution of cyclin antiserum for 1 h, washed, and incubated with a 1:20 dilution of fluorescein-conjugated rabbit antimouse immunoglobulin (Antibodies Inc.,

Davis, CA) for 30 min. After washing, the cells were incubated in 2 ml of 150 U RNAse A/ml PBS (Worthington, Freehold, NJ) for 20 min at 37°C, centrifuged, and resuspended in 50 ug/ml propidium iodide (Calbiochem-Behring, San Diego, CA) in PBS at least 1 h prior to flow cytometry. Unless otherwise specified, all procedures took place at 4°C.

Flow Cytometry

Fluorescence-activated cell sorter (FACS) analysis of the cell cycle was performed as described [16].

cDNA Library Construction

Total cellular RNA was isolated by lysis of L2 cells in guanidinium isothiocyanate and density gradient ultracentrifugation over CsCl [17]. Polyadenylated RNA was selected by two passages over an oligo (dT) column and the cDNA library was constructed according to described methods [17].

Screening

Fifty thousand phage from the murine cDNA library were plated on agar plates. DNA was transferred to Colony/Plaque Screen hybridization membranes (NEN, Dupont, Boston, MA) and probed with a ^{32}P-labeled rat cDNA encoding cyclin [kindly provided by P. Nakane, Tokai University School of Medicine, Japan; 18], as described previously [19]. cDNA probes were labeled using the random primer extension method [20] with deoxycytidine 5′-[alpha-^{32}P]triphosphate (3,000 Ci/mmol; Amersham, Arlington Heights, IL). Hybridization with the rat cDNA probe in a secondary screening was confirmed by Southern blot analysis.

Subcloning and Sequencing

Inserts from cDNA clones hybridizing with the rat cyclin cDNA were isolated by agarose gel electrophoresis and electroelution and were subcloned into M13mp18 [17]. Clones were partially sequenced by the dideoxy chain termination method using Sequenase, a modified form of T7 DNA polymerase [21; United States Biochemical Corp., Cleveland, OH]. Sequence analysis and comparison with the European Molecular Biology Laboratories database was carried out using the Intelligenetics Software Program (Mountain View, CA) on Bionet^{tm}.

Northern Blot Analysis

RNA was prepared from L2 cells and Northern blot analysis was performed as previously described [22]. Probes were prepared as for screening and included cyclin cDNA and a murine cDNA encoding the beta chain of the T-cell receptor (TcR) (pUC25, a gift of Y. Hashimoto, University of Pennsylvania). Radioactivity on autoradiographs was quantified by scanning laser densitometry, using an LKB Ultroscan XL densitometer.

RESULTS

Previous immunofluorescence and biosynthetic analyses of cyclin during cellular proliferation have demonstrated cyclin expression varies in the cell population as a function of time following rIL2 stimulation [12,13]. Progression through the cell cycle in these studies was based on a separate previous examination of DNA content in the total cell population with time following stimulation. However, a precise description of cyclin expression during the cell cycle requires simultaneous analyses of cyclin immunofluorescence and DNA content in individual cells. Figure 1 illustrates the results of simultaneous analyses of cyclin immunofluorescence and DNA content in IL2-stimulated cloned T cells during late G_1, G_1 to S phase transition, and active proliferation. L2 cells were maintained in culture by weekly stimulation with alloantigen and rIL2. The DNA content in 88% of the unstimulated cells, at the end of such a weekly cycle, was consistent with that of G_1 cells containing a normal diploid amount of DNA (Fig. 1). Twelve percent of quiescent L2 cells have apparent tetrapoloid DNA content, which is probably due to doublets for which no correction was made in these studies. Cyclin fluorescence was negligible in quiescent L2 cells. Previous experiments from this laboratory demonstrated synthesis of p36 protein [12], later shown to be cyclin [13], during the initial G_1 stage in IL2-stimulated L2 cells [13]. We therefore chose to begin our examination of cyclin expression at a period encompassing late G_1. Cyclin immunofluoresence was strikingly different at 21 h,

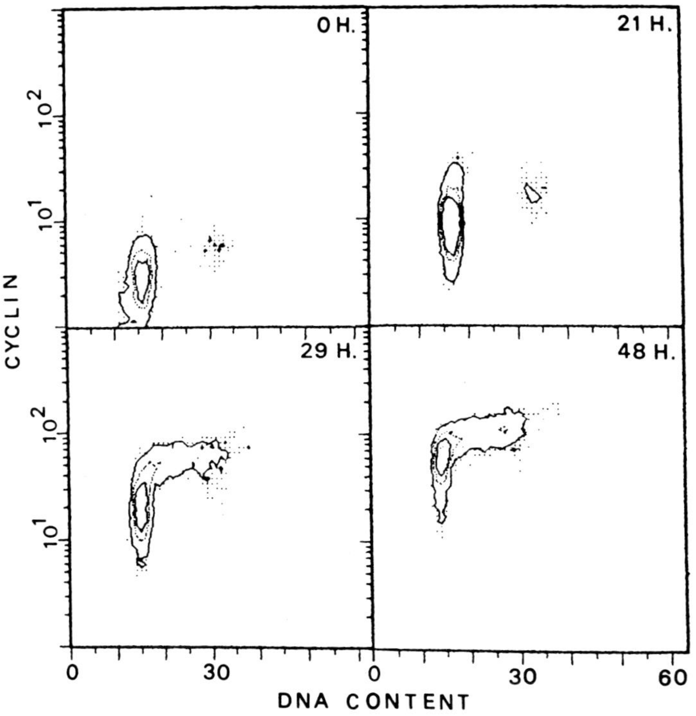

Fig. 1. Cell cycle analysis and cyclin immunofluorescence of rIL2-stimulated L2 cells. L2 cells were stimulated with 100 U/ml human recombinant IL2. At the indicated times, cells were harvested from culture, washed, and stained with human cyclin antiserum; this was followed by fluorescein-conjugated rabbit antihuman globulin and propidium iodide, as described [16]. The intensity of cyclin-associated fluorescence was plotted on a logarithmic scale on the ordinate. Propidium iodide fluorescence (DNA content) was plotted on a linear scale in arbitrary units on the abscissa. At 0 h, the cyclin fluorescence was identical to that obtained following control staining of cells with the rabbit antihuman globulin alone and thus represents background fluorescence. Each panel represents the analysis of 10,000 cells, and the contour lines represent the number of cells at each level (5, 10, 40, and 70).

compared to unstimulated cells. Approximately 66% of the cells containing a G_1 DNA content expressed cyclin and all of the cells at the G_1 to S phase transition or in S phase expressed cyclin. The mean increase in cyclin expression at 21 h compared to unstimulated cells was approximately fourfold. Eight hours later, at 29 h, 30% of the cells were in $S/G_2/M$ phases, and, again all of these cells expressed cyclin (Fig. 1). Further, the intensity of staining for cyclin was increased approximately fourfold over that seen at 21 h. At 48 h, the population was cycling asynchronously and cells could be identified in all phases of the cell cycle (Fig. 1). All cells at 48 h expressed cyclin, though the intensity of staining was greatest in those cells with the largest DNA content (Fig. 1), with a mean fluorescence intensity approximately 20-fold greater than that of unstimulated cells.

Human and rat cDNA encoding cyclin have recently been identified and their full sequences determined [18,25]. A murine cyclin cDNA would provide a tool for examining cyclin expression at the levels of mRNA accumulation, stability, and ultimately, transcription, as well as identification of cyclin clones from murine genomic libraries. In order to isolate a cDNA-encoding murine cyclin, we screened a cDNA library from IL2-stimulated L2 cells with the rat cDNA as a probe. Six of these clones (.01% of plated clones) hybridized with the rat cDNA in secondary screenings and Southern blot analysis. One of these clones, H5.3, was subcloned into M13mp18, and the 5′ end sequence was determined. This partial sequence, and its comparison to the relevant sequences of human and rat cyclin cDNA, is shown in Figure 2. The partial murine sequence was compared with all nucleotide sequences in the European Molecular Biology Laboratories (EMBL) and National Institutes of Health (NIH) databases and showed similarity only to the human and rat cyclin sequences. The 5′ partial murine sequence was slightly more similar to the rat than the human sequence (88% compared to 82% nucleotides matched, respectively). This clone was 1.1 kb in length, approximately the length of the rat and human clones (1.145 kb and 1.1 kb, respectively), and may therefore represent a full-length clone. Further sequence analysis will be required to determine the extent to which the 5′ end of the murine sequence is represented in this clone.

The murine cDNA clone was used as a probe for the determination of steady-state cyclin mRNA levels during IL2-driven L2 cell proliferation. Total RNA was isolated from L2 cells at various times following IL2 stimulation and subjected to Northern blot analysis. In unstimulated cells, cyclin mRNA was barely detectable (Fig. 3A,B). Cyclin message was increased as early as 1 h following stimulation and

```
                                                                          64
HUMAN    ATGTTCGAGGCGCGCCTGGTCCAGGGCTCCATCCTCAAGAAGGTGTTGGAGGCACTCAAGGACC
                  :    :::: :::::::::::::::::::::::::: ::::::: ::::: ::::
MOUSE        AGCTCGAATCCCTGATCCAGGGCTCCATCCTGAAGAAGGTGCTGGAGGCTCTCAAAGACC
                  : :  ::::::::::::::::::::::::::::::::::::::: ::::::::::
RAT      ATGTTTGAGGCACGCCTGATCCAGGGCTCCATCCTGAAGAAGGTGCTGGAGGCCCTCAAAGACC

                                                                         126
HUMAN      TCATCAACGAGGCCTGCTGGGATATTAGCTCCAGCGGTGTAAACCTGCAGAGCATGGACTCG
           ::::::: ::::::::::::::: :::::::: :::: :: :::::::::::::::::::::
MOUSE      TCATCAATGAGGCCTGCTGGGACGTCAGCTCGGGCGGCGTGAACCTGCAGAGCATGGACTCG
           ::::::::::::::::::::::: :::::::::::::::::::::::::::::::::: :::
RAT        TCATCAATGAGGCCTGCTGGGACATCAGTTCGGGCGGCGTGAACCTACAGAGCATGGATTCG
```

Fig. 2. Partial nucleotide sequence of the coding strand of the murine cDNA (H5.3) encoding cyclin and comparison with the human and rat cyclin sequences [18,25]. Nucleotides are numbered in the 5′ to 3′ direction and begin with the first residue of the initiation codon ATG (human and rat). Spaces represent areas where the nucleotides are different between species.

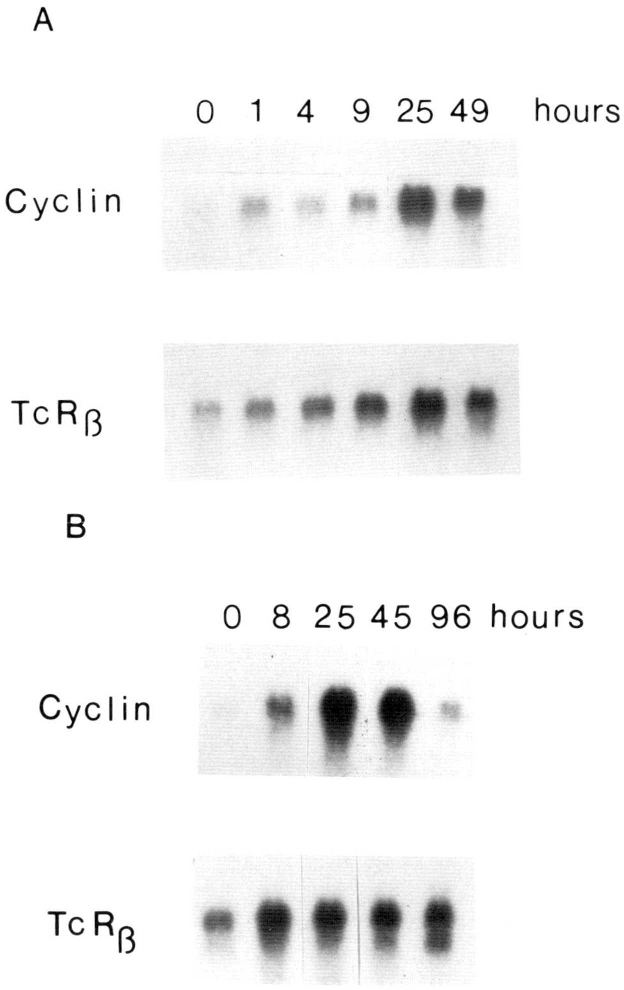

Fig. 3. Kinetics of accumulation of cyclin or the beta chain of the T cell receptor mRNA in L2 cells stimulated with 100 U/ml rIL2. Total RNA was isolated at 0, 1, 4, 9, 25, and 49 h **(A)** or 0, 8, 25, 45, and 96 h **(B)** after stimulation and subjected to Northern blot analysis as described in Materials and Methods.

increased with time reaching a maximum (24-fold stimulation compared to unstimulated cells) at 25 h after IL2 stimulation (Figs. 3A, 4A). We next examined cyclin mRNA accumulation at times corresponding to four discrete periods in the cell cycle: mid-G_1 (8 h), G_1 to S phase transition (25 h), active proliferation (45 h), and return to quiescence (96 h). At 8 h, cyclin mRNA levels were elevated nearly ten-fold above those seen in unstimulated cells (Figs. 3B, 4). Maximal cyclin mRNA accumulation (at 25 and 45 h) was greater than 30-fold above unstimulated levels. At 96 h, when the majority of cells had returned to a postmitotic G_1 state [26], cyclin mRNA levels dropped to approximately fivefold above unstimulated cells. The difference in the time of maximal cyclin message accumulation in the two experiments probably represents a difference in the degree to which the L2 cells were stimulated.

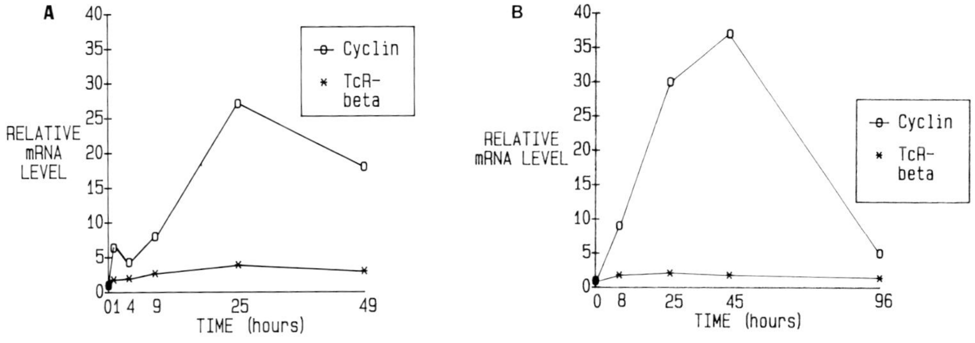

Fig. 4. Densitometric analysis of autoradiograms from Northern blots illustrated in Figure 3, probed with cDNA for cyclin or TcR-beta. Panels **A** and **B** are graphic representations of the data presented in Fig. 3, A and B, respectively. Data are presented as the relative increase in mRNA accumulation in stimulated compared to unstimulated cells.

To confirm that the amounts of RNA were relatively constant, the gels were stained with ethidium bromide and examined under ultraviolet illumination. The 28S and 18S ribosomal RNA bands were identified and appeared quantitatively similar on each lane (data not shown). To assess the relative change in cyclin RNA accumulation, levels of cyclin message were compared to the levels of message encoding the beta chain of the T-cell receptor (TcR-beta). The TcR-beta mRNA levels were only slightly elevated (2–3-fold) in stimulated compared to unstimulated cells (Figs. 3, 4), consistent with previous reports [27]. Thus, the increase in cyclin expression in Figure 3 represents a genuine response to rIL2 stimulation.

DISCUSSION

Cyclin, also called proliferating cell nuclear antigen, is a nuclear protein of 36,000-dalton molecular weight, first identified as an antigen recognized by autoantibodies produced in approximately 3% of patients with systemic lupus erythematosus [28]. These antibodies reacted preferentially with nuclei of proliferating normal and transformed cells [28]. Expression of cyclin, as identified by immunofluorescence, closely parallels ^{3}H-thymidine incorporation in proliferating cells [7]. In order to study the regulation of cyclin expression, and, conversely, how cyclin may possibly regulate DNA synthesis, the expression of cyclin must be examined at several levels. These levels may be considered to include the distribution of cyclin as identified by antibodies, cyclin protein synthesis and transcription of the cyclin gene.

Early studies suggested that cyclin expression was restricted to S phase [7,28]. However, cyclin can be detected by immunofluorescence at low levels in formaldehyde-fixed quiescent 3T3 cells [29]. Flow cytometric analysis of lectin-stimulated lymphocytes by Kurki et al. [23] suggested that cyclin was present later during G_1 activation but highest in S phase. Our previous studies indicated that cyclin biosynthesis was initiated as early as 4 h after stimulation with IL2 [12] but only suggested that cyclin was present in all phases of the cell cycle in actively proliferating L2 cells [13]. Thus, we chose a time course for analyzing cyclin immunofluorescence and DNA content to demonstrate that cyclin levels are increased after IL2 stimulation in essentially all proliferating cells during all phases of the cell cycle.

Figure 1 clearly demonstrates that cyclin protein expression is tightly associated with proliferation, and not merely S phase, of L2 cells stimulated with rIL2; that is, cyclin levels are high in all cell-cycle phases ($G_1/S/G_2/M$) of actively proliferating cells. Cyclin detected in these experiments may be considered to include newly synthesized protein as well as accumulated protein. The flow cytometric experiments are considerably more sensitive than experiments in which cyclin-associated immunofluorescence is evaluated by microscopy. Further, fixation in formaldehyde reveals cyclin immunofluorescence at times when cyclin is undetected in methanol-fixed cells (i.e., in unstimulated cells) [29]. These methodological differences provide an explanation for the contrasting results of Celis et al. [7] who detected extremely weak cyclin immunofluorescence in G_2 and M phases of the cell cycle. The idea that cyclin protein expression is proliferation specific rather than S phase specific is supported by the findings of Bravo and Bravo [29], who also detected cyclin during the transition of 3T3 cells from a growing to a quiescent state, by pulse-chase experiments and immunoblotting. Kurki et al. [23] also detected significant levels of cyclin immunofluorescence at G_2/M in mitogen-stimulated splenocytes, though these levels were reduced compared to S-phase; G_1- associated fluorescence was minimal in those studies, in contrast to our findings.

Northern blot analysis of cyclin mRNA levels (Fig. 3) demonstrates that cyclin expression is also highly regulated throughout the cell cycle in IL2-stimulated L2 cells at the level of mRNA accumulation. Matsumoto et al. [18] detected significant levels of cyclin mRNA at 48 h, corresponding to maximal DNA synthesis in PHA-stimulated human peripheral blood lymphocytes. However, in contrast to our results, they did not detect cyclin message in unstimulated cells or in cells that were incorporating thymidine at 72 h after stimulation. Similarly, Almendral et al. [25] did not detect cyclin message in unstimulated quiescent 3T3 cells or in 3T3 cells until 12 h following serum stimulation. In those studies, maximal accumulation was detected at 16–18 h and cyclin levels declined dramatically at 24 h after stimulation, though histone H3 levels remained high, providing evidence that the cells were in S phase [30]. Our data correlate closely with previous cyclin protein synthesis data in IL2-stimulated L2 cells [12] and the present observation on cyclin immunofluorescence, since significant cyclin protein synthesis, immunofluorescence and RNA accumulation were detected at all phases of the cell cycle. It is puzzling to us that the cyclin message could not be detected at the later time points analyzed by Matsumoto et al. [18] and Almendral et al. [25], since our studies clearly demonstrate that the message remains elevated after S phase in IL2-stimulated L2 cells. An examination of cyclin mRNA expression in several types of cells stimulated to proliferate with various ligands may provide insight into the contrast between these findings. The differences in kinetics of the onset of expression of cyclin mRNA in IL2-stimulated L2 cells compared to 3T3 cells and PHA-stimulated human peripheral blood lymphocytes may be related to the relatively greater degree of quiescence of the latter cell types compared to L2 cells; L2 cells are maintained at a G_1 stage prior to stimulation.

Cyclin protein synthesis [12] and cyclin mRNA accumulation could be detected during early G_1 (4 h following stimulation) and was reduced, though still detectable, during the return to quiescence (at 96 h). This observation provides evidence, for the first time, that cyclin protein expression, which also decreases during the return to G_1, is directly related to the mRNA levels. Thus, cyclin expression is likely to be regulated at the level of mRNA accumulation (i.e., message stability and/or transcriptional activation).

Maximal cyclin mRNA accumulation was observed at 24–48 h after stimulation. Previous experiments published by us demonstrate that in IL2-stimulated L2 cells, c-myc mRNA rises rapidly, reaching a maximum at 1–4 h after stimulation, and is almost undetectable at 28 h [22]. Thus, maximum cyclin mRNA accumulation follows c-myc expression and remains elevated for a longer period of time than c-myc in IL2-stimulated L2 cells. The difference in kinetics of induction of cyclin and c-myc mRNA suggests different mechanisms of regulation between genes activated early in G_1 compared to those activated at the G_1 to S phase transition. Such mechanisms may include differences in the stability of cyclin and c-myc mRNA. C-myc mRNA is relatively short-lived [$T_{1/2}$ of 10–60 min depending on the cell type; 31,32] and message stability is considered to play an important role in regulating c-myc mRNA levels since transcription of the c-myc gene may be only slightly induced in proliferating cells [33]. Cyclin mRNA stability and its contribution to regulation of cyclin expression remains to be explored.

The partial sequence of the murine cDNA encoding cyclin was similar to that of rat and human cDNA clones. The identification of a murine cDNA provides a tool for the identification of murine cyclin genomic clones for study of the organization of the cyclin gene and its regulation at the transcriptional level.

The present studies have demonstrated that cyclin protein and mRNA accumulation are highly regulated throughout the cell cycle in a cloned T-cell stimulated with IL2. Previous studies have identified cyclin as a proliferation-associated protein and have highlighted the changes in cyclin nuclear distribution and/or antigenicity. Identification and characterization of the regulation of cyclin expression at multiple levels will provide a framework within which to address questions concerning the molecular mechanisms regulating cyclin gene expression during cellular proliferation.

ACKNOWLEDGMENTS

These studies were supported by NIH grants AI21681 and GM36962 (M.B.P.). M.B.P. is a Hartford Foundation Fellow. P.M.S. was supported by NIH training grant CA09140-12. The authors acknowledge the support of the Bionet Resource, from the Biomedical Research Technology Program, Division of Research Resources, NIH.

REFERENCES

1. Baserga R: Proc Natl Acad Sci USA 54:1141–1148, 1965.
2. Brooks RF: Cell 12:311–317, 1977.
3. Smith JB, Rozengut E: Proc Natl Acad Sci USA 75:5560–5564, 1978.
4. Hirasawa K, Nishizuka Y: Annu Rev Pharmacol Toxicol 25:147–170, 1985.
5. Rozengurt E, Rodrigues-Pena M, Smith KA: Proc Natl Acad Sci USA 80:7244–7248, 1983.
6. Pardee AB: Cancer Res 47:1488–1491, 1987.
7. Celis JE, Madsen P, Nielson S, Celis A: Leuk Res 10:237–249, 1986.
8. Bravo R, MacDonald-Bravo R: EMBO J 4:655–661, 1985.

9. Bravo R, Frank R, Blundell PA, Macdonald-Bravo H: Nature 326:515–517, 1987.
10. Prelich G, Stillman B: Cell 53:117–126, 1988.
11. Prystowsky MB, Otten G, Pierce SK, Shay J, Olsham J, Fitch FW: Lymphokine Res 12:23–38, 1985.
12. Sabath DE, Monos DS, Lee SC, Deutsch C, Prystowsky MB: Proc Natl Acad Sci USA 83:4739–4743, 1986.
13. Moore KS, Sullivan K, Tan EM, Prystowsky MB: J Biol Chem 262:8447–8450, 1987.
14. Glasebrook AL, Fitch FW: J Exp Med 151:876–895, 1980.
15. Wang A, Lu SD, Mark DF: Science 224:1431–1433, 1984.
16. Clevenger G, Bauer KD, Epstein AL: Cytometry 6:208–214, 1985.
17. Davis LG, Dibner MD, Battey JF: "Basic Methods in Molecular Biology." New York: Elsevier, pp 129–275, 1986.
18. Matsumoto K, Moriuchi T, Kaji T, Nakane PK: EMBO J 6:637–642, 1987.
19. Benton W, Davis R: Science 196:180–182, 1977.
20. Feinberg AP, Vogelstein B: Anal Biochem 132:6–13, 1983.
21. Tabor S, Richardson C: Proc Natl Acad Sci USA 84:4767–4771, 1987.
22. Reed JC, Sabath DE, Hoover RG, Prystowsky MB: Mol Cell Biol 5:3361–3368, 1985.
23. Kurki P, Vanderlaan M, Dolbeare F, Gray J, Tan EM: Exp Cell Res 166:209–219, 1986.
24. Kurki P, Lotz M, Ogata K, Tan EM: J Immunol 138:4114–4120, 1987.
25. Almendral JM, Huebsch D, Blundell PA, MacDonald-Bravo H, Bravo K: Proc Natl Acad Sci USA 84:1575–1579, 1987.
26. Lee SC, Sabath DE, Deutsch C, Prystowsky MB: J Cell Biol 102:1200–1208, 1986.
27. Yoshikai Y, Yanagi Y, Suciu-Foca N, Mak TW: Nature 310:506–508, 1984.
28. Miyachi K, Futzler MJ, Tan EM: J Immunol 121:2228–2234, 1978.
29. Bravo R, MacDonald-Bravo H: J Cell Biol 105:1549–1554, 1987.
30. Plumb M, Stein J, Stein G: Nucleic Acids Res 11:2391–2401, 1983.
31. Dani Ch, Blanchard JM, Piechaczyk M, El Sabouty S, Marty L, Jeanteur P: Proc Natl Acad Sci USA 81:7046–7050, 1984.
32. Piechaczyk M, Yang J-Q, Blanchard J-M, Jeanteur P, Marcu KB: Cell 42:589–597, 1985.
33. Greenberg M, Ziff EB: Nature 311:433–438, 1984.

Journal of Cellular Biochemistry 38:229–236 (1988)
Growth Factors and Their Receptors: Genetic
Control and Rational Application 49–56

Evidence for Autocrine Activation of a Tyrosine Kinase in a Human Gastric Carcinoma Cell Line

Silvia Giordano, Maria Flavia Di Renzo, Radha P. Narsimhan, Luca Tamagnone, Elena V. Gerbaudo, Loredana Chiadó-Piat, and Paolo M. Comoglio

Department of Biomedical Sciences and Oncology, University of Torino Medical School, Torino, Italy

Phosphotyrosine (P-Tyr) antibodies have been used to identify the phosphorylated forms of growth factor receptors and oncogene-coded tyrosine kinases. Western blot analysis of a gastric carcinoma cell line with P-Tyr antibodies revealed a tyrosine-phosphorylated protein of M_r 145,000 (P145). In addition, in vitro phosphorylation with (γ-^{32}P)ATP of P-Tyr immunoprecipitates of the same cells resulted in labelling of this protein on tyrosine. P145 appears to be a transmembrane glycoprotein, with features suggestive of a growth factor receptor. However, the in vivo or in vitro addition of known growth factors did not affect P145 tyrosine phosphorylation. We now report that P145 is rapidly dephosphorylated in vivo when cells are exposed to low pH, a condition known to dissociate ligands from their receptors. The addition of serum-free medium, conditioned by the gastric carcinoma cells, fully restores the tyrosine phosphorylation lost with acid treatment. These data suggest that the activity responsible for P145 phosphorylation on tyrosine, whether intrinsic to P145 itself or due to an associated kinase, is stimulated by a factor secreted by the tumor cells themselves.

Key words: oncogenes, growth factors, phosphotyrosine, autocriny, stomach cancer

Several independent findings support the hypothesis that the malignant behavior of some tumors is sustained by an "autocrine" loop, wherein cancer cells produce growth factors that stimulate their own growth (1–5). Many growth factor receptors are associated with a protein tyrosine kinase activity, which acts—in concert with other biochemical events—as a signal transducer, triggering the mitogenic intracellular responses (6–10). Upon the binding of their specific growth factors, the receptors endowed with tyrosine kinase activity are transiently activated and phosphorylate themselves on tyrosine. In malignant cells, these receptors may be permanently activated, even in the absence of exogenous growth factors, due to structural altera-

Received February 1, 1988; accepted July 8, 1988.

tions of the receptors [11–12] or due to an autocrine stimulation [2]. This results in their constitutive phosphorylation.

Phosphotyrosine (P-Tyr) antibodies have been used successfully to identify the autophosphorylated form of receptors for known growth factors and the oncogene-coded proteins endowed with tyrosine kinase activity (10, 13–21). Using P-Tyr antibodies, it has been shown that platelet-derived growth factor (PDGF) and epidermal growth factor (EGF) receptors are phosphorylated only upon the addition of exogenous growth factor and are rapidly dephosphorylated (20–21). The same technique has also shown that kinases involved in autocrine loops are permanently activated and that phosphotyrosine-containing proteins accumulate within the cells (16, 18). We recently used P-Tyr antibodies to screen a number of human tumor cells; a few contained high levels of proteins phosphorylated on tyrosine (17). A tyrosine-phosphorylated protein with the M_r of 145,000 (P145) was unique to a gastric carcinoma line. After immunoprecipitation with P-Tyr antibodies, P145 displayed a strong associated protein kinase activity in vitro, becoming phosphorylated on tyrosine. Neither immunologic cross-reactions nor similarities in phosphopeptide maps were found between P145 and known tyrosine kinases (36). The present work provides evidence that the tyrosine phosphorylation of P145 in the gastric carcinoma cell line studied may be stimulated through an autocrine loop.

MATERIALS AND METHODS

Cells and Reagents

The GTL-16 cell line is a clonal line derived from a poorly differentiated gastric carcinoma cell line (22). Normal human gastric mucosa was obtained as described previously (17). SR-BALB cell line is a mouse fibroblast line transformed by the Schmidt-Ruppin D strain of Rous sarcoma virus. P-Tyr antibodies were raised against p-amino-benzene-phosphonate and affinity purified as previously described (13). They are defined as phosphotyrosine antibodies (α-P-Tyr).

Western Blotting

Cells were washed with phosphate-buffered saline (PBS), pH 7.4, and solubilized in boiling Laemmli buffer (23) with or without reducing agents. Samples were adjusted to a protein concentration of 200 μg/well, run in SDS-PAGE and transferred onto nitrocellulose sheets as previously described (24, 25). Blots were probed with 9 μg/ml of purified P-Tyr antibodies, followed by ^{125}I-labelled *S. aureus* protein-A (Amersham). The M_r of labelled proteins was estimated using cotransferred ^{14}C-methylated myosin (200,000), phosphorylase b (92,500), bovine serum albumin (69,000), egg albumin (46,000), and carbonic anhydrase (30,000) (Amersham).

Immunocomplex Kinase Assay

After immunoprecipitation with P-Tyr antibodies or control preimmune rabbit immunoglobulins, immunocomplexes were collected on Protein-A Sepharose, washed, and phosphorylated in the presence of 10 mM $MnCl_2$ and 10 μCi (γ-^{32}P)ATP (specific activity: 7,000 Ci/mM, Amersham) at 0°C for 5 min. Samples were eluted off the protein-A Sepharose by boiling in Laemmli buffer. These samples were subjected to SDS-PAGE followed by autoradiography for 2 h using an intensifying screen.

Acid Treatment and Stimulation by Conditioned Medium

Cells were incubated with an isotonic solution containing 20 mM acetic acid (pH 3.7), 150mM NaCl, and 0.25% BSA for 5 min at 4°C. Under these conditions, known growth factors such as EGF and PDGF are dissociated from their cell surface receptors without affecting cell viability (20, 26).

For stimulation with conditioned medium, exponentionally growing cultures were incubated in serum-free medium for 24 h and then trypsinized and allowed to recover in medium without serum for an additional 48 h. Cells were treated with the above-described isotonic acid solution for 5 min, washed twice with PBS, and incubated with conditioned serum-free medium. The latter was collected from cells previously grown in serum-free medium for 48 h and trypsinized as above. Negative controls included incubation with isotonic acid solution (pH adjusted to 7.0 with NaOH) and physiological saline.

RESULTS

Detection and Localization of Proteins Phosphorylated on Tyrosine in a Gastric Carcinoma Cell Line GTL-16

Whole-cell proteins were solubilized in Laemmli buffer (23), separated by PAGE, transferred to nitrocellulose sheets, and probed with purified antibodies against P-Tyr followed by radiolabelled protein-A, as described elsewhere (11). As previously reported (17), no labelled bands were observed in normal human fibroblasts grown in vitro nor in samples harvested from the gastric mucosa of donors affected by different pathologies. A protein of the approximate M_r of 145,000 was detected in samples solubilized from the gastric tumor cell line GTL-16 (Fig. 1A). Longer exposure of the gels or the vanadate treatment of the cells allowed the detection of other phosphotyrosine-containing bands of different M_r (data not shown). The presence of authentic P-Tyr in P145 was confirmed by immunoprecipitating it from cells metabolically labelled with ^{32}P-orthophosphate (36).

In Vitro Tyrosine Kinase Activity

It was previously shown (36) that P145 exhibited an associated tyrosine kinase activity, which can be demonstrated in an in vitro kinase assay. Proteins extracted from cells by nonionic detergent were immunoprecipitated with either P-Tyr antibodies or control rabbit immunoglobulins. When immunoprecipitates were exposed to (γ-^{32}P)ATP to assay the kinase activity, P145 became phosphorylated (Fig. 1C). In analogy with the known tyrosine kinases (7-8) and in view of the fact that P145 is virtually the only protein phosphorylated on tyrosine in the assay, it seems reasonable to propose that P145 phosphorylates itself on tyrosine (Fig. 1C).

The band seen on Western blot represents the same protein that becomes autophosphorylated, because as described elsewhere (36), both proteins show an unusual shift in their electophoretic mobility when separated in SDS-PAGE under nonreducing conditions, i.e., in the absence of beta-mercapto-ethanol. This shift has never been observed when known tyrosine kinases were analyzed under the same conditions. It had been attributed to the presence of a light chain covalently linked to P145, as previously shown (36). In cells lacking P145 on Western blot, in vitro phosphorylation of proteins with similar M_r was not observed (14, 17).

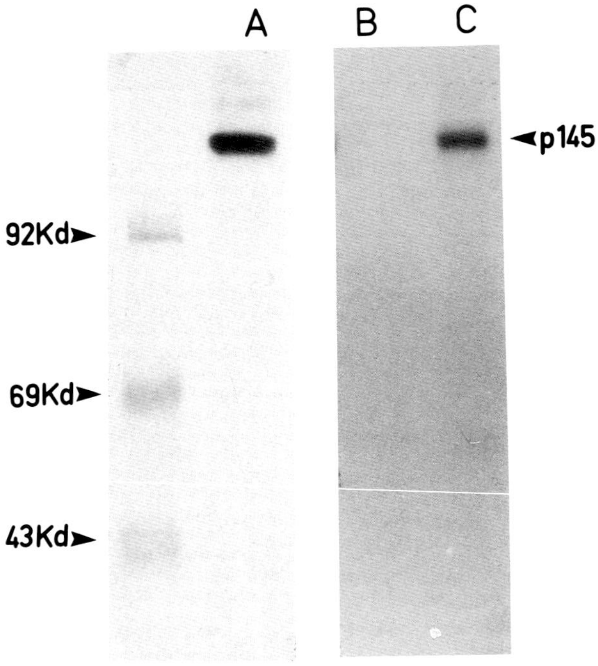

Fig. 1. Tyrosine phosphorylation of P145 in vivo and in vitro. **A**, Western blot analysis of proteins solubilized from GTL-16 cells, probed with P-Tyr antibodies. **B,C**, Immunocomplex kinase assay of P145 precipitated by P-Tyr antibodies (C) or by preimmune rabbit immunoglobulins (B).

Tyrosine Phosphorylation of P145 in Live Cells Is Sustained by an Autocrine Loop

Because it has been shown that the constitutive phosphorylation of receptors could be sustained by factors produced by the cells themselves, experiments were designed to interfere with such a putative autocrine stimulation. GTL-16 tumor cells were treated with mild acid solutions (see Methods). A short acid treatment rapidly removes bound EGF or PDGF molecules from their receptors and switches off growth-factor induced tyrosine kinase activity. The effect is rapid and reversible (20).

GTL-16 cells were incubated at pH 3.7 for different lengths of time. The extent of tyrosine phosphorylation of P145 was evaluated on Western blots probed with P-Tyr antibodies. Following acid treatment, dephosphorylation of P145 was almost complete with 5–15 min (Fig. 2).

To rule out the possibility that the drop of intracellular pH could be directly responsible for the fall of the tyrosine kinase activity, mouse fibroblasts transformed by v-src were treated with acid for the same lengths of time: the activity of $pp60^{src}$ was unaffected (Fig. 2). Thus, the acid treatment appears to interfere with events taking place at the cell surface, most likely the binding of an extracellular factor.

Evidence suggesting the presence of a factor secreted by the gastric carcinoma cells themselves was obtained by studying the effect of conditioned medium. As discussed above and shown in Figure 3, the phosphorylation of P145 is abolished by treating cells with acid. P145 is rephosphorylated if the cells are allowed to recover in serum-free medium at 4°C for a minimum of 15 min (Fig. 3). The addition of

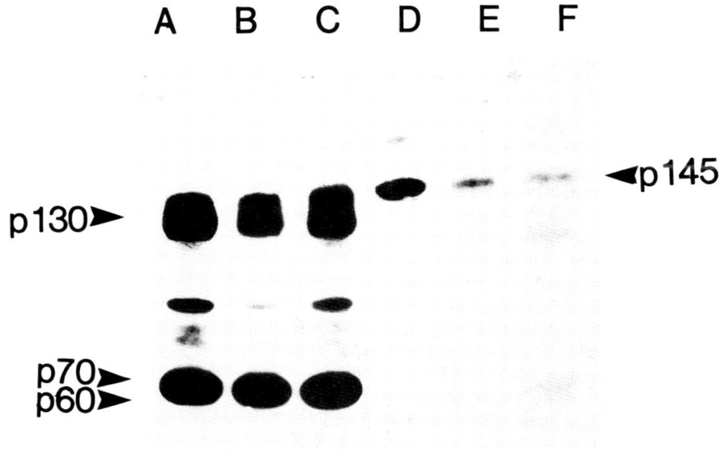

Fig. 2. Effects of mild acid treatment on P145 phosphorylation in intact cells. GTL-16 cells (**D,E,F**) or cells transformed by v-src (**A,B,C**) were treated by an isotonic acid solution pH 3.7 (see Methods) for 5 min (B,E) or 15 min (C,F). The tyrosine phosphorylation of P145 observed in untreated cells (D) is abolished by acid treatment (E,F). The treatment had no effect on the activity of $pp60^{src}$ (A,B,C). Intact cells were solubilized and protein phosphorylated on tyrosine were analyzed in Western blot with P-Tyr antibodies. In v-src transformed cells, the proteins phosphorylated on tyrosine other than $pp60^{src}$ have been described elsewhere (14).

conditioned medium restores the tyrosine phosphorylation of P145 within 5 min at the same temperature. P145 phosphorylation was not restored either by treating the cells with the same acid solution neutralized to pH 7.00 or by treating them with nonconditioned medium in presence or absence of serum (data not shown). Similarly, media conditioned by a number of cell lines other than GTL-16 did not stimulate P145 phosphorylation. These data suggest that the GTL-16 cells produce and secrete in the medium a factor responsible for stimulating the phosphorylation of P145.

DISCUSSION

The data reported in this paper show that in a gastric carcinoma cell line the tyrosine phosphorylation of a membrane protein of M_r 145,000 is sustained by an extracellular factor produced by the cells themselves.

P145 was previously identified as a glycoprotein exposed at the cell surface because it incorporated ^{3}H-D-Glucosamine, was bound by lentil lectin, was labelled by radioactive Iodine under nonpermeating conditions, and was cleaved by mild trypsin treatment of intact cells (36). After immunoprecipitation with P-Try antibodies, P145 displayed a strong associated tyrosine kinase activity in vitro. In immunocomplex-kinase assays, P145 itself was by far the major detectable substrate. In analogy with the known tyrosine kinases, which almost invariably phosphorylate themselves, it was speculated that P145 itself could be tyrosine kinase.

The transmembrane structure and the tyrosine kinase activity of P145 are features of growth-factor receptors. The possible identity of P145 and other known tyrosine kinase receptors (27–31) was disproved by data reported and discussed elsewhere (36).

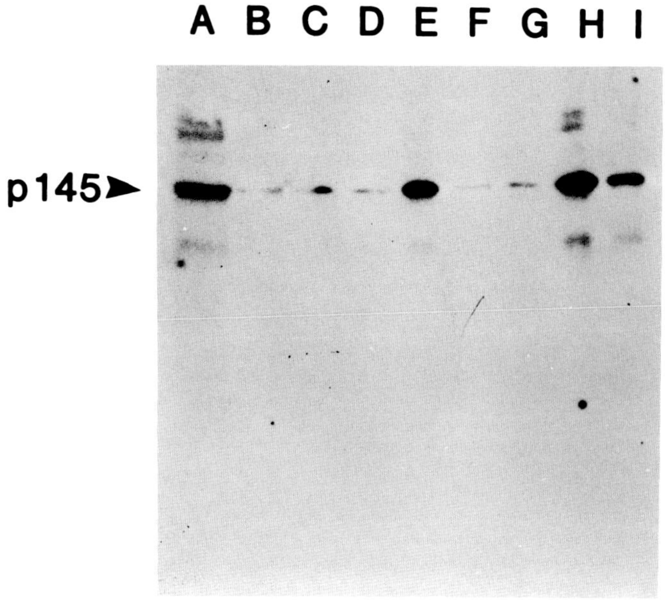

Fig. 3. Stimulation of P145 tyrosine phosphorylation by conditioned medium. Intact cells were solubilized by Laemmli buffer and analyzed by Western blot using P-Tyr antibodies. **A**, GTL-16 cells grown in serum free medium for 3 days. **B**, The same cells treated by mild acid solution as described in the Methods section. **C**, Cells treated as in B and incubated with neutralized acid solution (see Methods section) for 5 min at 4°C. **D**, Cells treated as in B and incubated with physiological saline for 5 min at 4°C. **E**, Cells treated as in B and stimulated with conditioned medium for 5 min at 4°C. **F**, Cells treated as in B and incubated in the neutralized acid solution for 10 min at 4°C. **G**, Cells treated as in B and incubated with physiological saline for 10 min at 4°C. **H**, Cells treated as in B and stimulated with conditioned medium for 10 min at 4°C. **I**, Recovery of cells as in B after 20 min in neutralized acid solution at 4°C.

The present study demonstrated that P145 is dephosphorylated when cells are exposed to a treatment known to dissociate ligands from their receptors and also that tyrosine phosphorylation of P145 is fully restored upon the addition of conditioned medium to the cells. This suggests that the kinase associated with P145 may be stimulated by binding a soluble factor secreted by the gastric carcinoma cells themselves. However, the possibility that P145 could be transphosphorylated by a different kinase representing the actual receptor for the factor present in the medium cannot be ruled out. A similar transphosphorylation mechanism has been shown to operate in the case of erbB-2 phosphorylation by the EGF-receptor (32, 33).

The existence of autocrine circuits of growth stimulation has been reported in physiologic as well as in pathologic conditions, including the autocrine loop triggered by IL-2 (34) in lymphocytes, by FGF (35) in gliomas, by PDGF (7), or by bombesin (2) in small-cell lung carcinomas. In some instances, autocrine activation of a receptor-associated tyrosine kinase also has been demonstrated (16, 18). Evidences for the existence in the cell line studied of one or more autocrine circuits include the following: (1) cell growth in the absence of serum; (2) the presence of a membrane protein (P145) with structural features common to growth-factor receptors; (3) loss of phosphorylation of P145 after acid wash, followed by spontaneous recovery in the absence of serum; (4) stimulation of P145 phosphorylation by conditioned medium. Whether or not tyrosine kinase activity sustained by the autocrine loop(s) is relevant to the altered cell growth of the line remains to be investigated.

ACKNOWLEDGMENTS

This work was supported by funds granted by the P.F. Oncologia of the Italian C.N.R. to P.M.C. (no. 87.01241.44) and to M.F.D.R. (87.01269.44) and by the Italian Association for Cancer Research (A.I.R.C.). S.G. is the recipient of A.I.R.C. fellowships, and L.C.P. is the recipient of Ghirotti fellowship.

We thank Dr. C. Ponzetto for helpful discussion and revision of the manuscript.

REFERENCES

1. Betsholtz C, Heldin C-H, Nister M, Ek B, Wasteson A, Westermark B: Cell 39:447–457, 1984.
2. Cuttitta F, Carney DN, Mulshine J, Moody TW, Fedorko J, Fischler A, Minna JD: Nature 316:823–826, 1985.
3. Rettenmier CW, Roussel MF, Ashmun RA, Ralph P, Price K, Sherr CJ: Mol Cell Biol 7:2378–2387, 1987.
4. Sporn MB, Roberts AB: Nature (London) 313:745–747, 1985.
5. Sporn MB, Todaro GJ: New Engl J Med 303:878–880, 1980.
6. Hunter T: Cell 50:823–829, 1987.
7. Heldin CH, Westermark B: Cell 37:9–20, 1984.
8. Hunter T, Cooper JA: Annu Rev Biochem 54:897–930, 1985.
9. Sherr CJ, Rettenmier CW, Sacca R, Roussel MF, Look AY, Stanley ER: Cell 41:665–676, 1985.
10. Cirillo D, Gaudino G, Naldini L, Comoglio PW: Mol Cell Biol 6:4641–4649, 1986.
11. Downward J, Yarden Y, Mayes E, Scrace G, Totty N, Stockwell P, Ullrich A, Schlessinger J, Waterfield M: Nature (London) 307:521–527, 1984.
12. Ullrich A, Coussens L, Hayflick J, Dull T, Gray A, Tam A, Lee J, Yarden Y, Libermann T, Schlessinger J, Downward J, Mayes E, Whittle N, Waterfield M, Seeburg P: Nature (London) 309:418–425, 1984.
13. Comoglio PM, Di Renzo MF, Tarone G, Giancotti FG, Naldini L, Marchisio PC: EMBO J 3:483–489, 1984.
14. Di Renzo MF, Ferracini R, Naldini L, Giordano S, Comoglio PM: Eur J Biochem 158:383–391, 1986.
15. Frackelton AR Jr, Tremble TM, Williams LT: J Biol Chem 259:7909–7915, 1984.
16. Gaudino G, Cirillo D, Naldini L, Rossino P, Comoglio PM: Proc Natl Acad Sci USA 85:2166–2170, 1987.
17. Giordano S, Di Renzo MF, Cirillo D, Naldini L, Chiadó-Piat L, Comoglio PM: Int J Cancer 39:482–487, 1987.
18. Leal F, Williams LT, Robbins KC, Aaronson S: Science 230:327–330, 1985.
19. Ross AH, Baltimore D, Heisen HN: Nature (London) 294:654–656, 1981.
20. Sturani E, Zippel R, Toschi L, Comoglio PM, Alberghina L: Mol Cell Biol 8:1345–1351, 1988.
21. Zippel R, Sturani E, Toschi L, Naldini L, Alberghina L, Comoglio PM: Biochim Biophys Acta 881:54–61, 1986.
22. Motoyama T, Hojo H, Suzuki T, Oboshi S: Acta Med Biol (Niigata University, Niigata, Japan) 27:49–52, 1984.
23. Laemmli JK: Nature (London) 227:680–685, 1970.
24. Burnette WN: Anal Biochem 112:195–204, 1981.
25. Towbin H, Staehelin T, Gordon J: Proc Natl Acad Sci USA 76:4350–4353, 1979.
26. Garrett JS, Coughlin SR, Niman HL, Tremble PM, Giels GM, Williams LT: Proc Natl Acad Sci USA. 81:7466–7470, 1984.
27. Park M, Dean M, Kaul K, Braun MJ, Gonda MA, Wande Woude G: Proc Natl Acad Sci USA 84:6379–6383, 1987.

28. Martin Zanca D, Hughes SH, Barbacid M: Nature (London) 319:743–748, 1986.
29. Martin Zanca D, Mitra G, Long LK, Barbicid M: In "Cold Spring Harbor Symposia on Quantitative Biology," 1986, pp 983–992.
30. Besmer P, Murphy JE, George PC, Qiu F, Bergold FP, Lederman L, Snyder HW, Brodeur D, Zuckerman EE, Hardy WD: Nature (London) 320:415–421, 1986.
31. Yarden Y, Kuang WJ, Yang-Feng T, Coussens L, Munemitsu S, Dull TJ, Chen E, Schlessinger J, Francke U, Ullrich A: EMBO J 6:3341–3352, 1987.
32. Kabowaki T, Kasuga M, Tobe K, Takaku F: Biochim Biophys Res Comm 144:699–704, 1987.
33. Akiyama T, Saito T, Ogawara H, Toyoshima K, Yamamoto T, Mol Cell Biol 8:1019–1026, 1988.
34. Linch DC, Wallace DL, O'Flynn K: Immunol Rev 95:137–159, 1987.
35. Libemann TA, Friesel R, Jaye M, Lyall RM, Westermark B, Drohan W, Schimdt A, Maciag A, Schlessinger J: EMBO J 6:1627–1632, 1987.
36. Giordano S, Di Renzo MF, Chiadó-Piat L, Ferracini R, Comoglio PM: Mol Cell Biol 8:3510–3518, 1988.

Journal of Cellular Biochemistry 38:237–249 (1988)
Growth Factors and Their Receptors: Genetic
Control and Rational Application 57–69

Characterization of a Bombesin Receptor on Swiss Mouse 3T3 Cells by Affinity Cross-Linking

James Sinnett-Smith, Ian Zachary, and Enrique Rozengurt

Imperial Cancer Research Fund, London WC2A 3PX, England

We have previously identified by chemical cross-linking a cell surface protein in Swiss 3T3 cells of apparent Mr 75,000-85,000, which may represent a major component of the receptor for peptides of the bombesin family in these cells [1]. Because bombesin-like peptides may interact with other cell surface molecules, it was important to establish the correlation between receptor binding and functions of this complex and further characterize the Mr 75,000-85,000 cross-linked protein. Detailed time courses carried out at different temperatures demonstrated that the Mr 75,000-85,000 affinity-labelled band was the earliest cross-linked complex detected in Swiss 3T3 cells incubated with ^{125}I-labelled gastrin-releasing peptide (^{125}I-GRP). Furthermore, the ability of various nonradioactive bombesin agonists and antagonists to block the formation of the Mr 75,000-85,000 cross-linked complex correlated extremely well (r = 0.994) with the relative capacity of these peptides to inhibit ^{125}I-GRP specific binding. Pretreatment with unlabelled GRP for up to 6 h caused only a slight decrease in both specific ^{125}I-GRP binding and the affinity labelling of the Mr 75,000-85,000 protein. We also show that the cross-linked complex is a glycoprotein. First, solubilized affinity labelled Mr 75,000-85,000 complex applied to wheat germ lectin-sepharose columns was eluted by addition of 0.3 M N-acetyl-D-glucosamine. Second, treatment with endo-β-N-acetylglucosaminidase F reduced the apparent molecular weight of the affinity-labelled band from 75,000-85,000 to 43,000, indicating the presence of N-linked oligosaccharide groups.

Key words: cell surface glycoprotein, peptide

Regulatory peptides, which act as local hormones or neurotransmitters in an autocrine or paracrine fashion on adjacent cells are increasingly implicated in the control of cell proliferation [reviewed in 1]. In particular, Rozengurt and Sinnett-

Abbreviations: SDS–PAGE, sodium dodecyl sulphate-polyacrylamide gel electrophoresis; EGS, ethylene glycolbis (succinimidyl succinate); DSS, disuccinimidyl suberate; DSP, dithiobis (succinimidyl propionate); BSOCOES, Bis[2-(succinimidooxycarbonyloxy) ethyl] sulfone; GRP, gastrin releasing peptide; endo F, endo-β-N-acetylglucosaminidase F; GlcNAc, N-acetyl-D-glucosamine; Hepes, 4-(2-hydroxyethyl)-1-piperazineethane sulphonic acid; EDTA, ethylene diaminetetraacetic acid; NP40, nonidet NP40; DMEM, Dulbecco's modified Eagle's medium; SP, substance P.

Received March 4, 1988; accepted July 8, 1988.

Smith [2] found that the amphibian tetradecapetide bombesin [3] is a potent mitogen for Swiss 3T3 cells and acts synergistically with insulin and other growth-promoting factors at subnanomolar concentrations. Mammalian peptides structurally related to bombesin, including gastrin-releasing peptide (GRP) and the neuromedins [4-7], also stimulate DNA synthesis in these cells [8]. These peptides bind to high-affinity receptors in Swiss 3T3 cells, which are distinct from those of other mitogens for these cells [8]. Structure–activity relationships demonstrate that the stereospecific requirement for receptor binding [8,9] and elicitation of biological responses, including mitogenesis [8], is satisfied by the highly conserved COOH-terminal heptapeptide of this family of neuropeptides.

After binding, bombesin and structurally related peptides elicit a complex array of early biological responses [reviewed in 10] including enhanced phosphoinositide metabolism [11-13], mobilization of Ca^{2+} from intracellular stores [12-14], activation of protein kinase C [9,15-17], inhibition of ^{125}I-epidermal growth factor–binding [9,15,18] stimulation of Na^+/H^+ antiport activity (14), and induction of the cellular oncogenes c-fos and c-myc [19-22]. In addition, bombesin-like peptides are present in high concentrations in small-cell lung carcinoma [23-25], where they could act as autocrine growth factors [2]. In spite of the potential importance of peptides of the bombesin family as modulators of animal cell growth [1], little is known about the molecular properties of the receptor for these peptides.

Recently, we have identified by chemical cross-linking a cell surface protein in Swiss 3T3 cells of apparent Mr 75,000–85,000 that specifically recognizes ^{125}I-GRP [28]. These findings suggest that this protein is a major component of the receptor for peptides of the bombesin family in these cells. Here we provide further evidence supporting this proposition and show that the affinity-labelled Mr 75,000–85,000 complex is a glycoprotein, as judged by chromatography on wheat germ lectin columns and by treatment with endo-β-N-acetylglucosaminidase (endo F), an enzyme that cleaves glycans from asparagine-linked glycoproteins [29].

MATERIALS AND METHODS

Cell Culture

Stock cultures of Swiss 3T3 cells [30] were maintained in Dulbecco-Vogt's modified Eagle's medium (DMEM) supplemented with 10% fetal bovine serum, penicillin (100 U/ml), and streptomycin (100 μg/ml) in a humidified atmosphere of 10% CO_2:90% air at 37°C. For experimental purposes, 10^5 cells were subcultured in 33 mm Nunc Petri dishes with 2 ml DMEM containing 10% fetal bovine serum and used at least 6 days after the last change of the medium. The cells were arrested in the G_1/G_o phase of the cell cycle, as judged by cytofluorometric analysis and by the fact that less than 1% of the cells were autoradiographically labelled after a 40-h exposure to (^{3}H)-thymidine [31].

Chemical Cross-Linking of ^{125}I-GRP

Chemical cross-linking was performed as described previously [28]. Briefly, confluent and quiescent cultures of Swiss 3T3 cells were incubated at 37°C in 1 ml of binding medium (0.14M NaCl, 10 mM Na_2HPO_4, 1.8 mM KH_2PO_4, 5 mM KCl, 1.8 mM $CaCl_2$, 1 mM $MgCl_2$, 25 mM HEPES at pH 7.0) containing ^{125}I-GRP at 1 nM. After 10 min of incubation (unless otherwise indicated) the cultures were washed and incubated in binding medium, pH 7.4, containing 6 mM ethylene glycolbis (succininimidyl succinate) (EGS) or 1 mM dithiobis (succinimidyl propionate) (DSP) at 15°C for 15 min. The cross-linking agents (EGS and DSP) were dissolved in dimethyl sulfoxide immediately prior to use and were added to binding medium to give a final solvent concentration of 1–2%. The cultures were then washed twice with phosphate buffered saline and solubilized in 0.1 ml of 2× sample buffer, which contained 0.2 M Tris-HCl pH 6.8, 10% (w/v) glycerol, 6% sodium dodecyl sulfate (SDS) (w/v), 4% β-mercaptoethanol (v/v) and 2 mM EDTA. Samples were immediately heated at 100°C for 3–5 min and analyzed by SDS-polyacrylamide gel electrophoresis.

SDS-Polyacrylamide Gel Electrophoresis

Slab gel electrophoresis was performed with 7.5% acrylamide in the separating gel and 5% in the stacking gel, and 0.1% SDS [32]. After electrophoresis, gels were stained, destained, and dried down onto paper for autoradiography with Fuji x-ray film. Dried gels were exposed to film for 4–8 days. The Mr 75,000–85,000 band from autoradiograms was scanned with a Joyce-Loebl double-beam densitometer, and the areas under specific peaks were measured with a Hewlett-Packard digitizer.

MATERIALS

EGS, DSS, DSP and BSOCOES were purchased from Pierce and Warriner (UK) Ltd. (Chester, England). Bombesin, 3-[(3-cholamidopropyl) dimethylammonia] 1-propanesulphonate (CHAPS), Triton X-100, deoxycholate, digitonin, and nonidet-NP40 (NP40) were obtained from Sigma chemical Co. (St Louis, MO). Wheat germ lectin sepharose 6 MB was obtained from Pharmacia (Uppsala, Sweden). Endo F was purchased from NEN Research Products (Boston, MA). ^{125}I-GRP was obtained from Amersham (England). GRP, Neuromedin B, bombesin [8-14], [DArg^{1}, DPro^{2}, DTrp^{7,9}, Leu^{11}]substance P, and [DArg^{1}, DPhe^{5}, DTrp^{7,9}, Leu^{11}]substance P were purchased from Pensinsula Laboratories Europe Ltd. (St. Helens, England).

RESULTS

Formation of the Mr 75,000–85,000 Cross-Linked Complex: Time and Temperature Dependence

Analysis by SDS–PAGE of extracts of Swiss 3T3 cells that had been preincubated with ^{125}I-GRP for various times at different temperatures and subsequently treated with the homobifunctional disuccinimidyl cross-linking agent EGS for 15 min at 15°C revealed the presence of a major broad band migrating with an apparent Mr 75,000–85,000. Figure 1 shows that the level of this band increased rapidly at 37°C and reached a maximum after 7.5 min of incubation with ^{125}I-GRP. The rate of formation of the cross-linked complex decreased in a progressive fashion when the incubations with ^{125}I-GRP were carried out at 24°C or at 15°C instead of at 37°C (Fig 1). In other experiments, cultures incubated with ^{125}I-GRP for 5 min were washed and subsequently treated with EGS for various times at 37°C. Maximum cross-linking was obtained after 5 min of exposure to EGS. Longer incubations at 37°C with the cross-linking agent resulted in the formation of additional complexes of higher molecular weight (results not shown). These kinetic studies demonstrate

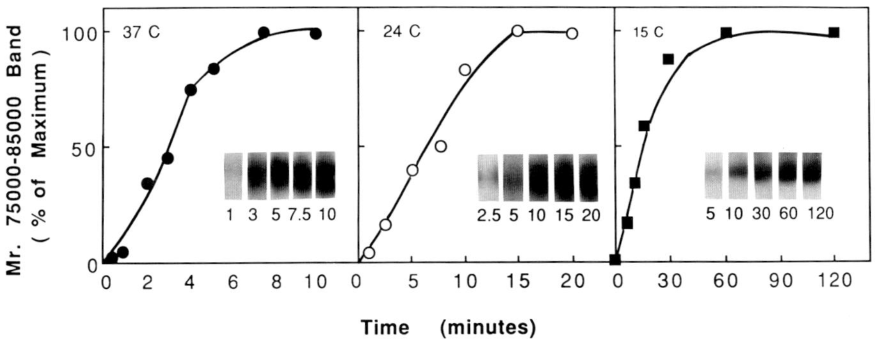

Fig. 1. Time course of the appearance of the Mr 75,000–85,000 affinity-labelled protein at various temperatures. Confluent and quiescent cultures of Swiss 3T3 cells were incubated with ^{125}I-GRP (1 nM) at either 37°C **(left-hand panel)**, 24°C **(centre panel)**, or 15°C **(right-hand panel)**. At the times indicated, chemical cross-linking was carried out and the level of Mr 75,000–85,000 quantified as described in Materials and Methods. The values shown are expressed as a percentage of the maximum level and are the mean of three independent experiments obtained at each temperature. The inserts show some representative autoradiograms.

that the Mr 75,000–85,000 affinity labelled band is the earliest cross-linked complex detected in Swiss 3T3 cells incubated with ^{125}I-GRP.

Covalent Nature of the Mr 75,000–85,000 Complex

Previous studies demonstrated that not only EGS but also the homobifunctional cross-linking agents DSS, DSP, and BSOCOES were effective in promoting covalent cross-linking of ^{125}I-GRP to Swiss 3T3 cells, resulting in the formation of a major band of Mr 75,000–85,000 [28]. In the present studies we confirmed these observations, using these cross-linking agents at 37°C instead of at 15°C. That the Mr 75,000–85,000 constitutes genuine covalent linking of ^{125}I-GRP to the putative bombesin receptor was verified with the thiol-cleavable cross-linking agent DSP. As shown in Figure 2, DSP promoted the cross-linking of ^{125}I-GRP to the Mr 75,000–85,000 protein. Reduction of the disulfide bond of DSP by incubating the extracts with various concentrations of 2-mercaptoethanol resulted in dose-dependent disappearance of the Mr 75,000-85,000 band.

Effect of Bombesin Agonists and Antagonists on the Formation of the Mr 75,000–85,000 Complex

The possibility that the Mr 75,000–85,000 band is a surface component of Swiss 3T3 cells closely related to the receptor for peptides of the bombesin family was substantiated by cross-linking ^{125}I-GRP to cultures incubated in the presence of different concentrations of unlabelled bombesin agonists and antagonists. Figure 3 shows that addition of bombesin, GRP, neuromedin B or the 8–14 amino acid fragment of bombesin (bombesin 8-14) decreased the level of the Mr 75,000–85,000 complex in a concentration-dependent fashion. The concentrations of bombesin, GRP, neuromedin B, and bombesin 8-14 required to produce 50% inhibition of cross-linking (IC_{50}, cross-linking) were 0.7 nM, 8 nM, 34 nM, and 900 nM, respectively. In contrast, the NH_2-terminal fragment of GRP (GRP, 1–16) did not prevent the formation of the cross-linked complex even at 10 μM.

The neuropeptide substance P (SP) neither inhibits the binding of ^{125}I-GRP to 3T3 cells [8] nor prevents the formation of the Mr 75,000–85,000 cross-linked

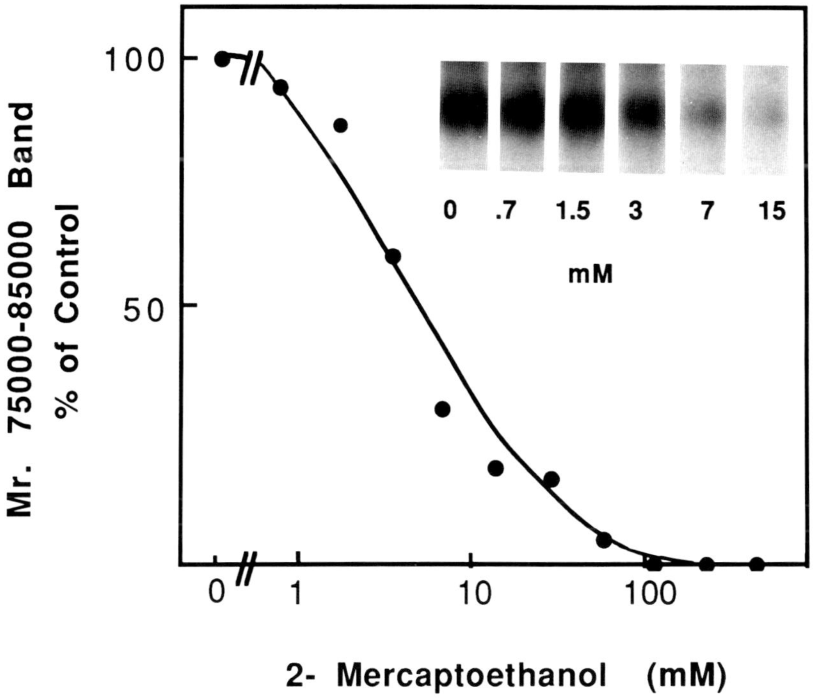

Fig. 2. Effect of 2-mercaptoethanol on DSP-mediated cross-linking of ^{125}I-GRP to the Mr 75,000–85,000 protein. Confluent and quiescent cultures of 3T3 cells were incubated with ^{125}I-GRP (1 nM) at 37°C for 10 min. Chemical cross-linking using DSP was performed at 15°C for 15 min, as described in Materials and Methods. The cell monolayers were then dissolved with 2× sample buffer containing varying concentrations of 2-mercaptoethanol, boiled for 5 min, and analysed by SDS–PAGE. The areas under the individual peak for the Mr 75,000–85,000 protein are shown as a function of the concentration of 2-mercaptoethanol and are expressed as a percentage of the control value. The inset shows the autoradiogram obtained at the lower concentrations of reducing agent.

complex [28]. SP also fails to stimulate DNA synthesis in quiescent cells [2]. However, SP antagonists are potent bombesin antagonists in 3T3 cells [8,33]. Figure 3 also includes the effect of various concentrations of [DArg^1, DPro^2, DTrp^{7,9}, Leu^{11}]SP and [DArg^1, DPhe^5, DTrp^{7,9}, Leu^{11}]SP [33] on the formation of the Mr 75,000–85,000 band. These antagonists decreased the level of the Mr 75,000–85,000 band in a concentration-dependent fashion. The concentrations of [DArg^1, DPhe^5, DTrp^{7,9}, Leu^{11}]SP and [DArg^1, DPro^2, DTrp^{7,9}, Leu^{11}]SP that exhibit a 50% inhibition of cross-linking (IC_{50}, cross-linking) were 3,700 nM and 17,000 nM, respectively.

The concentrations of agonist or antagonists required to produce 50% inhibition of the level of the Mr 75,000–85,000 complex (IC_{50}, cross-linking) were derived from Figure 3, whereas the concentrations of these peptides that cause 50% inhibition of ^{125}I-GRP specific binding to Swiss 3T3 cells were derived from previous published curves [8,33]. Figure 4 shows that the IC_{50} for cross-linking correlated extremely well with the values of IC_{50} for binding ($r = 0.994$). The striking parallelism between the ability of the peptides to inhibit binding and cross-linking considerably strengthens our conclusion that the Mr 75,000–85,000 complex is a major component of the receptor for peptides of the bombesin family.

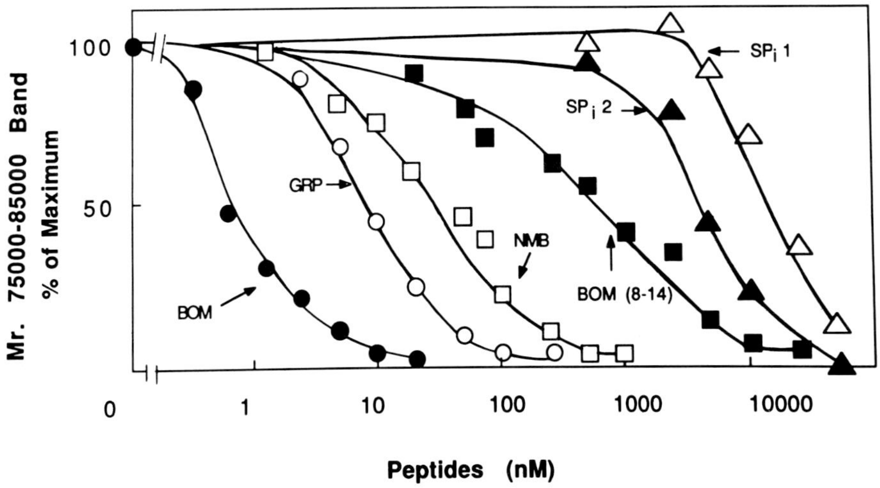

Fig. 3. Effect of various bombesin agonists and antagonists on the affinity labelling of the Mr 75,000–85,000 protein. Confluent and quiescent cultures of 3T3 cells were incubated at 37°C for 10 min with 1 nM ^{125}I-GRP either in the presence or absence of the following agonists and antagonists at the concentrations indicated: bombesin (●, Bom): GRP (○); neuromedin B (□, NMB); bombesin 8-14 (■, Bomb 8–14); [DArg^{1}, DPro^{2}, DTrp^{7,9}, Leu^{11}]substance P (△, SPi 1); and [DArg^{1}, DPhe^{5}, DTrp^{7,9}, Leu^{11}]substance P (▲, SPi 2). Chemical cross-linking was then performed with 6 mM EGS, as described in Materials and Methods. The results are the composite of three independent experiments and are expressed in each case as a percentage of the control value, obtained from scanning densitometry of the autoradiograms.

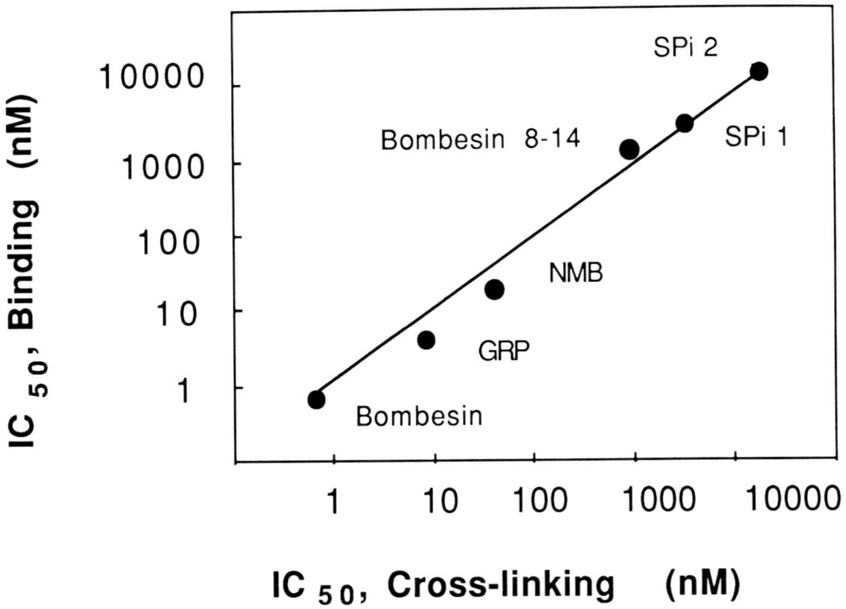

Fig. 4. The ability of various bombesin agonists and antagonists to inhibit ^{125}I-GRP binding correlates with the capacity of these peptides to block the formation of the Mr 75,000–85,000 affinity-labelled complex. The concentrations of peptides that gave 50% inhibition in the binding of ^{125}I-GRP to 3T3 cells (IC_{50} binding) from previously published data [8,33] were plotted against the concentrations of the same peptides that gave a 50% inhibition of chemical cross-linking (IC_{50}, cross-linking) obtained from Figure 3.

Pretreatment With GRP Does Not Alter the Level of the Mr 75,000–85,000 Cross-Linked Complex

Zachary and Rozengurt [34] reported that exposure of Swiss 3T3 cells to either bombesin or GRP causes only a slight decrease in the number of cell-surface binding sites for these peptides. Furthermore, pretreatment with GRP did not reduce the level of the Mr 75,000–85,000 affinity-labelled band. These findings led to the conclusion that peptides of the bombesin family, in contrast to other mitogens, do not cause down regulation of their specific high affinity receptors in Swiss 3T3 cells. Recently Kris et al. [35] reported that treatment for 30 min with 1 nM GRP markedly reduced the level of the Mr 75,000–85,000 affinity-labelled band; they concluded that bombesin strongly down-regulates its receptor. In view of this discrepancy, we decided to reassess whether pretreatment with GRP alters the specific binding of ^{125}I-GRP and/or the level of the affinity-labelled Mr 75,000–85,000 protein. Figure 5 shows that treatment of Swiss 3T3 cells with either 3.6 or 36 nM GRP for up to 6 h, followed by extensive washing at 37°C to remove surface-bound ligand, caused only a slight decrease in the level of specific binding. As a control, other cultures were pretreated with EGF at 3.3 nM for various times and then incubated with ^{125}I-EGF. Pretreatment with EGF caused the expected down regulation of surface receptors (Fig. 5, left). To determine whether exposure of the cells to GRP changed the level of the Mr 75,000–85,000 protein, parallel cultures were treated with GRP for various times, washed extensively, incubated with ^{125}I-GRP, and then treated with EGS. As shown in Figure 5 (right), pretreatment with 3.6 or 36 nM GRP did not cause any significant alteration of the Mr 75,000–85,000 protein. In other experiments we pretreated the cultures with 3.6 nM GRP and washed them at 4°C, as reported by Kris et al. [35]. We found

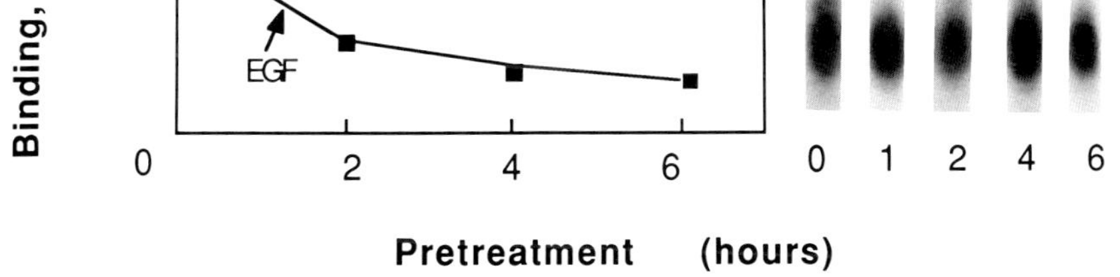

Fig. 5. Effect of pretreatment with GRP on the binding of ^{125}I-GRP to Swiss 3T3 cells and on the ^{125}I-GRP affinity labelling of the Mr 75,000–85,000 protein in Swiss 3T3 cells. **Left panel:** Confluent cultures of 3T3 cells were incubated at 37°C in either the absence or presence of 3.6 nM GRP (●), 36 nM GRP (○) or 3.3 nM EGF (■). After the times indicated, all cultures were washed 5× at 37°C, with a 15-min incubation after the third wash with DMEM, and incubated at 37°C with either 1 nM ^{125}I-GRP for 30 min (○,●) or 2 nM ^{125}I-EGF for 1 h (■). Values are expressed as a percentage of the binding to control, unpretreated cultures and represent the mean of duplicate determinations. **Right panel:** Parallel cultures of 3T3 cells were incubated with 1 nM ^{125}I-GRP for 5 min at 37°C, and chemical cross-linking was carried out as described in Materials and Methods. The autoradiograms obtained from the chemical cross-linking are shown in the right-hand panel.

a marked decrease in ^{125}I-GRP binding and cross-linking, suggesting that the apparent decrease in the level of the Mr 75,000–85,000 affinity-labelled band noted by Kris et al. [35] may be the result of incomplete removal of nonradioactive surface-bound ligand rather than genuine down regulation.

Characterization of the Mr 75,000–85,000 Cross-Linked Complex as a Glycoprotein

Treatment with the nonionic detergents NP40 and Triton X-100 for 2 h at 4°C promoted effective solubilization of the Mr 75,000–85,000 affinity labelled band, whereas other detergents, including deoxycholate, digitonin, and CHAPS, were less effective (Fig. 6). Solubilization by NP40 depended on the amount of detergent; maximum solubilization by NP40 was achieved at a concentration of 1.25% (Fig. 6). The Mr 75,000–85,000 affinity-labelled band solubilized by NP40 or Triton X-100 was recovered in the supernatant after centrifugation at 100,000*g* for 1 h, as revealed by SDS–PAGE followed by autoradiography of the supernatant and pellet after treatment with each detergent.

To determine whether the affinity-labelled Mr 75,000-85,000 complex is a glycoprotein, solubilized preparations were subjected to affinity chromatography in lectin columns [36]. After incubation with ^{125}I-GRP and cross-linking with EGS, 3T3 cells were treated with 1% NP40. The solubilized material was chromatographed on a wheat germ lectin sepharose 6MB column. After extensive washing with the equilibrating buffer and subsequently with this buffer containing 0.4 M NaCl, bound

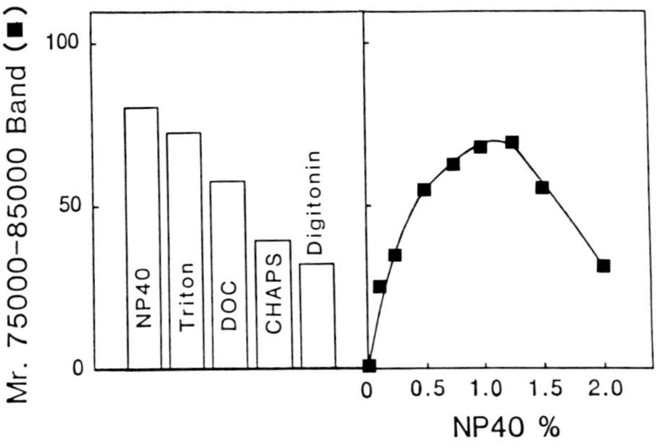

Fig. 6. Solubilization of the 75,000–85,000 affinity-labelled protein. Confluent cultures of 3T3 cells were incubated with ^{125}I-GRP (1 nM) at 37°C for 10 min and chemically cross-linked with EGS as described in Materials and Methods. The individual cultures were solubilized and scraped at 4°C into a solubilization buffer consisting of 50 mM Hepes, 100 mM NaCl, 5 mM EDTA, phenylmethylsulfonyl fluoride (50 μM), aprotonin (3.5 μg/ml), pepstatin A (1 μg/ml), leupeptin (1 μg/ml), and bacitracin (1 μg/ml), plus the following detergents: 1% NP40, 1% Triton X-100 (TRITON), 5 mM deoxycholate (DOC), 6 mM CHAPS, and 2% digitonin (**left panel**). Other cultures were treated with solubilization buffer containing varying percentages of NP40 (**right panel**). In all cases, the cell suspensions were incubated at 4°C with gentle shaking for 2 h, and then passed through a 23-g needle before centrifugation at 100,000*g* for 1 h. The supernatants obtained were analyzed by SDS–PAGE. As a control, parallel cultures chemically cross-linked to ^{125}I-GRP were scraped into the solubilization buffer in the absence of any detergent. Following centrifugation at 100,000*g* the pellet was solubilized with 2× sample buffer and analyzed by SDS–PAGE. The values shown are expressed as a percentage of the level of Mr 75,000–85,000 protein obtained from the pellet. The supernatant showed a complete absence of the Mr 75,000–85,000 protein.

glycoproteins were eluted from the column with buffer containing 0.3 M N-acetyl-N-glucosamine (GlcNAc). Individual fractions were examined by SDS–PAGE. As shown in Figure 7, the Mr 75,000–85,000 band was tightly bound to the lectin column and it was eluted by GlcNAc. Similar results were obtained when 3T3 cells were solubilized with 1% Triton X-100 instead of NP4O. In contrast, little retardation was observed when the solubilized material was applied to a lentil sepharose 4B column. These findings suggest that the Mr 75,000–85,000 contains carbohydrate chains with GlcNAc residues.

In order to substantiate the conclusion that the Mr 75,000–85,000 band is a glycoprotein, we examined the effect of endo-β-N-acetyl-glucosaminidase F (endo F) treatment on this complex. This enzyme cleaves N-linked oligosaccharides from glycoproteins and is therefore a useful probe for the further elucidation of the nature of putative cell-surface glycoproteins [29]. The affinity-labelled Mr 75,00–85,000 protein was located on dried, unfixed polyacrylamide gels by autoradiography and excised. Individual gel slices were then rehydrated and incubated with different amounts of endo F. As shown in Figure 8, treatment with endo F decreased the apparent molecular weight of the affinity-labelled band from 75,000–85,000 to 43,000. The appearance of this band was not due to a nonspecific proteolytic degradation of the Mr 75,000–85,000 protein, because neither the molecular weight nor the broad, fuzzy appearance of the higher-molecular-weight band was affected when gel slices

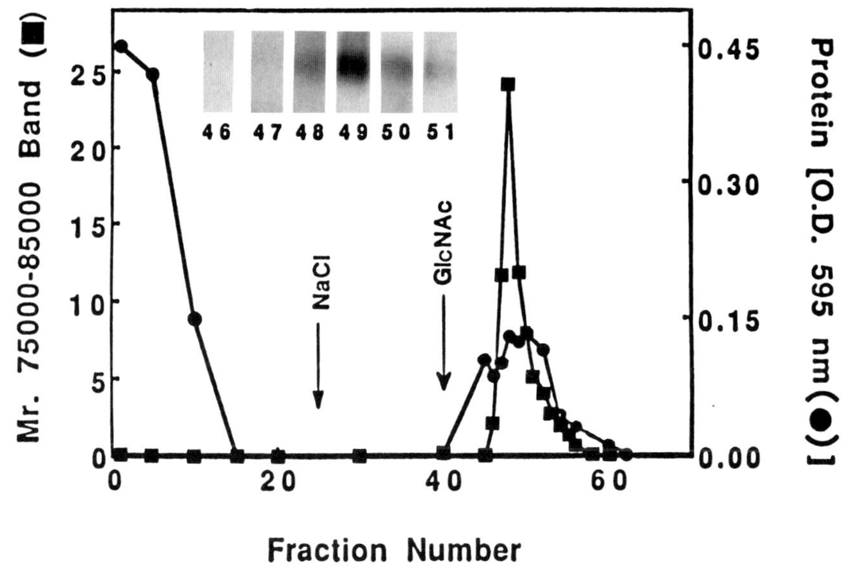

Fig. 7. Chromatography of the affinity-labelled Mr 75,000–85,000 protein on wheat germ lectin sepharose. ^{125}I-GRP (1 nM) was cross-linked to confluent Swiss 3T3 cells (10^7) and the cultures were then solubilized with 1% NP40 as described in Figure 1. The supernatant from the 100,000*g* spin was diluted with solubilization buffer to give a final concentration of 0.2 % NP40, and loaded onto a 1-ml wheat germ lectin-sepharose 6MB column. The column was washed sequentially with solubilization buffer containing 0.1% NP40 then with solubilization buffer containing 0.1% NP40 and 0.4 M NaCl. Bound glycoproteins were eluted with the same buffer containing 0.1% NP40, 0.4 M NaCl, and 0.3 M GlcNAc. Aliquots (0.1 ml) of the fractions (0.3 ml) were analyzed by SDS–PAGE. The results represent the integration values in arbitrary units obtained from the autoradiogram shown in the inset. Proteins were measured by the method of Bradford with Coomassie blue G-250.

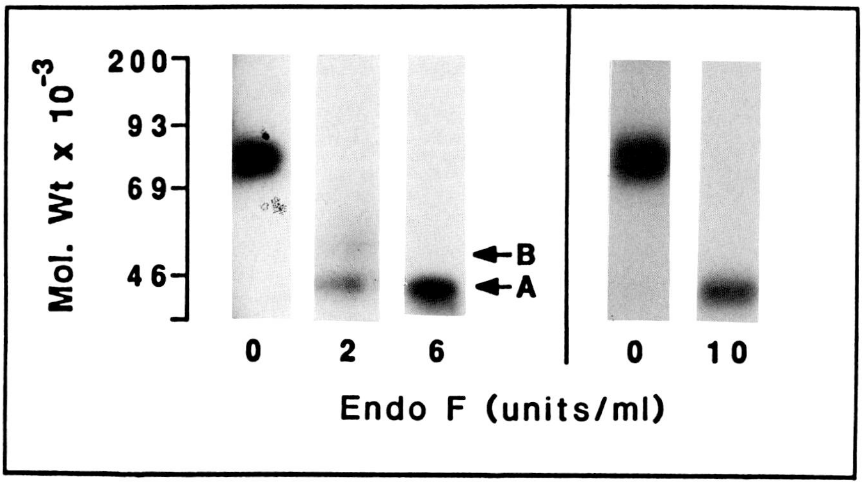

Fig. 8. Endo F treatment of the affinity-labelled Mr 75,000–85,000 protein. **Left:** Following chemical cross-linking of ^{125}I-GRP to Swiss 3T3 cells, the affinity-labelled protein was resolved on a 10% acrylamide gel under reducing conditions, located by autoradiography of the dried unfixed gels, and excised. Each gel slice was placed in an individual tube and rehydrated in 0.3 ml of endo F digestion buffer (0.1 M Na phosphate, pH 6.1, 50 mM EDTA, 1% NP40, 0.1% SDS, and 1% β-mercaptoethanol). The accessibility of enzyme was then increased by sequentially freeze-thawing the tubes through three cycles of alternate incubations in dry ice and 37°C for 10 min each. Endo F was then added to each tube at the indicated concentrations for 24 h at 37°C. **Right:** Fractions containing the Mr 75,000–85,000 protein were pooled, and endo F (10 units/ml) was added directly to a 100-μl aliquot and incubated at 37°C for 24 h. The reactions were terminated by the addition of 10× sample buffer and heating for 5 min at 100°C. The gel slices and incubation mixtures were resolved on a second 10% SDS–PAGE.

were treated under identical conditions in the absence of the enzyme (Fig. 8). The effect of endo F on the molecular weight of the Mr 75,000–85,000 band was concentration-dependent, and a discrete intermediate band could be observed (Fig. 8), suggesting the presence of at least two N-linked oligosaccharide groups. Incubation with endo-β-acetylglucosaminidase H (1 μg/ml), which cleaves specifically high-mannose oligosaccharides [37], did not alter the level of the Mr 75,000–85,000 band or cause the appearance of any lower-molecular-weight species. In other experiments, fractions eluted from wheat germ lectin sepharose columns containing the Mr 75,000–85,000 affinity-labelled protein were treated with endo F. In agreement with the results shown in Figure 8 (left), the molecular weight of the complex decreased to 43,000 (Fig. 8, right).

DISCUSSION

To gain information concerning some of the molecular properties of the bombesin/GRP receptor, we used affinity-labelling methodology to identify surface components of Swiss 3T3 cells that specifically recognize ^{125}I-GRP. Homobifunctional disuccinimidyl cross-linking agents have been successfully employed to identify receptors for other peptide mitogens, including those for EGF and PDGF. Analysis of extracts of cells that had been preincubated with radiolabelled GRP and subsequently treated with a cross-linking reagent revealed the presence of a major band migrating with apparent Mr 75,000–85,000. Several lines of evidence support the conclusion that this protein is the receptor or a major component of the receptor for peptides of the bombesin family in these cells: (a) The formation of the Mr 75,000–85,000 affinity labelled complex is promoted by a variety of disuccinimidyl cross-linking agents, including EGS, DSS, BSOCOES, and the thiol-clearable agent DSP [28] (Fig. 2). (b) The inhibition of ^{125}I-GRP affinity-labelling of this band with unlabelled GRP corresponds closely with the ability of GRP to inhibit the binding of the labelled ligand in parallel cultures [28]. Indeed, we have demonstrated that the ability of various nonradioactive peptide agonists and antagonists to inhibit the formation of the Mr 75,000–85,000 affinity labelled complex (Fig. 3) correlates extremely well ($r = 0.994$) with the relative capacity of these peptide to inhibit ^{125}I-GRP specific binding (Fig. 4). (c) Time courses carried out at different temperatures show that the Mr 75,000–85,000 band is the earliest cross-linked complex detected in cells incubated with ^{125}I-GRP (Fig. 1). (d) ^{125}I-GRP affinity-labelling of the Mr 75,000–85,000 band is specific, as shown by the lack of effect of a panel of other mitogens and neuropeptides on cross-linking [28]. (e) The Mr 75,000–85,000 protein was not found in other cell lines that do not exhibit receptors for bombesin-like peptides [8,28]. (f) the dependence of affinity labelling of the Mr 75,000–85,000 protein on the concentration of ^{125}I-GRP closely paralleled the ability of the unlabelled peptide to stimulate DNA synthesis and a variety of other biological responses in Swiss 3T3 cells [8,9,14,15,21]. All these findings taken together strongly suggest that an Mr 75,000–85,000 surface protein is the receptor or a major component of the receptor for peptides of the bombesin family in Swiss 3T3 cells.

It is well known that the binding of polypeptide growth factors such as EGF and platelet-derived growth factor (PDGF) to their receptors is followed by rapid internalization and intracellular degradation of the ligand as well as the receptor. This process results in a marked reduction in the number of surface binding sites in the target cell (down regulation). Receptor down regulation was originally demonstrated by ligand-binding studies with ^{125}I-EGF [38-40] and ^{125}I-PDGF [41-43] and subsequently verified with specific antibodies directed against the receptor molecule [44,45]. We reported that exposure of Swiss 3T3 to mitogenic concentrations of bombesin or GRP caused only a small decrease in ^{125}I-GRP binding and no significant alteration in the level of the Mr 75,000–85,000 cross-linked complex [34]. These findings indicated that mitogenic peptides of the bombesin family do not cause down regulation of their receptors in Swiss 3T3 cells. In contrast, Kris et al. [35] reported that treatment with 1 nM GRP for 30 min was sufficient to completely down-regulate the Mr 75,000–85,000 complex. In view of this discrepancy, here we have reassessed the effect of GRP pretreatment on the binding and cross-linking of ^{125}I-GRP to Swiss 3T3 cells. As shown in Figure 5, pretreatment with 3.6 or 36 nM nonradioactive GRP for up to 6 h caused only a slight decrease both in specific binding and in the level of the Mr 75,000–85,000, in accord with our previous results. The abolition of ^{125}I-GRP cross-linking by prior exposure to GRP reported by Kris et al. [35] may represent incomplete removal of surface-bound GRP before the incubation with radiolabelled GRP. The absence of ligand-induced down regulation [34] (Fig. 5) taken together with the rapid internalization and degradation of ^{125}I-GRP shown previously [34] raises the possibility that receptors for bombesin-like peptides in Swiss 3T3 cells return functionally intact to the plasma membrane, i.e., they recycle in the same fashion as receptors involved in nutrient uptake [46]. Further studies using additional reagents, such as specific antibodies, will be necessary to test this hypothesis directly.

Regardless of the precise mechanism, the results shown in Figure 5 provide an additional example of the tight correlation between specific ^{125}I-GRP binding and formation of the Mr 75,000-85,000 cross-linked complex.

The broad, diffuse, appearance of the Mr 75,000–85,000 affinity-labelled band raised the possibility that the cross-linked complex is a cell surface glycoprotein. Various detergents, particularly NP40 and Triton X-100, effectively solubilize the affinity labelled Mr 75,000–85,000 complex. Solubilized complex binds to wheat germ lectin-sepharose columns and can be eluted with N-acetyl-D-glucosamine (Fig. 7). This indicates that the complex contains carbohydrate chains with GlcNAc residues [36]. Furthermore, treatment of the affinity-labelled band with endo F, an enzyme that cleaves N-linked oligosaccharides from glycoproteins [29], reduced the apparent molecular weight of the labelled cross-linked complex to 43,000. The effect of endo F on the apparent molecular weight of the Mr 75,000–85,000 complex was concentration dependent, and a discrete intermediate band could be detected, suggesting the presence of at least two N-linked oligosaccharide chains. The striking reduction in apparent molecular weight seen after endo-F treatment suggests that the cross-linked complex is heavily glycosylated and/or that it migrates anomalously in SDS gels. Kris et al. [35] also found that treatment of the Mr 75,000–85,000 cross-linked complex with N-glycanase caused a marked reduction in its apparent molecular weight. Recently, Masu et al. [47] reported cDNA cloning of a receptor for the tachykinin neuropeptide substance K using an oocyte expression system. It is noteworthy that the deduced molecular weight of the core polypeptide is 43,066 and that it possesses two potential N-glycosylation sites [47]. Thus, the small size of the core polypeptide noted here is not unique for a bombesin-family peptide receptor; it is shared with other neuropeptide receptors.

Addition of bombesin-like peptides to quiescent Swiss 3T3 cells stimulates a complex array of early and late molecular events [1,2,9-22]. The mechanism by which a small-size receptor (in comparison with EGF or PDGF receptors) can initiate such an array of responses remains an important and challenging question. A single protein or various G proteins [48] may be implicated in the signal transduction process [19,49]. It has also been reported that an Mr 115,000 protein in Swiss 3T3 cells was phosphorylated in tyrosine in response to bombesin [49], an observation that remains controversial [16]. Further experimental work is warranted to determine the relationship, if any, between the Mr 75,000–85,000 affinity-labelled glycoprotein and bombesin-stimulated tyrosine kinase activity.

In conclusion, we have identified further by chemical cross-linking a cell-surface protein in Swiss 3T3 cells of apparent Mr 75,000–85,000 that may represent the receptor or a major component of the receptor for bombesin-like peptides. The molecule is a glycoprotein with a core polypeptide component of Mr 43,000, which contains at least two N-linked oligosaccharide side chains. This knowledge should prove useful in future studies directed at the purification and molecular characterization of the bombesin receptor.

REFERENCES

1. Zachary I, Woll, PJ, Rozengurt E: Dev Biol 124:295–308, 1987.
2. Rozengurt E, Sinnett-Smith J: Proc Natl Acad Sci USA 80:2936–2940, 1983.
3. Anastasi A, Erpsamer V, Bucci M: Experientia 27:166–167, 1971.

4. McDonald TJ, Jornvall H, Nilsson G, Vagne M, Ghatei M, Bloom SR, Mutt V: Biochem Biophys Res Commun 90:227–233, 1979.
5. Minamino N, Kangawa K, Matsuo H: Biochem Biophys Res Commun 114:541–548, 1983.
6. Minamino N, Kangawa K, Matsuo H: Biochem Biophys Res Commun 119:14–20, 1984.
7. Minamino N, Sudoh T, Kangawa K, Matsuo H: Biochem Biophys Res Commun 130:685–691, 1985.
8. Zachary I, Rozengurt E: Proc Natl Acad Sci USA 82:7616–7620, 1985.
9. Zachary I, Rozengurt E: Cancer Surv 4:729–765, 1985.
10. Rozengurt E: Science 234:161–166, 1986.
11. Heslop JP, Blakeley DM, Brown KD, Irvine RF, Berridge MJ: Cell 47:703–709, 1986.
12. Takuwa N, Takuwa Y, Bollag WE, Rasmussen H: J Biol Chem 262:182–188, 1987.
13. Lopez-Rivas A, Mendoza SA, Nånberg E, Sinnett-Smith J, Rozengurt E: Proc Natl Acad Sci USA 84:5768–5772, 1987.
14. Mendoza SM, Schneider JA, Lopez-Rivas A, Sinnett-Smith JW, Rozengurt E: J Cell Biol 102:2223–2233, 1986.
15. Zachary I, Sinnett-Smith JW, Rozengurt E: J Cell Biol 102:2211–2222, 1986.
16. Isacke CM, Meisenhelder J, Brown KD, Gould KJ, Gould SJ, Hunter T: EMBO J 5:2889–2898, 1986.
17. Rodriguez-Pena A, Zachary I, Rozengurt E: Biochem Biophys Res Commun 140:379-385, 1986.
18. Brown KD, Blay J, Irvine RF, Heslop JP, Berridge MJ: Biochem Biophys Res Commun 123:377–384, 1984.
19. Letterio JJ, Coughlin SR, Williams LT: Science 234:1117–1119, 1986.
20. Palumbo AP, Rossino P, Commoglio PM: Exp Cell Res 167:276–280, 1986.
21. Rozengurt E, Sinnett-Smith JW: J Cell Physiol 131:218–225, 1987.
22. Bravo R, Macdonald-Bravo H, Muller R, Hubsch D, Almendral JM: Exp Cell Res 170:103–109, 1987.
23. Moody TW, Pert CB, Gazdar AF, Carney DN, Minna JD: Science 214:1246–1248, 1981.
24. Wood SM, Wood JR, Ghatei MA, Lee YC, O'Shaughnessy D, Bloom SR: J Clin Endocrin Metabol 53:1310–1312, 1981.
25. Erisman MD, Linnoila RI, Hernandez O, Diaugustine RP, Lazarus LH: Proc Natl Acad Sci USA 79:2379–2383, 1982.
26. Cuttitta F, Carney DN, Mulshine J, Moody TW, Fedorko J, Fischler A, Minna JD: Nature 316:823–826, 1985.
27. Carney DN, Cuttitta F, Moody TW, Minna JD: Cancer Res 47:821–825, 1987.
28. Zachary I, Rozengurt E: J Biol Chem 262:3947–3950, 1987.
29. Elder JH, Alexander S: Proc Natl Acad Sci USA 79:4540–4544, 1982.
30. Todaro GJ, Green H: J Cell Biol 17:299–313, 1963.
31. Dicker P, Rozengurt E: Nature 287:607–612, 1980.
32. Laemmli UK: Nature 227:680–685, 1970.
33. Woll PJ, Rozengurt E: Proc Natl Acad Sci USA (in press), 1988.
34. Zachary I, Rozengurt E: EMBO J 6:2233–2239, 1987.
35. Kris RM, Hazan R, Villines J, Moody TW, Schlessinger J: J Biol Chem 262:11215–11220, 1987.
36. Osawa T, Tsuji T: Ann Rev Biochem 56:21–42, 1987.
37. Koide N, Muramatsu T: J Biol Chem 249:4897–4904, 1974.
38. Carpenter G, Cohen S: J Cell Biol 71:159–171, 1976.
39. Das M, Fox CF: Proc Natl Acad Sci USA 75:2644–2648, 1978.
40. Knipp MN, Connolly DT, Lane MD: J Biol Chem 257:11489–11496, 1982.
41. Bowen-Pope DF, Ross R: J Biol Chem 257:5161–5171, 1982.
42. Heldin C-H, Wasteson A, Westermark B: J Biol Chem 257:4216–4221, 1982.
43. Pike LJ, Bowen-Pope DF, Ross R, Krebs EG: J Biol Chem 258:9383–9390, 1983.
44. Stoscheck CN, Carpenter G: J Cell Biol 98:1048–1053, 1984.
45. Beguinot L, Lyall RM, Willingham MC, Pastan I: Proc Natl Acad Sci USA 81:2384–2388, 1984.
46. Goldstein JL, Brown MS, Anderson RGW, Russell DW, Scheider WJ: Ann Rev Cell Biol 1:1–39, 1985.
47. Masu Y, Nakayama K, Tamaki H, Herada Y, Kuno M, Nakanishi S: Nature 329:836–838. 1987.
48. Gilman A: Ann Rev Biochem 56:615–650, 1987.
49. Zachary I, Millar J, Nanberg E, Higgins T, Rozengurt E: Biochem Biophys Res Commun 146:456–463, 1987.
50. Cirillo DM, Gaudino G, Naldini L, Comoglio PM: Mol Cell Biol 6:4641–4649, 1986.

Journal of Cellular Biochemistry 39:1–11 (1989)
Growth Factors and Their Receptors: Genetic
Control and Rational Application 71–81

Novel Method for Studying mRNA Phenotypes in Single or Small Numbers of Cells

Daniel A. Rappolee, Alice Wang, David Mark, and Zena Werb

Laboratory of Radiobiology and Environmental Health, University of California, San Francisco, California 94143 (D.A.R., Z.W.); Department of Molecular Biology, Cetus Corporation, Emeryville, California 94608 (A.W., D.M.).

Biological processes, such as growth control, are often governed by biochemical steps involving mRNA transcripts that are short-lived and have a low copy number. Furthermore, the cells involved in these processes are often available in low numbers from in vivo sources. We now report a method that is superior to in situ hybridization, RNA blot analysis, and the nuclease protection assay for the study of short-lived, low-copy-number mRNA transcripts. The method consists of a microprocedure for isolating RNA from one to a few thousand cells and two coupled enzymatic steps: reverse transcription of whole cellular RNA, followed by amplification of the cDNA by a specifically primed polymerase chain reaction to give specific cDNA fragments that can be visualized on agarose gels by ethidium bromide staining. With this method we have detected actin mRNA from a single cell, or < 100 cRNA molecules, and have quantified differences in RNA concentrations of less than threefold. The reverse transcription reaction products can be divided for the polymerase chain reaction, and several mRNA species can be assayed simultaneously. Therefore, we call the method single-cell mRNA phenotyping. This technique is applicable to the analysis of low-copy-number growth factor transcripts in cells in culture and in vivo.

Key words: cDNA, mRNA, reverse transcription, polymerase chain reaction, single-cell mRNA detection, actin, growth-factor transcripts

The study of growth factor transcription has been slowed by a number of obstacles. Growth factors are generally transcribed in low copy numbers and have a short half-life [1]. In addition, often low numbers of primary cells can be obtained from in vivo sources. In situ hybridization analysis for mRNA in cells in tissue is

Abbreviations used: GuSCN, guanidine thiocyanate; IL-1α, interleukin-1α; MMLV–RT, Moloney murine leukemia virus reverse transcriptase; NGF, nerve growth factor; PCR, polymerase chain reaction; RT, reverse transcription; Taq, *Thermus aquaticus*.

Received March 22, 1988; accepted July 6, 1988.

difficult and insensitive [2]. One approach has been to use continuous cell lines to obtain sufficient material for growth factor studies, but immortalization of cell lines is often produced by ectopic expression of the growth factors in question [3].

Two developments have now permitted development of a novel method for RNA analysis. First, a method for cell-free DNA replication was reported [4]. This method, the polymerase chain reaction (PCR), involves two oligonucleotide primers that flank the DNA sequence to be amplified, repeated cycles of thermal denaturation of the DNA, annealing of primers to their complementary sequences, and primer extension to give an exponential accumulation of the target fragments. PCR has been used for selective enrichment of DNA sequences and sequence variations and for high-efficiency cloning of genomic sequences and has been improved by the use of a newly isolated thermostable DNA polymerase from *Thermus aquaticus* (Taq) [5]. In addition, the reverse transcriptase from Moloney murine leukemia virus (MMLV—RT), which produces cDNA from whole cellular RNA mixtures, was recently isolated [6].

We now report that a microadaptation of the guanidine thiocyanate/cesium chloride (GuSCN/CsCl) gradient ultracentrifugation technique [7], followed by MMLV–RT production of cDNA and PCR amplification of specific cDNA subfragments, allows the quantitative assay of multiple transcript types from a low number of total transcripts. We have found that a) a single cell yields a signal for β-actin mRNA; b) as few as 100 cRNA transcripts can be detected; c) each cDNA molecule can generate 10^9 to 10^{10} PCR fragments; and d) $<$ threefold differences in mRNA can be resolved after PCR amplification of cDNA.

MATERIALS AND METHODS

Materials

Cells were thioglycollate-elicited peritoneal macrophages [8] harvested on day 4 after stimulation from CF1 mice (Charles River Laboratories). Hep-SK cells were obtained from the American Type Culture Collection. Restriction endonucleases were obtained from BRL and Promega Biotech and used according to the manufacturer's instructions. Taq DNA polymerase and DNA Thermocycler programmable heating block were obtained from Perkin Elmer-Cetus Corporation. MMLV–RT was purchased from BRL. cDNA probe for chick actin was a gift of M. Kirschner [9]. Primers were purchased from UCSF Biomolecular Resource Center or were provided by Cetus Corporation. [^{32}P]dCTP (3,000 Ci/mmol) was purchased from Amersham. SeaKem ME and NuSieve GTG agaroses were purchased from FMC Corporation. DNA 1-kb molecular weight standards were purchased from BRL, and Zeta-bind nylon membranes were purchased from Bio-Rad. Apparatus for agarose gel electrophoresis was purchased from American Bionetics. CsCl was purchased from Pharmacia, and GuSCN was purchased from Fluka. Oligo-dT primers and dNTPs (dATP, dCTP, TTP, dGTP) were purchased from Pharmacia. RNAsin and T7 polymerase were purchased from Promega Biotech. Acetylated bovine serum albumin was purchased from Sigma. Nerve growth factor (NGF) cRNA, produced from an SP6 clone, was a gift of Dr. Dennis Clegg. The poly A^+-pBR322-IL-1α cRNA was synthesized from a plasmid template (D. Mark, unpublished results) by T7 polymerase, purified by oligo-dT chromatography, and quantified by absorbance at 260 nm.

The mouse β-actin [10] primers were (5′-primer) 5′-GTGGGCCGCTCTAGGCACCA-3′, 5′ = residue 125, and (3′-primer) 5′-TGGCCTTAGGGTGCAGGGGG-3′, 3′ = residue 264. The mouse NGF [11] primers were (5′ primer) 5′-CCAAGGACGCAGCTTTCTAT-3′, 5′ = residue 246, and (3′ primer) 5′-CTCCGGTGAGTCCTGTTGAA-3′, 3′ = residue 649.

RNA Preparation

RNA was prepared by a microadaptation of the GuSCN/CsCl gradient ultracentrifugation method [7]. Briefly, macrophages (as few as one cell per microtiter well, as determined by microscopic examination) were cultured, washed with 0.9% NaCl, lysed in 100 μl of 4 M GuSCN containing 10–20 μg *Escherichia coli* ribosomal RNA as carrier, layered over 100 μl 5.7 M CsCl, and centrifuged for approximately 20×10^6 *g*-min/cm of gradient length, in a Beckman TL-100 tabletop ultracentrifuge or in a Beckman Airfuge. RNA was ethanol precipitated, centrifuged, and then washed in 80% ethanol. The dried RNA pellet was redissolved directly in reverse transcription (RT) buffer (described below) or in diethyl pyrocarbonate-treated H_2O containing RNAsin.

Reverse Transcription

RNA was reverse transcribed into cDNA essentially as previously described [6]. RNA (from one cell equivalent to 1 μg) was incubated at 37°C for 60 min with a mixture of 100 U of MMLV–RT and the following reagents: 0.2 μg oligo-dT primer, 3 mM $MgCl_2$, 10 mM Tris-HCl buffer, pH 8.3, 75 mM KCl, 1 μg acetylated bovine serum albumin, 0.5 mM dNTP, and 5 U of RNAsin in 10 μl volume. The RT can be repeated by addition of 50 U of fresh MMLV–RT after a 93°C, 5-min denaturing step, followed by flash cooling to 4°C.

Polymerase Chain Reaction

PCR was performed essentially as previously described [5]. A small portion of RT products (1 μl) was mixed with 1 U of Taq DNA polymerase, 50 pmol of 20–25 base-long oligonucleotide 5′ and 3′ sequence-specific primers, in a buffer containing 10 mM Tris-HCl, 2.5 mM $MgCl_2$, 50 mM KCl, 5 μg acetylated bovine serum albumin, pH 8.3, in 50 μl volume. The mixture was overlaid with mineral oil to prevent evaporation and then amplified by PCR in a repeated 3-temperature cycle on the Thermocycler programmable heating block. The temperature used in the annealing cycle was varied because groups of primers had different melting temperatures; temperatures were also lowered when primers were used in other species for which the DNA sequence was not known. For samples to be amplified for >60 cycles of PCR, 1 U of Taq polymerase was added at 60 cycles.

Agarose Gel Electrophoresis

A portion of the PCR mixture (5 μl) was added to 4 μl of loading dye mix and electrophoresed in an 80-V constant-voltage field in 3% NuSieve GTG/1% SeaKem ME agaroses until the bromphenol blue dye front had migrated 6 cm. Gels were stained for 10 min in ethidium bromide and destained in H_2O for at least 30 min.

DNA Blot Analysis

DNA separated on agarose gels was transferred to Zeta-bind nylon membranes as previously described [12]. Gels were incubated for two 10-min periods at ambient temperature with 0.25 M HCl, incubated overnight in 0.4 M NaOH, and rinsed in 2 × SSC (1 × SSC = 0.15 M sodium chloride, 0.015 M sodium citrate). cDNA probes were prepared by random priming [13]. Zeta-bind transfers were hybridized by standard techniques [13] and washed at 60°C for 60 min in 0.2 × SSC containing 0.1% SDS before autoradiography.

Restriction Enzyme Analysis

PCR products were ethanol precipitated and resuspended in the appropriate restriction enzyme buffer. Three-quarters of the resuspended products were digested, and 5 μl of this digest and 5 μl of the undigested products were then separated on a 4% agarose gel as described above.

Quantitative Analysis

Two procedures were used to obtain quantitative data: a) An RT mixture from thioglycollate-elicited macrophages was mixed with PCR reagents (those previously described plus 100 μCi of [^{32}P]dCTP, spec. act. 3,000 Ci/mmol) in 2-μl aliquots after 1/100, 1/10, or no dilution. Samples of this mixture (5 μl) were removed after 35, 40, 45, 50, and 55 cycles of the PCR, mixed with loading dye, and then separated on a 1.5-mm-thick 5% polyacrylamide/7 M urea gel [13] and run at 20 mA constant current until the bromphenol blue front had migrated 9 cm into the separating gel. The gel was autoradiographed, appropriate bands were cut from the gel, and radioactivity was determined by scintillation fluorography. Data were plotted by exponential curve fitting with an Apple Cricket Graph® program. b) Samples of pBR322-IL-1α cRNA were serially diluted threefold and mixed with 1 μg of Hep-SK RNA so that the RT mixture contained 3×10^2 to 1.3×10^7 copies of cRNA. After RT, the cDNA was amplified by PCR for 40 cycles in the presence of [^{32}P]dCTP to follow incorporation into the specific cDNA fragment band.

RESULTS AND DISCUSSION

A diagram outlining the steps in the RNA phenotyping analysis is shown in Figure 1. The first step was to obtain the mRNA, which can be done by any conventional procedure for obtaining total or cytoplasmic RNA from tissue or large numbers of cells [13] or by a micromethod for obtaining total RNA from 1 to 10^5 cells (Fig. 1a). The RNA was then reverse transcribed to prepare cDNA (Fig. 1b). This cDNA preparation was divided as desired for analyzing specific mRNA transcripts. For each mRNA species to be analyzed, a designated sequence (Fig. 1b, arrowheads) was selected, and 5′ and 3′ sequence oligonucleotides were prepared according to the criteria outlined in Table I. The primers were added to the cDNA preparation along with Taq polymerase and dNTPs. The cDNA was then thermally denatured and afterward cooled to allow annealing of primers, and the temperature was then increased to the optimum for primer extension by Taq polymerase (Fig. 2). In the first PCR cycle only the 5′ primers annealed to the cDNA from the RT (Fig. 1c); after primer extension a cDNA sequence extending from the 5′ primer for various lengths, up to and including the 3′ end of the cDNA, was obtained. In the second cycle, the 3′ primers annealed to the DNA synthesized in the first cycle, and the 5′ primers annealed to the original cDNA; after primer extension the first copies

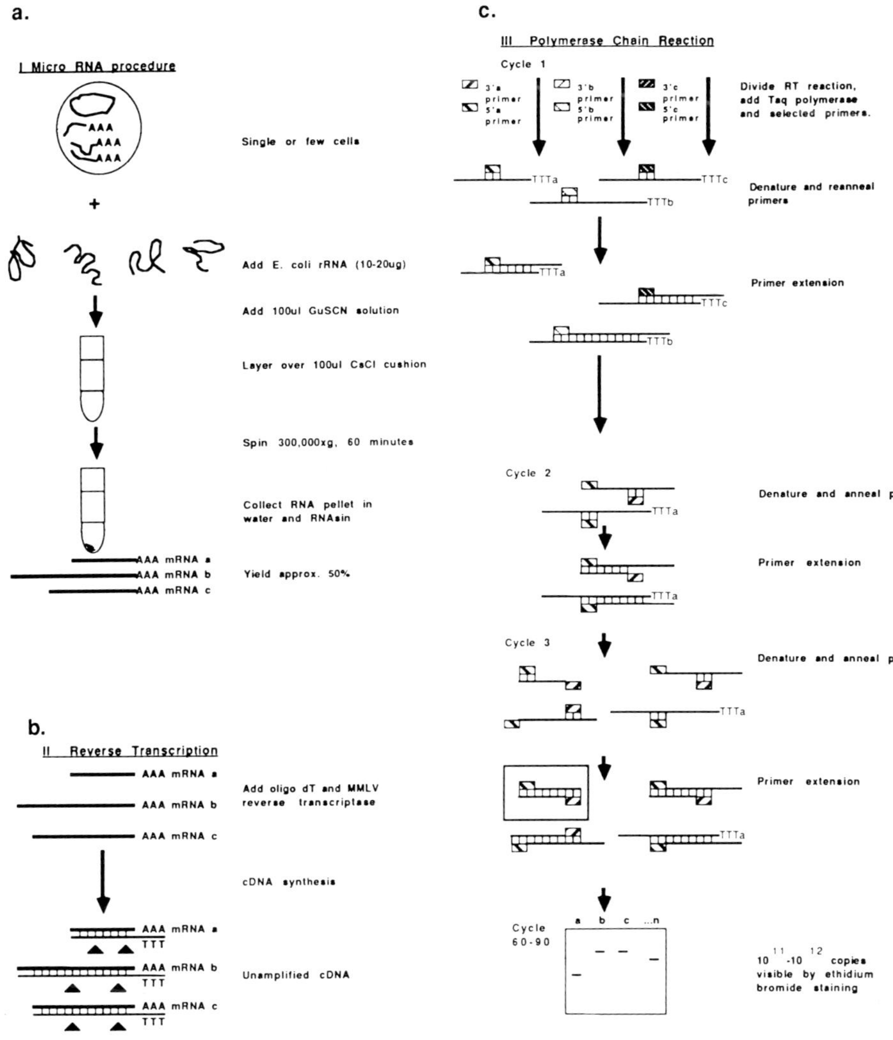

Fig. 1. Steps involved in mRNA phenotyping in low numbers of cells. **a:** Purification of whole cellular RNA from small numbers of cells. **b:** Reverse transcription using MMLV–RT. The arrowheads indicate the designated sequences for amplification by PCR. **c:** Early PCR cycles and the visualization of PCR products in an agarose gel. Note that cycles 2 and 3 focus on amplification of cDNA a only, but the schematic of the agarose gel at bottom represents cDNA a, b, and c.

of DNA of the exact size of the designated sequence flanked by the primers were produced. In subsequent cycles, this exact-size DNA became dominant, so that by 30 to 90 cycles, the 10^{11} to 10^{12} copies of cDNA, visible by ethidium bromide staining, contained only this specific sequence (Fig. 1c). The intensity of the ethidium bromide signal was related to the number of fragments in each well over a ~20-fold dilution range (data not shown). The ideal equation describing the generation of fragments in

TABLE I. Criteria for Choosing Oligonucleotide Primer Pairs for RNA Phenotyping Analysis

1. Target mRNA sequences should be ~0.2 to 0.5 kb in length.
2. For most efficient reverse transcription, target sequences should be toward the 3′ end of the coding region of the mRNA sequence; in general the 3′ untranslated region is not used because of the lack of conservation of these between species.
3. Target sequences should contain an intron/exon border so that genomic DNA and mRNA can be distinguished.
4. Target sequences should contain a diagnostic restriction endonuclease cleavage site to permit analytical validation of fragments.
5. Melting temperature of the two oligonucleotide sequences should be similar and close to the extension temperature in the assay.
6. Target sequences should be within available cDNA probes for validation by DNA blot analysis.

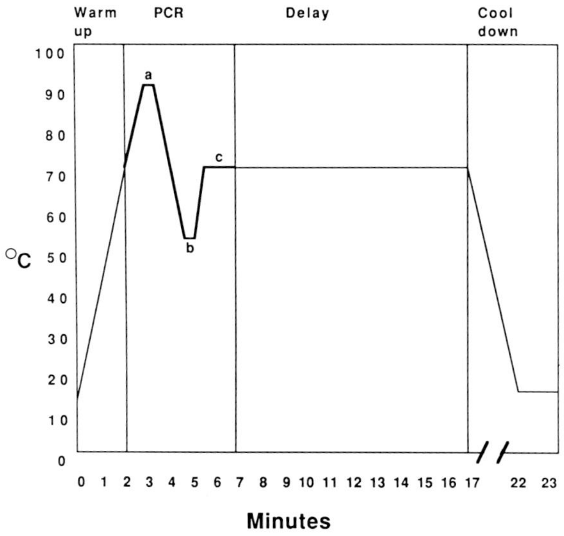

Fig. 2. The four stages in the operation of the programmable heating block for PCR. The three-temperature PCR cycle is repeated 30 to 90 times. PCR kinetics: **a** = dimer denaturing (95°C, 30 s); **b** = primer-template reannealing (50–70°C, 30 s); **c** = primer extension (72°C, 90 s).

the PCR is exponential, similar to equations for describing bacterial growth and compound interest:

$$N = N_0(\text{eff.})^n$$

where N is number of amplified fragments, N_0 is number of input cDNA molecules, eff. is efficiency (1–2 is equivalent to 0–100%), and n is cycle number. However, empirically we have found that cDNA generates fragments efficiently and exponentially only during the first 10–20 cycles of the PCR. PCR cycles 30–60 are often only 10% efficient (Fig. 3). These data also show that the theoretical equation does not describe the generation of the fragments. Moreover, extrapolation of the data by curve fitting does not work and shows that small changes in the efficiency of reaction make it impossible to predict N_0 from late cycles, except with the use of a standard, as shown below.

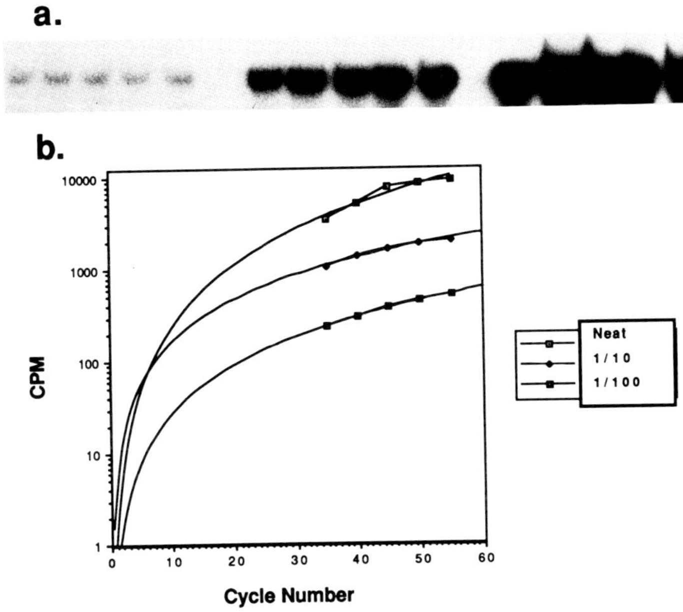

Fig. 3. Quantification of differences in input actin mRNA by sampling at various cycles. The efficiency of the generation of PCR products declines after the first 10 to 20 cycles. Note that tenfold differences in input cDNA result in resolvable differences in output specific signal at all time points measured. These data were obtained by polyacrylamide gel electrophoresis of ^{32}P-labeled cDNA products of PCR as described in Materials and Methods. **a:** Autoradiogram of gel. The three sections are 1/100, 1/10, and undiluted mRNA preparations, respectively, sampled at 35, 40, 45, 50, and 55 cycles; **b:** cpm of each band of the gel shown in **a**.

A single cell with approximately 1,000 β-actin transcripts [14] generated a signal after RT-PCR that was detectable by ethidium bromide staining after 65 PCR cycles (Fig. 4). With increasing actin mRNA input from 10^2 and 10^4 macrophages, the signal appeared at earlier cycles. At large cycle numbers, there was some variability in the strength of the ethidium bromide signal for individual samples; therefore the incorporation of [^{32}P]dCTP into the PCR-generated fragments was the preferred method for quantitative analysis. Signals were validated by four means: a) the size of the fragment; b) the presence of diagnostic restriction sites (Fig. 4a); c) detection by cDNA probes; and d) direct or indirect sequencing. The results from analysis of a single cell suggested that PCR is less efficient when cDNA mixtures, compared with genomic DNA mixtures, are used. Because the input number of cDNA molecules must be in the range of 1–1,000, we estimated (using the equation given earlier and a 10^{12}-fragment detection limit) that the PCR is 35–55% efficient when cDNA mixtures are used as templates. The PCR has previously been reported to be 67–90% efficient when genomic mixtures are used [4,5]. This discrepancy in efficiency is currently under investigation.

To test the sensitivity of the RT-PCR procedure, we used synthetic NGF cRNA produced from a plasmid containing the SP6 polymerase promoter. We have found that 100 copies of input cRNA can be detected by this method (Fig. 5). This finding agrees with the result from the single-cell experiment shown in Figure 4. In fact, because the input cRNA was not polyadenylated, the 3′ primer was used for priming

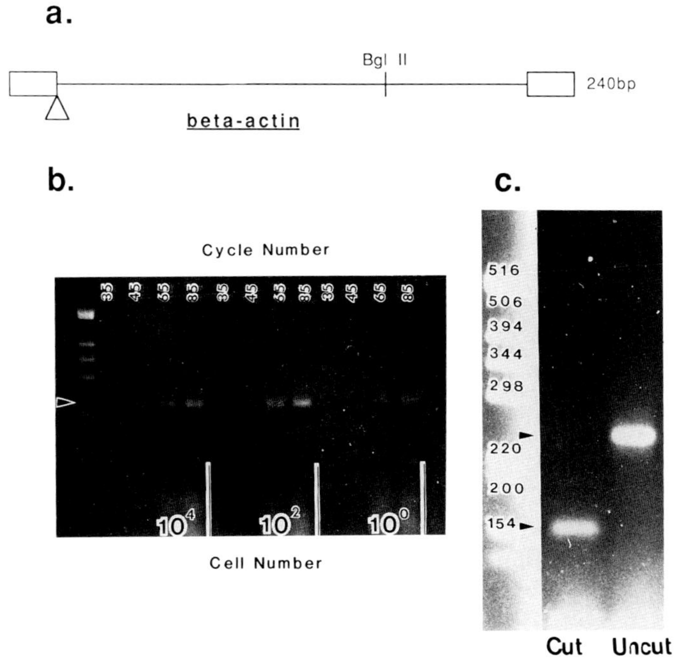

Fig. 4. Generation of a signal detectable by ethidium bromide staining from a single cell. RNA from 1, 10^2, and 10^4 macrophages was reverse transcribed, and the PCR analysis was performed with actin oligonucleotide primers. PCR products were separated by agarose gel electrophoresis, and the gel was stained with ethidium bromide. **a:** Diagram of target β-actin sequence between primers (**boxes**). The position of an intron (**triangle**) and the position of the *Bgl*II restriction cleavage site are shown. **b:** Initial signal validation by size (**arrowhead**). Numbers at the top are PCR cycles. **c:** Further signal validation by restriction enzyme cleavage of β-actin fragment with *Bgl*II (**Cut**). DNA molecular weight standards (in bp) are shown at left. **Arrowheads** indicate uncut (**top**) and cut (**bottom**) β-actin fragments.

the RT. Because the latter reaction is known to have a lower efficiency than oligo-dT priming, the sensitivity of the procedure may approach 1 copy of input mRNA with oligo-dT primer. Under ideal conditions the MMLV–RT is 10–50% efficient [13].

A difficulty in quantification is suggested by the equation we have been using ($N = N_0(\text{eff.})^n$). A small difference in efficiency could lead to large differences in N. Ideally, small input (N_0) mRNA could be quantified most accurately by using an exogenous cRNA with the same sequence (from the 5′ primer to the poly A tail) as the endogenous mRNA in question. The exogenous cRNA would be diluted serially as an internal standard, and then the amount of endogenous mRNA could be extrapolated from the equation of the line. For this endogenous standard to succeed, we first needed to analyze the resolving power of the RT–PCR procedure. In the experiment shown in Figure 6, 6.5×10^3 to 1.3×10^7 copies of specific cRNA were used in the RT-PCR in threefold dilutions. The results demonstrate that the RT–PCR can resolve < threefold differences in N_0 cRNA, with the greatest sensitivity at low N_0. These data will allow us to construct standard curves in an attempt to extrapolate a copy number for the low-copy N_0 endogenous mRNA. Input RNA present in samples of tissue in which input cell number cannot be determined accurately can be measured by RNA blot analysis of the cellular ribosomal RNA with cDNA probes

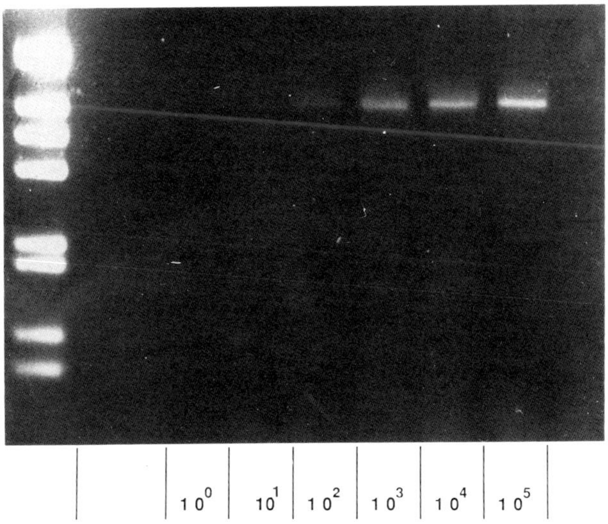

Fig. 5. Detection of a threshold of 100 molecules of NGF cRNA after RT-PCR, followed by agarose gel electrophoresis and ethidium bromide staining after 60 cycles of PCR.

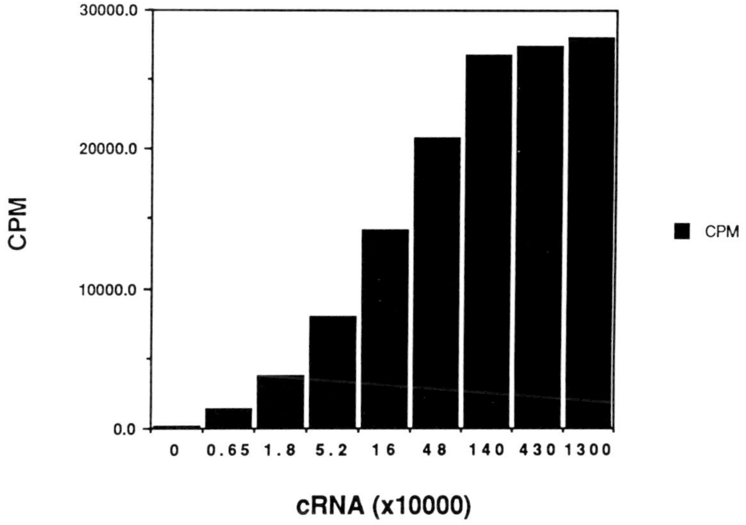

Fig. 6. Resolution of threefold differences in input (N_0) pBR322-IL-1α cRNA after PCR amplification of cDNA. Samples were serially diluted, then subjected to RT–PCR. The ^{32}P-labeled cDNA products were separated by gel electrophoresis and the radioactivity was determined.

for vertebrate 28S and/or 18S rRNA. This method can easily detect the rRNA in 2–10 cells (unpublished observation).

The combination of techniques for reverse transcribing mRNA into cDNA by using MMLV–RT and amplifying specific cDNA segments > 10^9-fold (using Taq DNA polymerase in the PCR) has allowed us to develop a method for phenotyping mRNA in low numbers of cells or with low amounts of mRNA. Some of the advantages and disadvantages of the method are listed in Tables II and III. The major

TABLE II. Advantages of RNA Phenotyping Method

1.	Rapid: results in < 1 day
2.	Sensitive: < 100 mRNA molecules detected
3.	Low numbers of cells: RNA from 1 to a few thousand cells can be analyzed
4.	mRNA phenotyping: > 10 different mRNA species within a single sample can be analyzed
5.	Rare mRNA: specific detection is possible
6.	Resolution: < threefold differences can be determined
7.	Alternative splicing: easy analysis
8.	Multiple species: primers often work across species
9.	Rapid cDNA cloning: preparation of species-specific probes is possible
10.	Rapid sequencing: amplified sequences can be sequenced directly
11.	Safe: analysis does not require ^{32}P
12.	Probes: rapid preparation of high specific activity probes
13.	Easy: many samples can be analyzed simultaneously
14.	Related molecules: easy discrimination between molecules

TABLE III. Disadvantages of RNA Phenotyping Method

1.	Does not give mRNA sizes directly
2.	Sensitive to mRNA secondary structure
3.	Does not give three-dimensional information about transcript distribution in cells and tissues
4.	Quantification takes effort
5.	Less resolution (3:1) than other methods
6.	Some sequence information needed
7.	Error rate is high (1/6000) for Taq polymerase

advantages of this novel RNA phenotyping procedure are the ease and rapidity with which mRNA transcripts from limited material can be unambiguously demonstrated. The simultaneous analysis of up to ten different mRNA species in a single small sample of RNA is possible [15]. Specific detection of rare mRNA species, such as those for growth factors, is possible [15,16]. By specific primer design, alternatively spliced mRNAs can be determined. With careful design (e.g., use of conserved amino acids with low degeneracy of codons), primers often work for several species (e.g., human PDGF-A in mouse) [15], even when the species-specific cDNA sequences have not been determined. The PCR fragments so generated can be sequenced and also used as homologous cDNA probes for other types of RNA analysis [17]. Specific primer design or specific restriction endonuclease analysis allows rapid analysis and discrimination of closely related molecules in very small samples [17]. Simultaneous analysis of several dozen samples differing in either input RNA and/or primer pairs is possible.

The major disadvantage of this method is that the size of mRNA species is not a direct result of the analysis. Knowledge about alternative splicing patterns is needed to circumvent this problem. As is true for any reaction involving reverse transcription, the technique is sensitive to secondary structure in mRNA. Of > 30 primer pairs designed, a few have been unsuccessful even for abundant mRNA species. Not all primer pairs work with the same efficiency or reliability. Quantitative analysis is cumbersome and difficult, although semiquantitative data can be obtained readily with dilution curves. Although this RT–PCR procedure can be more sensitive than in situ hybridization and has shown the presence of mRNA transcripts not found previously by the latter procedure, it does not give three-dimensional information about transcript distribution in cells and tissues. Good sequence homology is needed to cross species. Although the error rate for Taq polymerase has been estimated at about 1/6,000, only one base-pair error has been detected in sequencing about 5,000 bp of known sequence [17]. However, this could present a problem in analysis of point mutations in transcripts.

The potential exists for obtaining these data quantitatively. The RNA phenotyping procedure is superior in ease, speed, sensitivity, and resolution to mRNA analysis by in situ hybridization, RNA (Northern) blots, and nuclease protection assays. In other work, we have used this procedure to demonstrate growth-factor transcripts in macrophages stimulated by inflammatory agents in culture and in macrophages obtained directly from healing wounds in vivo [15]. We have also used this method to analyze the regulation of maternal and embryonic transcripts for growth factors [16] and the regulation of proteolytic enzymes [17] in single eggs and preimplantation embryos.

ACKNOWLEDGMENTS

This work was supported by the Office of Health and Environmental Research, U.S. Department of Energy (contract DE-AC03-76-SF01012). D.A.R. was the recipient of a U.C. Regents Fellowship and a U.C. Patent Fund Award. We thank Mike Doyle for providing the Hep-SK cells, Mary McKenney for editing the manuscript, and Renee Horiuchi for typing it.

REFERENCES

1. Shaw G, Kamen R: Cell 46:659, 1986.
2. Abraham J: Personal communication.
3. Ross R, Raines EW, Bowen-Pope DF: Cell 46:155, 1986.
4. Saiki RK, Scharf S, Faloona F, Mullis KB, Horn GT, Erlich HA, Arnheim N: Science 230:1350, 1985.
5. Saiki RK, Gelfand DH, Stoffel S, Scharf SJ, Higuchi R, Horn GT, Mullis KB, Erlich HA: Science 239:487, 1988.
6. Gerard GF: Focus (Bethesda Research Labs) 9:5, 1987.
7. Chirgwin J, Przybyla R, MacDonald R, Rutter W: Biochemistry 18:5294, 1979.
8. Werb Z, Chin JR: J Exp Med 158:1272–1293, 1983.
9. Cleveland DW, Lopata MA, MacDonald RJ, Cowan NJ, Rutter WJ, Kirschner MW: Cell 20:95, 1980.
10. Alonso S, Minty A, Bourlet Y, Buckingham M: J Mol Evol 23:11, 1986.
11. Scott J, Selly M, Urdea M, Quiroga M, Bell GI, Rutter WJ: Nature 302:538, 1983.
12. Reed KC, Mann DA: Nucl Acids Res 13:7207, 1985.
13. Maniatis T, Fritsch E, Sambrook J: "Molecular Cloning, A Laboratory Manual." Cold Spring Harbor, NY, Cold Spring Harbor Laboratory, 1982.
14. Werb Z: Unpublished results.
15. Rappolee DA, Mark D, Banda MJ, Werb Z: Science, 241:708, 1988.
16. Rappolee DA, Brenner CA, Schultz R, Mark D, Werb Z: Science, 241:1823, 1988.
17. Brenner CA, Adler RR, Rappolee DA, Pedersen RA, Behrendtsen O, Werb Z: Manuscript submitted for publication.

Journal of Cellular Biochemistry 39:13–23 (1989)
Growth Factors and Their Receptors: Genetic
Control and Rational Application 83–93

Characterization of Tumors Produced by Signal Peptide-Basic Fibroblast Growth Factor-Transformed Cells

Snezna Rogelj, Robert A. Weinberg, Paul Fanning, and Michael Klagsbrun

The Whitehead Institute for Biomedical Research, Cambridge, Massachusetts 02142 (S.R., R.A.W.); Children's Hospital, Harvard Medical School, Boston, Massachusetts 02115 (P.F., M.K.)

Basic fibroblast growth factor (bFGF) is found in a variety of cells and tissues. We have previously shown that bFGF is a transforming growth factor, but only when fused to a signal peptide (sp-bFGF). Cells expressing the native bFGF are tumorigenic in nude mice only, where the tumors form at a low frequency and grow very slowly as compared to sp-bFGF tumors. The cells transformed by the sp-bFGF growth factor gene cause rapidly growing tumors within 10 days in 100% of syngeneic and nude mice. In nude mice, the tumors are highly vascularized, while the vascularization in immunocompetent syngeneic mice is not as prominent. The syngeneic mice have a characteristic humoral immune response to sp-bFGF tumors, which differs from that mounted against *ras*-induced tumors. The ability of bFGF to induce tumorigenicity is significant in view of the recent discoveries of three new oncogenes: *hst*, *int-2,* and an oncogene from a human colon cancer. In addition to homology with FGF, the proteins encoded by these oncogenes all have a potential signal peptide at the protein's amino terminus, suggesting a mode of action analogous to that of our artificial signal peptide-bFGF (sp-bFGF) transforming growth factor model system.

Key words: autocrine transformation, angiogenesis, bFGF, heparin, oncogene, tumorigenicity, signal peptide

A number of distinct growth factors have been found associated with tumor cells. These growth factors have been associated with several properties of the tumor mass including angiogenesis—the generation of new capillary beds. Although the natural triggers of this process are unknown, neovascularization can be induced in certain in vivo models by several mitogens, including the basic fibroblast growth factor (bFGF) [1]. bFGF, originally isolated from brain and pituitary, is found in richly vascularized tissues such as placenta, retina, kidney, corpus luteum, and numerous solid tissues, as well as in vascular endothelial cells in culture. Analysis of bFGF synthesis in cell culture indicates that bFGF is primarily a cell-associated

Received March 3, 1988; accepted August 18, 1988.

growth factor. This lack of secretion is consistent with the apparent lack of a consensus signal peptide domain in the bovine and human bFGF genes. It is conceivable that this absence of a well-defined signal peptide is responsible for its failure to be secreted.

A family of new oncogenes encoding bFGF-like proteins was recently isolated from a Kaposi sarcoma cell line (*hst*), several human stomach cancers (*hst*), a human bladder cancer, and mouse mammary gland tumors caused by MMTV virus (*int-2*)[2–6]. The proteins encoded by these oncogenes share about 45% amino acid homology with the basic and acidic FGF. In addition, they all code for potential amino terminus signal peptide sequences.

bFGF is mitogenic for a broad spectrum of cells; probably all cells of mesodermal and neuroectodermal origin express FGF receptors on their surfaces [1,7]. The autocrine model of transformation would predict that bFGF can act as a transforming agent in any of these cell lines as long as it has access to the receptor, activation of which would then lead to constitutive proliferation. Presumably, access to the cell surface receptor can be achieved via secretion of the growth factor, or by co-compartmentalization of the growth factor and its receptor [8–11].

We have recently generated a series of NIH-3T3-derived cell lines that stably express various levels of bFGF with and without an artificial signal peptide. Only the cell lines expressing the chimeric signal peptide-bFGF protein were morphologically transformed and tumorigenic in syngeneic mice [12].

The tumors formed by our chimeric sp-bFGF expressing cell lines may be potential models for those tumors from which the above-mentioned FGF-homologous oncogenes were isolated. In this report we characterize the model tumors in terms of bFGF production, vascularization, growth in syngeneic and nude mice, and the humoral immune response of the animal to the tumor.

MATERIALS AND METHODS

Cell Lines Expressing the Native and the Chimeric bFGF [12]

Cell lines expressing various amounts of the native or the chimeric bovine brain bFGF were generated as described previously [12]. In brief, mammalian expression vectors directing expression of the normal bFGF protein or of a chimera consisting of bFGF fused to the immunoglobulin heavy chain signal peptide were transfected into NIH-3T3 cells, and stable clonal cell lines were isolated. Control cell lines (NIH-CONT) transfected only with the dominant selectable marker pSV2-neo were simultaneously generated. The bFGF protein production by these cell lines was measured by the ability of cell lysates or conditioned media to stimulate DNA synthesis in Balb-c/3T3 cell cultures [13,14]. In parallel, a *ras*-transformed NIH-3T3 cell line, EJ 62 Bam 6a [15], was transfected with the above expression plasmids, and stable, neomycin-resistant native or chimeric bFGF-expressing cell lines were selected.

Cell line nomenclature is summarized as follows: NIH, derived by transfection of NIH-3T3 mouse fibroblasts; B, expresses the native bovine brain bFGF; IgB, expresses the chimeric sp-bFGF protein where the immunoglobulin signal peptide has been fused to the native bFGF; N, expresses the dominant selectable marker for neomycin resistance; #, number assigned to that particular clone; CONT, control cell line expressing neomycin resistance; EJ, transformed with the Ha-*ras* derived from the human EJ/T24bladder carcinoma cell line.

Heparin-Sepharose Affinity Chromatography [13,14]

The cells were grown until they reached confluence, and the conditioned medium was removed. The cells (about 10^8 per experiment) were resuspended at 10^7 per ml in 1 M NaCl, 10 mM Tris-HCl, pH 7.0, and lysed by sonication. The cell debris was removed by centrifugation, and the supernate was diluted to 0.1 M NaCl. This cell lysate was applied to 3 ml heparin-Sepharose columns, and bFGF was eluted with a 0.1 to 3 M NaCl gradient (80 ml). The collected fractions (4 ml) were assayed for stimulation of DNA synthesis in BALB-c/3T3 cells [13,14].

Western Blotting

Protein samples were resolved by electrophoresis on sodium dodecyl sulfate-polyacrylamide gel electrophoresis (SDS-PAGE) and electrophoretically transferred to nitrocellulose paper. The nitrocellulose paper was probed with anti-bFGF anti-serum, and the bound antibodies were visualized with ^{125}I-protein A. The anti-bFGF antisera were prepared by immunizing rabbits with synthetic peptides corresponding to positions 1–12 (amino-terminal) and positions 33–43 (internal) of bFGF [13].

Tumorigenicity Studies

Four- to 6-week-old, male NIH-3T3 syngeneic NIH/NSF mice were obtained from NCI, Frederick, MD. They were injected subcutaneously with 2×10^6 cells within a week of arrival. The athymic nude mice (NUDE/NUDE) were also obtained from NCI, Frederick, MD. The day prior to injection with 2×10^6 cells, the nude mice were irradiated with 145 rad/min for 4 minutes from a cobalt source. The animals were observed for tumor formation for 3–4 months.

Histology

The tumors were harvested when they reached the size of about 1 cm^3 and were fixed in 2.5% glutaraldehyde with 2% paraformaldehyde in 0.14 M cacodylate buffer, pH 7.4, for 2 hours at room temperature. The tumors were dehydrated in a graded series of alcohols and embedded in JB4 Plastic (Polysciences). Sections (2 μm thick) were cut on a Reichert 2050 Supercut microtome and stained with hematoxylin and eosin. The histological sections were photographed on a Zeiss photomicroscope.

Analysis of the Humoral Immune Response to bFGF Tumors in Syngeneic Mice

Mice were bled from their tails 2–4 weeks after injection with tumorigenic cells. The blood was allowed to coagulate on ice for 30 minutes, and the sample was centrifuged for 10 minutes at 4°C in an Eppendorf centrifuge. The serum was removed into a fresh container and stored at 4°C, where it has been stable for the last 10 months.

Immunoprecipitation (IP) was used to analyze the serum for specific antibodies to proteins in the producing tumor, as well as in the parental NIH-3T3 cells. IP was carried out as follows: NIH-IgBNM6–1 or NIH-3T3 cells were grown in Dulbecco's minimum essential medium (DMEM) (Gibco) supplemented with 10% calf serum, penicillin and strep tomycin to about 3×10^6 cells per 100-mm dish. The cells were radiolabeled for 12 hours with 5 ml cysteine-free DMEM (Gibco) supplemented with 5% dialyzed calf serum, penicillin, streptomycin, and 50–100 μCi of ^{35}S-cysteine (specific activity 1,000 Ci/mmol, New England Nuclear) per 100-mm dish. After removal of media, the cells were washed twice with phosphate buffered saline (PBS), resuspended into 1 ml of cell lysis buffer (50 mM Tris-HCl, pH 8.0, 100 mM NaCl, 5 mM $MgCl_2$, and 0.5% (v/v) Nonidet P40 detergent), and incubated on ice for 5 minutes. The nuclei were removed with a 5-minute centrifugation in an Eppendorf microfuge. The supernatant was incubated for 4–24 hours with 4 μl of either normal mouse sera or sera collected from a tumor-bearing mouse. Twenty microliters of protein A-Sepharose was added and incubated for 30–60 minutes. The pellet was washed four times with cell lysis buffer containing 1% Triton X-100, 1% sodium deoxycholate, and 1% aprotinin solution (Sigma). Twenty microliters of SDS-PAGE sample buffer was added to the washed pellet, and the sample was resolved by electrophoresis on SDS-PAGE [16].

RESULTS

Cell Lines Expressing Either the Native or the Chimeric bFGF Protein

NIH-3T3 fibroblasts express very low levels of endogenous bFGF, about 0.04 units/10^4cells [12]. (One unit is the concentration of bFGF required to stimulate half-maximal DNA synthesis in Balb-c/3T3 cells [13].) However, they are known to possess bFGF receptors on their cell surfaces and to respond mitotically to exogenous bFGF [13,14]. As described previously, expression plasmids directing production of either the native or the chimeric (signal peptide-containing [sp-]) bFGF were transfected into these cells [12]. This enabled us to generate a series of NIH-3T3-derived cell lines that express various levels of either the native bovine brain bFGF (0.4–4.6 units/10^4 cells), or the chimeric sp-bFGF in which the mouse heavy chain immunoglobulin signal peptide is fused to the amino terminus of the native bovine brain bFGF molecule. More than 20 independent monoclonal cell lines were analyzed for expression of the chimeric sp-bFGF. The highest level of growth factor activity was found to be about 0.8 units/10^4 cells. The chimeric sp-bFGF appears to be toxic to NIH-3T3 cells when produced in higher amounts [12].

Control cell lines and cell lines expressing equivalent levels of the native and the chimeric protein were chosen for further characterization by heparin-Sepharose chromatography and Western blot analysis. Figure 1A shows heparin-Sepharose affinity elution profile for the cell lysates of three selected cell lines. bFGF elutes from heparin-affinity column at a characteristic 1.5 M NaCl [13]. As expected, very little growth factor activity elutes from a column of NIH-CONT control cell lysate. In contrast, lysates of cell lines expressing either the native or the chimeric sp-bFGF both contain heparin-binding, biologically active growth factor that elutes from the column with 1.5 M NaCl. The fractions eluting at 1.5 M NaCl were probed in an immunoblot [14] using an antibody raised against an internal region of bFGF [13]. Purified recombinant bovine brain bFGF was analyzed in parallel for positive control. The results are shown in Figure 1B. Two transfected cell lines, NIH-BNM7 (transfected with the vector directing expression of the native bFGF) and NIH-IgBNM6–1 (transfected with the vector directing expression of the chimeric sp-bFGF), express the 18-kD immunoreactive species that comigrates with bFGF standard. Additional higher molecular weight species (~20, 24, 30, and 35 kD) are present in the cell line expressing the chimeric sp-bFGF. These species may represent the uncleaved form of sp-bFGF molecule as well as some bFGF molecules that were modified post-translationally upon entering the endoplasmic reticulum.

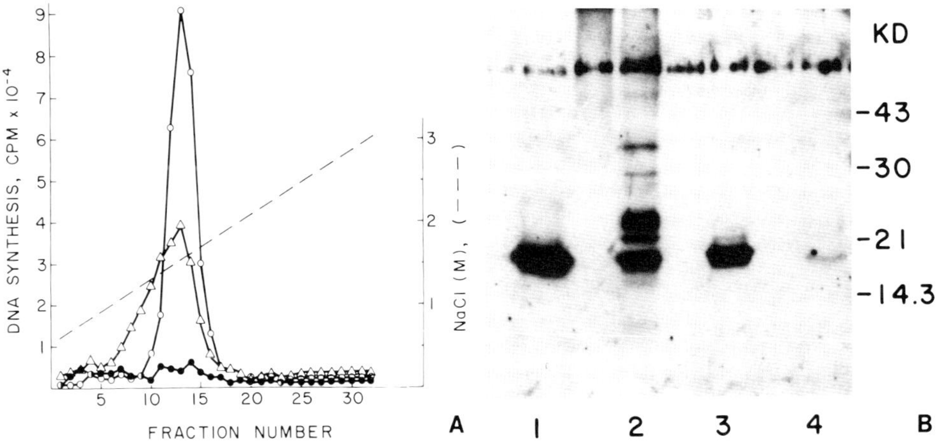

Fig. 1. **A:** Heparin-Sepharose Affinity chromatography elution profiles. Solid circles represent the control NIH-CONT cell lysate. Open circles represent the cell lysate of a bFGF-expressing cell line NIH-BNM7. Open triangles represent the cell lysate of a chimeric sp-bFGF-expressing cell line NIH-IgBNM6-1. **B:** Western blot analysis of the 1.5 M NaCl fractions from the above heparin-Sepharose column. **Lane 1:** Control recombinant bFGF. **Lane 2:** NIH-IgBNM6-I cell lysate. **Lane 3:** NIH-BNM7 cell lysate. **Lane 4:** NIH-CONT cell lysate.

No bFGF growth factor activity could be detected in the conditioned medium of any of the recombinant cell lines. Even when the media of the cell line expressing the chimeric sp-bFGF protein was conditioned in the presence of heparin (45 μg/ml), which stabilized free bFGF [6], no bFGF was detected in the medium. These results were confirmed using anti-bFGF amino-terminus peptide antibodies in an immunoprecipitation analysis (not shown).

Tumorigenicity of Cell Lines Expressing the Native or the Chimeric sp-bFGF

The tumorigenic potential of these and other cell lines was determined in both the immunocompetent syngeneic NIH/NSF and the athymic, immunodeficient nude mice. The results are summarized in Table I. The control cell lines were not tumorigenic in either the immunocompetent or the immunodeficient mouse type. However, biosynthesis of even very low levels of the chimeric sp-bFGF endows the cell with the ability to form tumors in 100% of mice of either type.

The response of the syngeneic mice to the cell lines elaborating the various levels of the native bFGF is different from that of the nude mice. These cell lines are not tumorigenic in immunocompetent mice even at levels as high as 4.6 units/10^4 cells, almost 10 times as much bFGF as is found in the endothelial cells. On the other hand, the tumors in nude mice do form with higher frequency when the cell lines produce elevated levels of the native bFGF. However, unlike the sp-bFGF tumors which appear at 100% frequency within 1–2 weeks of injection, these tumors have a late onset at around 5 weeks postinjection and occur with a frequency of about 50%.

Tumors Derive From the Injected NIH-3T3 Cells That Express bFGF

Neomycin-resistant cell lines were cultured from sp-bFGF tumors in both nude and syngeneic animal models and were analyzed for bFGF production using heparin affinity chromatography and immunoprecipitation. Figure 2 shows the heparin-Se-

TABLE I. Tumorigenicity of bFGF- and sp-bFGF-Expressing NIH-3T3 Cell Lines in the Syngeneic NIH/NSH Mice and the Immunodeficient NUDE/NUDE Mice

Cell lines injected	Units bFGF/ 10^4 cells	Tumors/ NSF mice	Time of appearance	Tumors/ NUDE mice	Time of appearance
NIH-CONT	0.06	0/12	0/4 mo[a]	0/6	0/3 mo[a]
NIH-BNMpc[c]	0.45	0/6	0/4 mo	0/6	0/3 mo
NIH-BNM6	0.6	0/6	0/3 mo	0/6	0/3 mo
NIH-BNM7	0.8	1/6	5 wk/3 mo	4/6	3 wk/3 mo
NIH-BNM46	1.4	0/3	0/5 mo	2/4	5 wk/3 mo
NIH-BNM48	1.4	0/3	0/5 mo	1/4	5 wk/3 mo
NIH-BNM39	3.8	0/3	0/5 mo	3/4	5 wk/3 mo
NIH-BNM35	4.6	0/3	0/5 mo	3/4	5 wk/3 mo
NIH-IgBNM6-1	0.8	38/38	1–2 wk	6/6	1–2 wk
NIH-IgBNM4-2	0.7	6/6	1–2 wk	4/4	1–2 wk
NIH-IgBNM29-1	0.7	6/6	1–2 wk	4/4	1–2 wk
NIH-EJ1	0.1	6/6	1–2 wk	3/3	1–2 wk
NIH-EJB10	2.2	6/6	1–2 wk	—	—
NIH-EJIgB13	0.8	3/3	1–2 wk	—	—

[a]0/4 mo means that no tumors in 4 months of observation.
[b]5 wk/3 mo means that the only tumors observed reached the size of ~0.5 cm^3 after about 5 weeks. The rest were tumor free for 3 months.
[c]NIH-BNMpc is a polyclonal neomycin-resistant cell line that expresses bFGF.

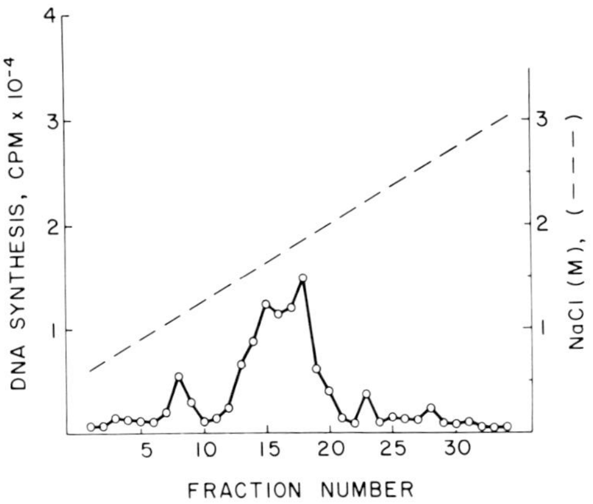

Fig. 2. Heparin-Sepharose affinity chromatography. Elution profile of a cell line derived from NIH-IgBNM6-1 tumor in an NIH/NSF mouse.

pharose elution profile of a cell line derived from a tumor that arose in an NIH/NSF mouse upon injection with the NIH-IgBNM6-1 cells. This neomycin-resistant cell line still expresses growth factor activity that elutes from heparin with 1.5 M NaCl. Also, a major 18-kD protein and several minor, higher molecular weight (~20 and 24 kD) proteins could be precipitated with the antisera raised against the amino-terminus peptide of bFGF (data not shown). These results indicate that the tumors did in fact arise from the injected, neomycin-resistant, chimeric sp-bFGF producing cell lines rather than from the host mouse cells.

Histological Analysis of the sp-bFGF Tumors

Histological analysis was performed on the sp-bFGF tumors from nude and immunocompetent syngeneic mice (Fig. 3A,B). Tumors in both animal types grow rapidly and are poorly differentiated. The tumors in the nude mice are far more vascularized than those from the same cell line growing in immunocompetent mice.

Specific Humoral Antibodies in Mice Bearing sp-bFGF Tumors

The immunocompetent NIH/NSF mice were checked for production of any specific antibodies that might be directed against bFGF or other novel antigens that may be present on the surface of the transformed cells. Control sera were obtained from mice injected with an equal number of the nontumorigenic parental NIH-3T3 cells.

The antibody fingerprint of sp-bFGF tumor-bearing sera is shown in Figure 4. While no proteins are immunoprecipitated by the control sera (Lane 1), a single strong 30-kD protein is specifically recognized by sera from two mice, each bearing a tumor that formed from an independently generated sp-bFGF- transformed cell line (Lanes 2, 3). This protein is uniquely recognized by sera from more than 40 mice injected with four independently isolated sp-bFGF clonal cell lines (data not shown). The antibody fingerprint of sp-bFGF tumors differs dramatically from the antibody fingerprint that forms in animals bearing tumors arising from *ras*-transformed NIH-3T3 cells. As seen in Lane 4, *ras* tumors also solicit a very specific antibody response in mice. This antibody fingerprint, as seen through the immunoprecipitating proteins, is also highly reproducible and was seen in more than ten mice bearing tumors formed from tumorigenic cells transformed with the *ras* oncogene.

To test for antigenic additivity of these tumors, cell lines stably expressing both sp-bFGF and *ras* were selected, injected into NIH/NSF mice, and the sera of the tumor-bearing animal were analyzed for presence of the *ras* and sp-bFGF antibody fingerprint in the IP assay. Indeed, in some cases, both sets of antibodies were found to be present in the same sera (Lane 6): the 30-kd protein characteristic of sp-bFGF tumors (as in Lanes 2, 3) and all the *ras* antibodies seen to be induced by *ras* tumors alone (as in Lane 4). A parallel set of cell lines expressing the native bFGF in addition to the *ras* proteins was also generated. Lane 5 represents the antibody fingerprint of one such *ras* tumor expressing the native bFGF.

The antigens to which the tumor-bearing mice developed antibodies are present not only on the tumor cells but are already expressed on the parental NIH-3T3 cells. In fact, additional data not shown here indicate that the 30-kD protein recognized by the sera of the NIH-IgBNM6–1 tumor-bearing mice is no more abundant in the immunizing NIH-IgBNM6–1 than it is in the nontumorigenic, parental NIH-3T3 cell line. In general, this observation is also true for most of the antigens recognized by the sera elicited by the *ras* tumors.

DISCUSSION

Work from a number of laboratories over the past years has indicated that autocrine growth stimulation can transform cells. This autocrine transformation requires the cellular expression of receptors capable of recognizing and responding to growth factors secreted by the cell itself [8–11].

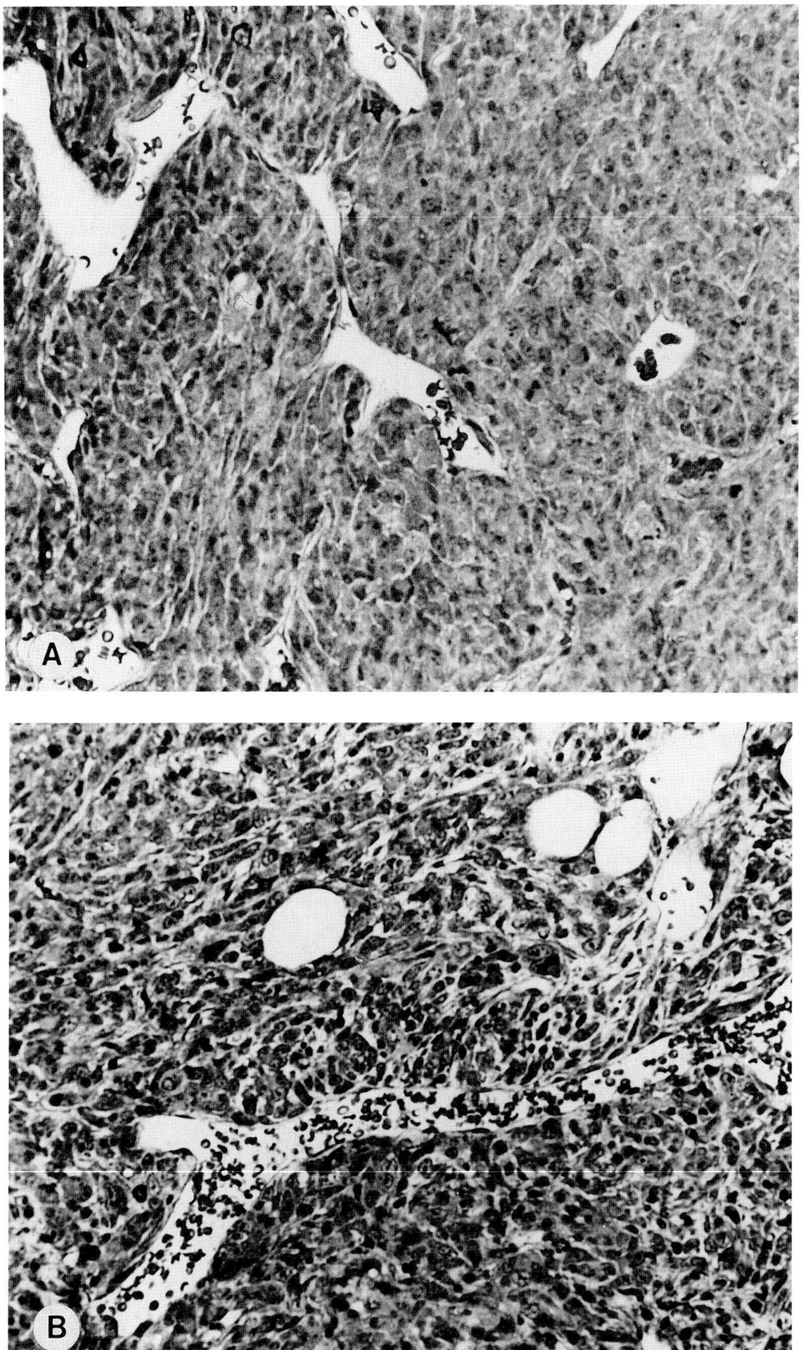

Fig. 3. Histology of NIH-IgBNM6–1 tumors. **A:** NUDE/NUDE mice. **B:** NIH-NSF mice.

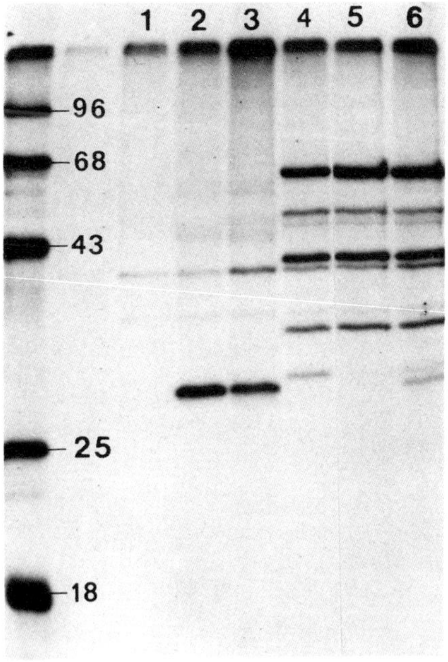

Fig. 4. Antibody fingerprinting of tumor-bearing sera. **Lane 1:** Control sera from a mouse injected with NIH-CONT cells. **Lanes 2, 3:** Sera from two mice injected with two independent sp-bFGF-producing clonal cell lines NIH-IgBNM6–1 and NIH-IgBNM29–1. **Lane 4:** Sera from an animal bearing an NIH-EJ1 tumor. **Lane 5:** Sera from an animal bearing tumor caused by an NIH-EJ cell line that also expresses the native bFGF (NIH-EJB10). **Lane 6:** Sera from an animal bearing a tumor caused by an NIH-EJ cell line also expressing the chimeric sp-bFGF (NIH-EJIgBI3). NIH-3T3 cell lysate was used as a substrate for this IP.

bFGF is a ubiquitous growth factor that is mitogenic for a wide variety of cell types, some of which also produce this factor. Yet, mere production of bFGF does not cause growth or transformation of such bFGF responsive cells. For example, bovine aortic endothelial and human umbilical vein endothelial cells synthesize high levels of intracellular bFGF [14,17], yet they require exogenous bFGF in order to proliferate. We have recently shown that bFGF is capable of tumorigenic transformation of NIH-3T3 fibroblasts only when a signal peptide is fused to bFGF. We speculate that in these cells, bFGF is localized in a subcellular compartment in which it has access to its cognate receptor.

Since bFGF is an angiogenic factor [18], the tumors induced by its autocrine action might be expected to be typified by characteristics such as unusually rich vascularization. The sp-bFGF tumors in nude mice indeed have prolific vasculature. In contrast, the tumors of the same origin in NIH/NSF mice are quite similar to *ras* induced sarcomas, and have inconspicuous vasculature. The reason for this is unclear. Perhaps some bFGF is secreted from the cells or simply leaks out of the necrosing tumor tissue. While the nude mice may respond to the angiogenic effect of externalized bFGF, the syngeneic mice may mount an immune defense against this new protein and neutralize it, thereby developing less vascularized tumors. However none of the tumor sera tested had detectable antibody titer to bFGF, so that the leakage of bFGF into the extracellular space in the tumor is probably not great. Thus, this discrepant vascularization of nude and syngeneic tumors is not explained.

In striking contrast with the immunocompetent mice, the nude mice do develop tumors from cells that synthesize large amounts of the native bFGF. However, in these cells the required bFGF levels are considerably higher (2–8 times), the tumors take much longer to form (5 weeks vs. 1–2 weeks), and they arise at a lower frequency (50% ± 25% vs. 100%) as compared to tumors produced by sp-bFGF-expressing cells. Again it is not clear why tumors form more readily in the nude mice. These cell lines that express very high levels of the native bFGF, though neither focus-forming nor morphologically transformed, do have lower serum requirements and do grow to higher saturation densities *in vitro* [12]. This growth may be dependent on leakage of bFGF from dying cells and may be responsible for tumor formation in the immunoincompetent nude mice, while it is blocked by the immune system in the syngeneic mice. If leakage is solely resposible for tumor formation in nude mice, then, in addition to the vascular tissues, some of the tumor mass may be made up of host cells. We have not yet looked for host cells in bFGF tumors in nude mice.

Although sp-bFGF tumors in syngeneic mice do not cause an identifiable antibody response to the bFGF molecule itself, they do evoke a typical humoral response that is characteristic for this type of tumor and is the same for all mice regardless of which sp-bFGF clonal cell line was injected. The mice develop a high titer of humoral antibodies directed specifically against a 30-kD protein of as yet undetermined subcellular localization. This 30-kD protein is not a tumor-specific antigen since it is found in equal abundance in the nontransformed parental cell line. This antigen may be presented in a new, immunogenic way in the sp-bFGF producing cells as compared to the parental NIH-3T3 cells. Although unlikely, it is also conceivable that the tumor formation removes immune self-tolerance for this particular antigen. Similarly, antibodies against a different set of self-antigens arise in mice injected with *ras* tumors. At least is some cases, both sets of antibodies are developed in response to sp-bFGF-producing *ras* tumors.

The antibody fingerprint as visualized in our IP assay enbles us to determine the transforming lesion within 1 week of tumor detection. Perhaps similar analysis of sera collected from patients with *hst* tumors (Kaposi sarcoma, stomach cancers) and others may detect an antibody fingerprint that will aid in tumor diagnosis and treatment.

ACKNOWLEDGMENTS

We wish to thank Jennifer Epp for techical help, Sandra Dethlefson for preparing histological sections, and Dr. Judah Folkman for indispensable advice and useful discussions.

REFERENCES

1. Folkman J, Klagsbrun M: Science 235:442–447, 1987.
2. Dickson C, Peters G: Nature 326:833, 1987.
3. Taira M, Yoshida T, Miyagawa K, Sakamoto H, Terada M, Sugimura T: Proc Natl Acad Sci USA 84:2980–2984, 1987.
4. Marx JL: Science 237:602–603, 1987.

5. Yoshida T, Miyagawa K, Odagiri H, Sakamoto H, Little PFR, Terada M, Sugimura T: Proc Natl Acad Sci USA 84:7305–7309, 1987.
6. Delli Bovi P, Curatola AM, Kern FG, Greco A, Ittman M, Basilico C: Cell 50:729–737, 1987.
7. Gospodarowicz D, Neufeld G, Schweigerer L: Mol Cell Endocrinol 46:187–204, 1986.
8. Leal F, Williams LT, Robbins KC, Aaronson SA: Science 230:327–330, 1985.
9. Robbins KC, Leal F, Pierce JH, Aaronson SA: The EMBO J 4:(7):1783–1792, 1985.
10. Betsholtz C, Westermark B, Ek B, Heldin C-H: Cell 39:447–457, 1984.
11. Keating MT, Williams LT: Science 239:914–916, 1988.
12. Rogelj S, Weinberg RA, Fanning P, Klagsbrun M: Nature 331:173–175, 1988.
13. Klagsbrun M, Sasse J, Sullivan R, Smith JA: Proc Natl Acad Sci USA 83:2448–2452, 1986.
14. Vlodavsky I, Folkman J, Sullivan R, Fridman R, Ishai-Michaeli R, Sasse J, Klagsbrun M: Proc Natl Acad Sci USA 84:2292–2296, 1987.
15. Bernstein SC, Weinberg RA: Proc Natl Acad Sci USA 82:1726–1730, 1985.
16. Laemmli UK: Nature (London) 227:680–685, 1970
17. Hannan R, Kourembans S, Flanders KC, Rogelj S, Roberts AB, Faller DV, Klagsbrun M: Accepted for publication in Growth Factors.
18. Shing Y, Folkman J, Sullivan R, Butterfield C, Murray J, Klagsbrun M: Science 223:1296–1299, 1984.

Journal of Cellular Biochemistry 39:117–128 (1989)
Growth Factors and Their Receptors: Genetic
Control and Rational Application 95–106

Immunological Study of Acidic Fibroblast Growth Factor (aFGF) Distribution in the Eye

D. Caruelle, B. Groux-Muscatelli, A. Gaudric, C. Sestier, G. Coscas, J.P. Caruelle, and D. Barritault

Universite Paris XII, Laboratoire de Biotechnologie des Cellules Eucaryotes, 94010 Creteil Cedex, France

During the last ten years, several groups, including the present authors, have detected growth factor activities in various ocular tissues, and the presence of a ubiquitous Eye-Derived Growth Factor (EDGF) has been described. More recently, isolation and characterization of this growth factor activity from the retina led to the identification of two molecules. These molecules were shown to be identical to other growth factors isolated from neuronal and non-neuronal tissues and are now designated as acidic and basic fibroblast growth factor (aFGF, bFGF). The biological function and the reason for the ubiquitous distribution of these factors remain unclear. Understanding may be improved by quantification of this distribution in various tissues during development. In the present study, specific polyclonal antibodies were raised against acidic FGF, aFGF was determined in various ocular tissues by enzyme immunoassay, and the localization of immunoreactive aFGF by immunohistological staining with fluorescent antibodies or with enzyme- or gold-labeled antibodies was studied.

In almost all tissues tested aFGF was found; but the retina, cornea, and vitreous body contained the highest levels of aFGF per gram of tissue. In the retina, aFGF was associated primarily with the nerve fiber layer and the inner and outer segments of the photoreceptors, whereas corneal aFGF was detected in the cytoplasma of the basal layer of epithelial cells.

Key words: growth factor, aFGF, immunoassay, eye, vitreous body, cornea, retina, lens

Growth-factor activity has been reported in various ocular tissues by several groups in the last ten years [1–4]. Although the original aim of the present authors was to characterize growth factors from the retina and to study their role in the control of lens growth and differentiation, the presence of growth factor activity in most ocular tissues—including the iris, vitreous body, ciliary processes, retina, and choroid—has been demonstrated [2]. These factors have been designated *eye-derived growth factors* (EDGF). Two (EDGF I and II), [5–7] which have a high affinity for

Received February 1, 1988; accepted August 8, 1988.

heparin, were identical to acidic and basic fibroblast growth factor (FGF) found in other nonocular tissues, and a third factor (EDGF III) was characterized [8]. A retina-derived growth factor has been also shown to be identical to aFGF [9], and both forms have been characterized by other groups.

The biological and physiological significance of these growth factors is unknown, but they are believed to have key roles in tissue growth and homeostasis throughout the whole life span.

High levels of bFGF have been reported in the vitreous bodies of patients with vasoproliferative disorders of the retina [10], suggesting pathological involvement. However, these data are subject to debate [11] (Clément et al., manuscript in preparation).

Accurate knowledge of the distribution of growth factors in various tissues of healthy and pathological eyes may improve understanding of their physiological roles and physiopathological implications.

Specific antibodies were raised against aFGF, an enzyme immunoassay was developed, and immunohistological studies were done to quantify and localize aFGF in normal bovine and human eyes.

MATERIALS AND METHODS

Antigen Purification and Immunization Procedure

As already described, aFGF was extracted from bovine brain [6]. Using this procedure, aFGF was shown to be completely pure (99%) after heparin-Sepharose chromatography. Over a 6-week period, female New Zealand albino rabbits were given weekly intradermal injections of 50 to 100 μg of aFGF emulsified in complete Freund's adjuvant (Sigma). Rabbits subsequently received monthly booster injections of 50 μg of aFGF emulsified in Freund's incomplete adjuvant. Rabbits were bled monthly ten days after each booster injection. Antibody production was monitored by enzyme linked immunosorbent assay (ELISA) [12].

Characterization of Antibodies

Immunoassays and immunohistological studies were carried out with sera from two different rabbits.

The characterization, specificity, and properties of the immune serum used for immunoassay have been reported elsewhere [13]. The criteria presented include ELISA, immunoblot, and inhibition of aFGF-induced biological activity by these anti-aFGF antibodies. No cross-reactivity with bFGF could be detected by any of these methods (fewer than 8/1,000).

Immunohistological studies were performed with another polyclonal antibody, which also showed no cross-reactivity with bFGF, as judged by ELISA.

Protein content was determined using the method of Bradford (14) with bovine serum albumin as the standard.

Immunoassay Procedure

All reagents used in the assays were diluted in 0.1 M phosphate buffer, pH 7.4, containing 0.4 M NaCl, 10^{-3} M EDTA, 0.1% bovine serum albumin (BSA), 0.01% NaN^3 (EIA buffer).

Enzyme Immunoassay (EIA) was performed as previously described according to the second antibody solid-phase EIA protocol [15].

Briefly, mouse antirabbit immunoglobulin G (IgG) monoclonal antibodies were adsorbed on 96-well microtiter plates and 1% BSA was added to saturate protein-binding sites. Plates were washed with 10^{-2} M phosphate buffer, pH 7.4, containing 0.05% Tween 20, each well was filled with 300 μl of EIA buffer, and the plates were stored at 4°C until use. Assays were performed in a total volume of 150 μl, with each component added in a 50-μl volume: standard aFGF solution, aFGF enzyme conjugate, and diluted anti-aFGF antiserum. Enzymatic tracer (aFGF-AChE) was obtained by coupling aFGF with the tetrameric form of acetylcholineesterase from the electric organs of the electric eel *Electrophorus electricus* [16].

After overnight incubation at 22°C, the plates were washed as described above, and 200 μl of Ellman's reagent [17] were added to each well. After 30 min to 2 h, the absorbance at 414 nm was measured using a Dynatech spectrophotometer plate reader.

Ocular Tissue Extracts

Ocular tissues were dissected from freshly slaughtered bovines and kept at 4°C. Homogeneous batches of specific tissues were crushed in a Moulinex blender and then homogenized in a tight-fitting potter homogenizer after the addition of 100 mM phosphate buffer pH 6.7, 2 M NaCl. After 1 h of continuous vigorous shaking at 4°C, samples were centrifuged (30 min, 20,000 g) and the ionic strength of the supernatant was adjusted to 0.65 M before application to a heparin-Sepharose chromatography column (50 mg of heparin-Sepharose were routinely used for extracts).

Preparation of Tissue Section

Tissues were fixed, within 30 min of death for bovines and within 4 h of death for humans in 4% paraformaldehyde in 0.1 M phosphate buffer, pH 7.4, for 4 h at 4°C. The tissues were dehydrated in a graded series of ethanol, cleared in toluene, and embedded in paraffin. Sections were cut to a thickness of 5 μm, mounted on slides, dried for 24 h at 37°C, and then stored at 4°C. Fixed frozen sections were compared with sections of paraffin.

Immunohistological Technique

The distribution of aFGF protein was analyzed by indirect immunofluorescence and immunoperoxidase or immunogold methods, enhanced by silver staining. Sections were preincubated in normal goat serum (10% in phosphate-buffered saline (PBS) 0.4 M NaCl, 0.1% BSA, 0.3% Triton x 100) for 30 min at 22°C. Preincubation medium then was discarded and, without washing, the sections were incubated in specific antiserum diluted 1/50 and 1/100 in a moist chamber at 22°C for 2 h. After incubation, sections were rinsed in PBS and incubated with goat antirabbit IgG antibodies (1/100 dilution) for 1 h at 22°C. This second antibody was goat antirabbit IgG antibody, which was conjugated with fluorescein isothiocyanate (FITC) (Tago) or with peroxidase (Pasteur Production) revealed by diaminobenzidine (Sigma) or labeled with colloidal gold [18] (Jansen) and treated by silver enhancement [19].

Several controls were performed without primary antibodies to control nonspecific immunoreactivity of reagents, with aFGF-absorbed primary antibody and preimmune serum.

RESULTS

Quantification of aFGF in Ocular Tissues

The sensitivity and specificity of EIA are presented in Figure 1. The minimal detection level was 252 pg/ml, with 50% sensitivity at 1.48 ng/ml of aFGF. At this level of sensitivity, the aFGF content of one eye was adequate for quantification.

The quantification of affinity-purified immunoreactive aFGF in ocular tissues is presented in Table I (sclera, choroid, and retinal pigment epithelium were extracted whole).

Immunohistological Studies

Immunolocalization was achieved using a second antibody labeled with a fluorochrome (FITC), an enzyme (peroxidase), or colloidal gold (silver enhancement as

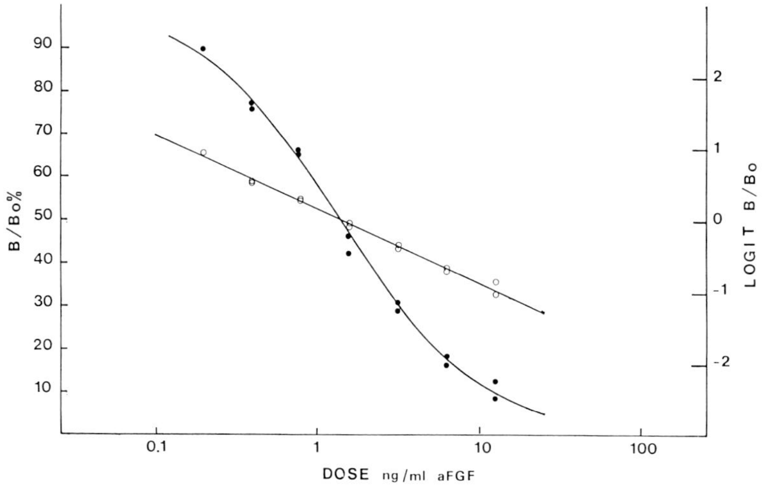

Fig. 1. Standard curve of AchE-aFGF (EIA). Displacement curve of AchE-aFGF in EIA (●) generated by standard aFGF [13] and corresponding linear log-log scale representation (○). The maximal sensitivity (defined as 50% B/BO) was 1.48 ng/ml and the minimal detection level (defined as 80% B/BO) was 252 pg/ml.

TABLE I. aFGF Distribution in Ocular Tissues*

Ocular tissue	Total content in one eye (ng)	ng of aFGF per g of tissue
Cornea	928	412
Aqueous humor	80	35
Lens	249	115
Vitreous body	4,790	295
Retina	243	479
Pigment epithelium choroid and sclera	1,376	15
Optic nerve		818

*Enzyme Immunoassays were performed with polyclonal anti-aFGF. Sclera, retinal pigment epithelium, and choroid were dissected whole. Weight of aqueous humor was calculated from volume measurements assuming a density of 1. The mean values from three independent experiments are given. Decreasing amounts of aFGF were noted from the optic nerve to the aqueous humor through retina, vitreous body, lens. aFGF was present in comparable amounts in the cornea and retina.

described in methods). Positive immunoreactions were noted at the same sites with these different techniques, whatever the treatment of sections.

Vitreous body and retina. Immunoperoxidase staining of a whole section of human retina is shown in Figure 2a–c. Generalized labeling of most cells can be seen at two magnifications (Fig. 2a, b), whereas no labeling was noted in the control with aFGF-absorbed primary antibodies (Fig. 2c).

A more detailed study of the different layers of the retina is presented in Figure 3a–c. Figure 3a shows marked immunoreaction of axon fibers; Figure 3b shows diffuse labeling of ganglion cells and intense labeling of some isolated cells in the plexiform layers. The intense labeling seen in Figure 3c is located in the inner and outer segments of the photoreceptors. Immunoperoxidase staining of the inner part of the retina (Fig. 4) confirmed the staining of the internal limiting membrane, nerve fibers, ganglion cells, stout columns, and footplates of the Müller cells (glial). Gold staining of the same region (Fig. 5) was noted in the nerve fiber layer and detached vitreous body.

aFGF in cornea and lens. Immunofluorescence of a healthy human cornea is presented in Figure 6a, b. Intense labeling was clearly located in the cytoplasm of the basal epithelial cell layer. Diffuse labeling was also seen throughout the epithelium. No significant aFGF immunoreaction was observed in the stroma, basal membranes (Bowman and Descemet), and endothelium (not shown).

Immunoperoxidase localization of aFGF in the lens (not shown) indicated that only the lens epithelium (i.e., the anterior cell layer beneath the anterior capsula, including the equatorial zone) contained significant amounts of immunoreactive material.

DISCUSSION

The availability of specific aFGF antibodies enabled us to perform rapid quantitative and qualitative studies in normal and pathological tissues.

The results presented here illustrate the complementarity of two techniques: the almost ubiquitous distribution of aFGF in ocular tissues illustrated by EIA and by histological staining.

The distribution of FGF in various ocular tissues has been reported and quantified by biological assay in previous studies. However, at that time, EDGF/FGF was partially purified and the existence of two biologically and chemically distinct forms of FGF was not suspected.

More recently, the purification of both forms of factors from the retina have enabled us to quantify the FGF content in the retina by biochemical and biological assays [8]. Results obtained with these assays indicated a content of 0.25 μg of pure, biologically active aFGF per 1,000 retinas [3, 8]. Several other groups reported similar values of FGF activity in the retina and vitreous body [9, 11, 20]. The value obtained for retina with EIA was almost the same (Table I). It is worth noting that this high correlation between biochemical and immunological quantification was obtained only after heparin-Sepharose chromatography of the homogenized tissue extracts. EIA of crude extracts routinely yielded higher and less reproducible values (not shown), although no cross-reactivity (below 8/1,000) with bFGF could be detected.

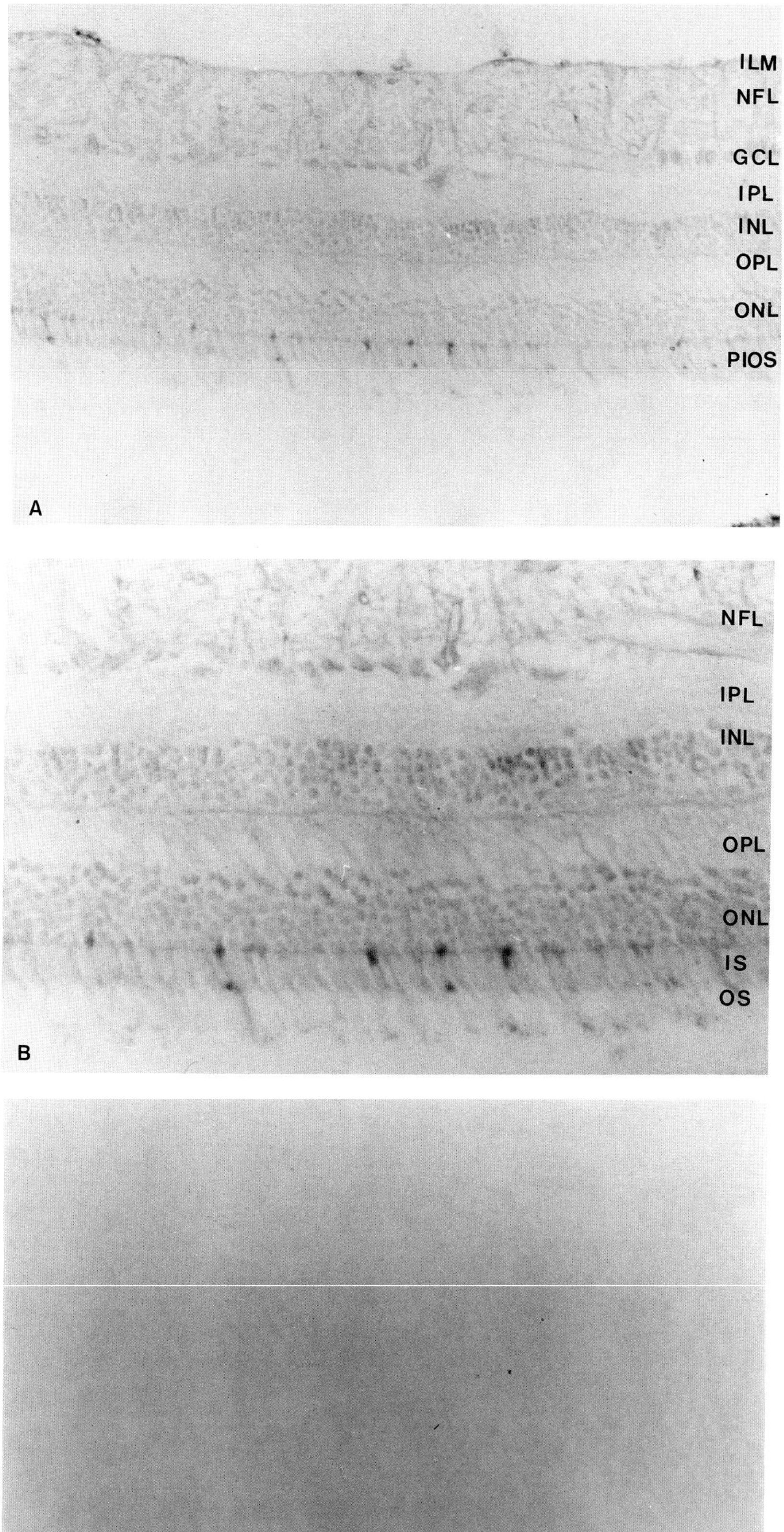

Figure 2.

Vitreous body and retina contained high levels of aFGF and immunohistological studies indicated extensive distribution, with 3 regions of higher concentration in the retina. The diffused distribution in the vitreous body could be due to postmortem or artefactual diffusion from the retina or from intravitreous regions containing higher concentrations. This is unlikely, however, as no variation of immunostaining could be detected in the vitreous body and also because aFGF and bFGF for biochemical analysis were purified from vitreous body collected within a few minutes of death (Mereau A., unpublished results).

The ubiquitous presence of aFGF immunoreactive material in the retina and the association of high concentrations of this material with specific regions suggest a possible physiological function for aFGF.

In agreement with previous reports of the presence of aFGF in brain neurons, aFGF was associated with the axon fibers of the neuroretina [21, 22, 23]. This neuronal localization of both acidic and basic FGF, and the in vitro studies [24, 25] indicating that these factors are necessary for growth and survival of neurons, suggest a similar role in vivo. More intriguing is the very intense staining of the inner and outer segments of the photoreceptors.

The presence of bFGF associated with photoreceptors purified by differential centrifugation has already been reported in the bovine retina [26]. However, disruption of the retina could induce the release of bFGF from other regions and nonspecific association with photoreceptors [27]. The presence of bFGF in retinoblastoma has also been reported [28]. Because retinoblastoma is considered to be derived from photoreceptor cells, it has been suggested that photoreceptors could be the source of bFGF in the normal retina. However, these authors [28] clearly ruled out the presence of aFGF in retinoblastoma. Therefore, our observations also indicate either that retinoblastoma cells derive only from a subset of non-aFGF-producing photoreceptor cells or that aFGF gene expression is lost through transformation or during tissue culture.

In the studies presented here, cross-reactivity between aFGF and bFGF has been excluded by ELISA. However, bFGF may have in situ cross-reactive epitopes that are undetected by ELISA. This possibility is unlikely, however, because absorbed primary aFGF antibodies were nonreactive, and retinal pigment epithelium (RPE) could not be immunostained with aFGF antibodies. This absence of the immunodetection of RPE is in good agreement with reports in which nontransformed RPE cells in tissue culture have been shown to synthesize bFGF only [29].

In previous studies, no significant amounts of growth factor activity were found in the lens [2]. The fact that immunoreactive aFGF material is associated with the lens epithelium in the equatorial zone and beneath the anterior capsula, but not with lens fibers, correlates well with what is known of the capacity of these cells to proliferate during lens growth.

Fig. 2. Immunoperoxidase localization of aFGF in human retina. **A,B:** Widespread positive immunoreaction in the different layers of the retina: ILM, inner limiting membrane; NFL, nerve fiber layer; GCL, ganglion cell layer; IPL, inner plexiform layer; INL, inner nuclear layer; OPL, outer plexiform layer; ONL, outer nuclear layer; PIOS, photoreceptor inner and outer segments; IS, inner segment; OS, outer segment. A, ×250; B, ×400. **C:** Negative control: aFGF-absorbed primary antibody. ×250.

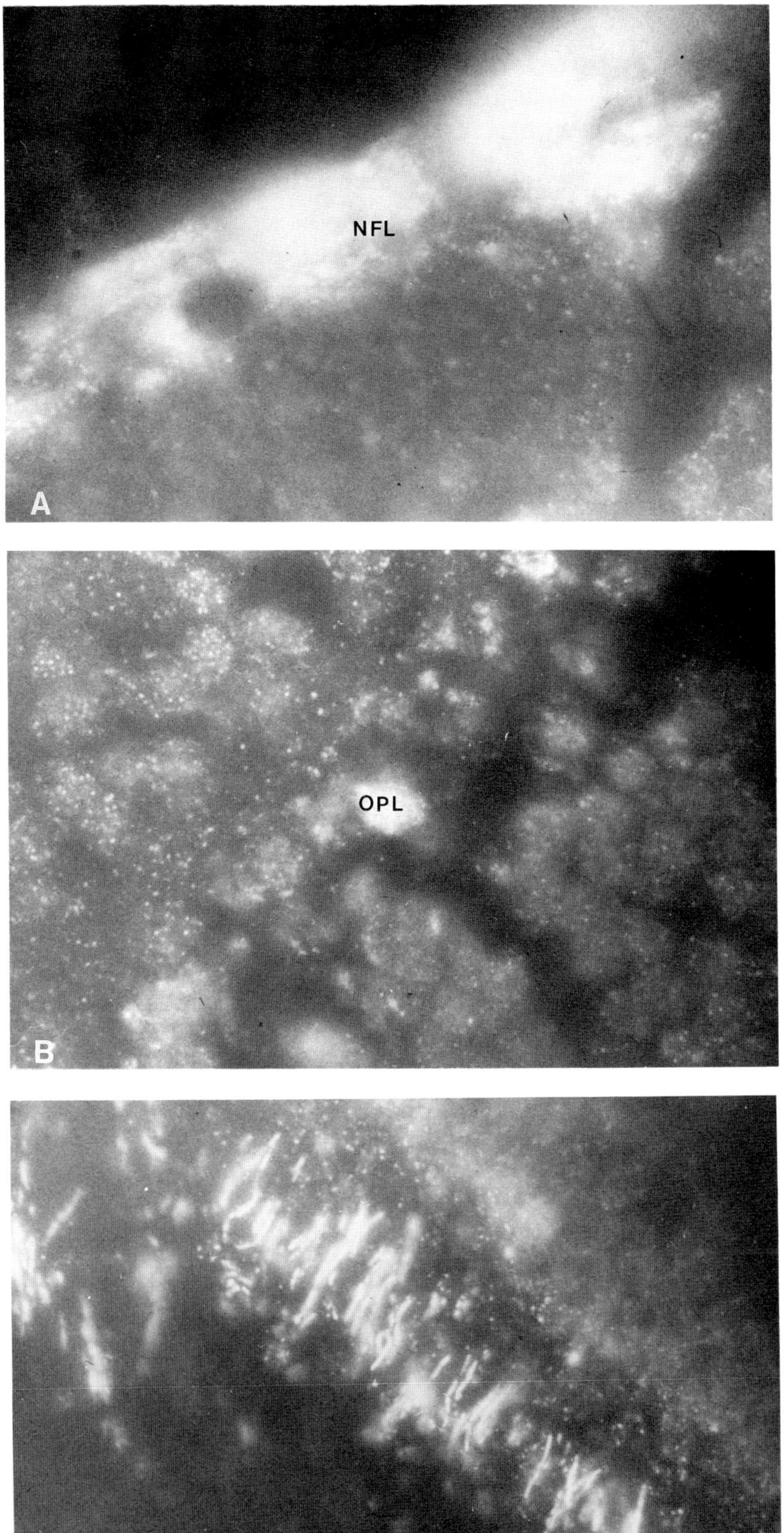

Figure 3.

Interestingly, the addition of labeled aFGF or bFGF to frozen sections of cornea or lens showed that lens capsula and Bowman and Descemet membranes contain receptors to FGF [30]. Under the present experimental conditions, we were unable to visualize significant amounts of FGF associated with these membranes, which suggests either that the antigen (aFGF) is inaccessible to the antibody or that antigen concentration is inadequate. Because it is known that FGF can be removed from Descemet membrane by heparinase treatment [30], it is likely that immunoreactive epitopes hidden in these basement membranes are inaccessible to aFGF antibodies.

Most of the aFGF immunoreactive material found in corneal epithelial lens cells seems to be associated with the basal cell layer, although positive staining was also detected throughout the epithelium and, to a lesser extent, in the corneal stroma.

Corneal epithelial cells are among those with the highest rate of renewal. EDGF (partially purified aFGF and bFGF) has been shown to accelerate reepithelialization of cornea in experimentally induced corneal ulcers [32] as well as in pathological ulcers [33].

The same results also have been obtained with purified aFGF and bFGF [34] and with epidermal growth factor (EGF) [32]. However, in studying EGF cross-reacting material in corneal epithelium, no staining was detected (unpublished results).

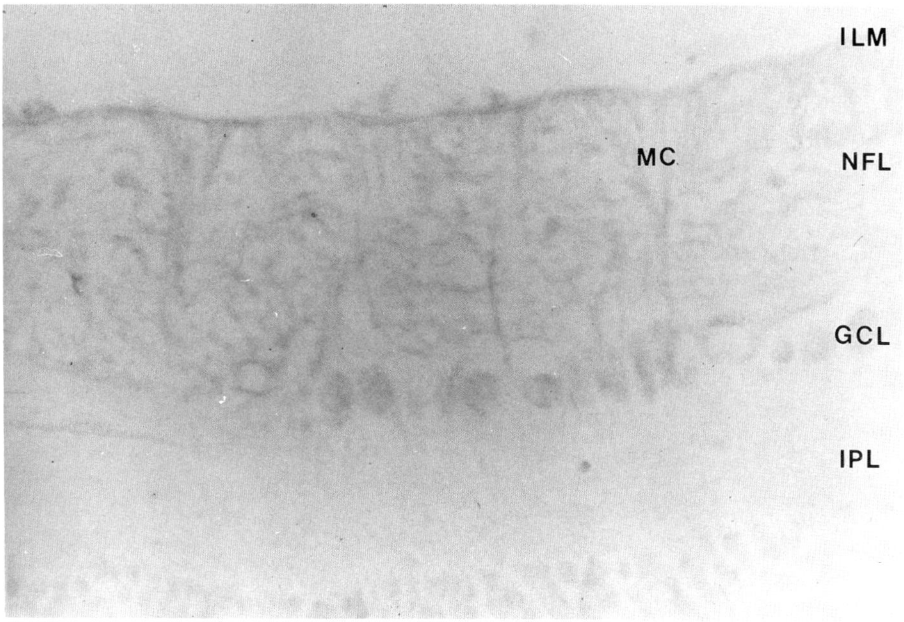

Fig. 4. Immunoperoxidase localization of aFGF in human retina. ×800. Positive immunoreaction in inner limiting membrane (ILM), axons of nerve fiber layer (NFL), ganglion cells (GCL), and the footplates of Müller cells (MC); IPL, inner plexiform layer.

Fig. 3. Immunogold (with silver enhancement) localization of aFGF in bovine retina. **A:** Labeling is apparent over the nerve fiber layer (NFL). **B:** Labeling is apparent in the outer plexiform layer (OPL). **C:** Labeling is apparent in the inner and outer segments of photoreceptors. ×1,000.

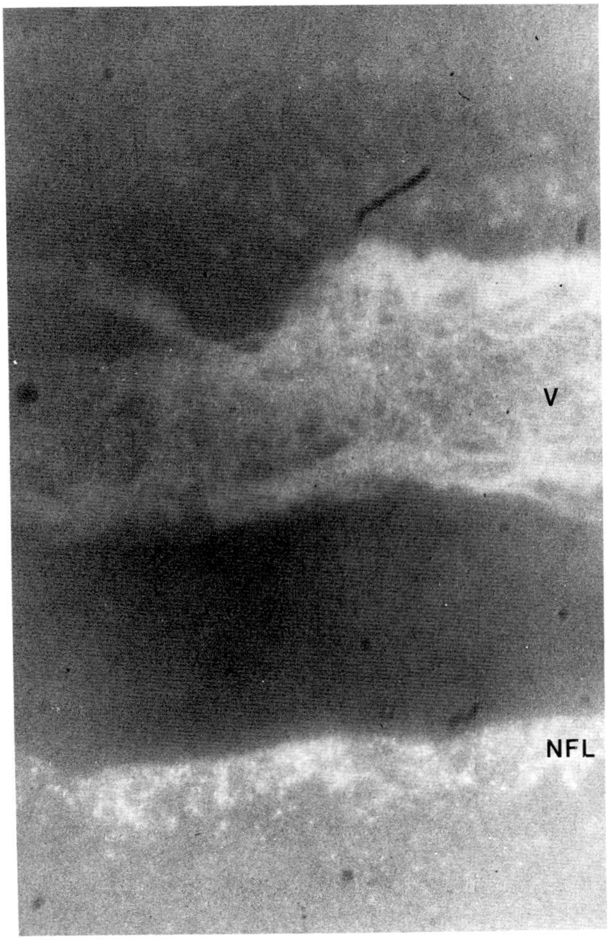

Fig. 5. Immunogold (with silver enhancement) localization of aFGF in bovine retina. Labeling of nerve fiber layer (NFL) and detached vitreous body (V). ×400.

It is therefore tempting to suggest that aFGF (and perhaps bFGF) is the growth factor implicated in vivo in the control of epithelial corneal cell proliferation.

ACKNOWLEDGMENTS

This work was supported by the Ministère de l'Education Nationale (Equipe recommandée), the Faculté de Médecine Créteil, Paris XII, and INSERM contrat No. 872002 and contrat CNAM No. 86254E to D.B.

The authors wish to thank Margaret Sterkers for providing histological preparations.

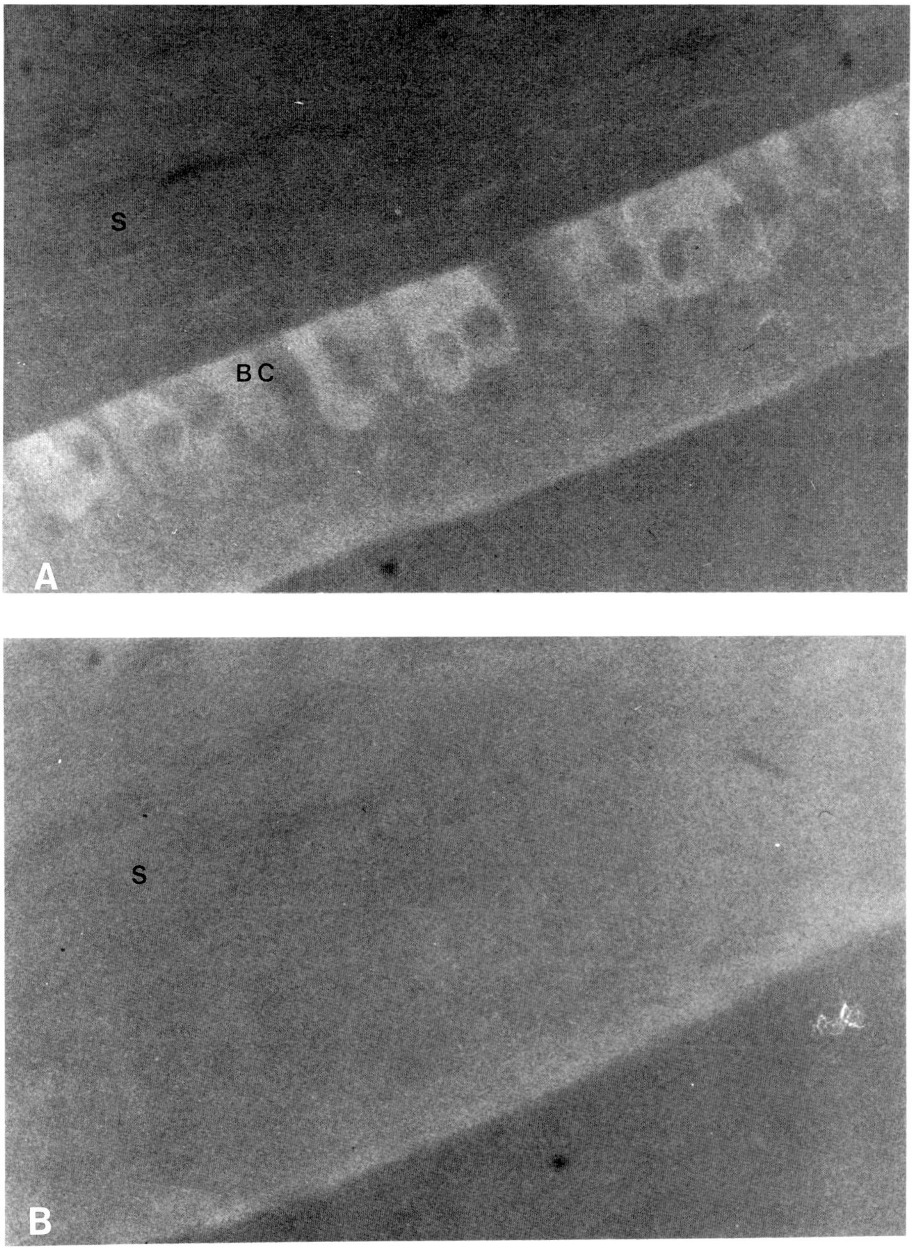

Fig. 6. Immunofluorescence localization of aFGF in human cornea. Tissue fixed in paraformaldehyde and processed for immunofluorescence as described in Materials and Methods. **A:** Intense labeling in the cytoplasma of epithelium basal cells (BC), stroma (S). **B:** negative control; aFGF absorbed primary antibody. ×400.

REFERENCES

1. Arruti C, Courtois Y: Exp Cell Res 117:283, 1978.
2. Barritault D, Arruti C, Courtois Y: Differentiation, 18:29, 1981.
3. D'Amore PA, Glaser BM, Brunson SK, Fenselau AH: Proc Natl Acad Sci USA 78:3068, 1981.
4. Chen CH, Chen SC: Invest Ophtalmol Vis Sci 19:592, 1980.
5. Schreiber AB, Kenney S, Kowalski J, Thomas KA, Gimenez-Gallego G, Rios-Candelore M, Di Salvo J, Barritault D, Courty J, Moenner M, Loret C, Burgess WH, Mehlman T, Friesel R, Johnson W, Maciag T: J Cell Biol 101:1623, 1985.
6. Courty J, Chevallier B, Moenner M, Loret C, Lagente O, Bohlen P, Courtois Y, Barritault D: Biochem Biophys Res Commun 136:102, 1986.
7. Courty J, Loret C, Chevallier B, Moenner M, Barritault D: Biochimie 69:511, 1987.
8. Courty J, Loret C, Moenner M, Chevallier B, Lagente O, Courtois Y, Barritault D: Biochimie 67:265, 1985.
9. D'Amore PA, Klagsbrun M: J Cell Biol 99:1545, 1984.
10. Baird A, Culler F, Jones KL, Guillemin P: Lancet 7:563, 1985.
11. Lutty GA, Chandler C, Behnett A, Fait C, Patz A: Curr Eye Res 5:9, 1986.
12. Parker PJ, Young S, Gullick WJ, Mayes ELV, Bennet P, Waterfield MD: J Biol Chem 259:9906, 1984.
13. Caruelle D, Grassi J, Courty J, Groux-Muscatelli B, Pradelles P, Barritault D, Caruelle JP: Anal Biochem 173:328, 1988.
14. Bradford MM: Anal Biochem 72:248, 1976.
15. Pradelles P, Grassi J, Maclouf J: Anal Chem 57:1170, 1985.
16. McLauglin LL, Wei Y, Leahy KM, Needlemen P, Grassi J, Pradelles P: Biochem Biophys Res Commun 144:469, 1987.
17. Ellman GL, Courtney KD, Andres V, Featherstone RM: Biochem Pharmacol 7:88, 1961.
18. De Mey J: In Polak J, Van Noorden S (eds): London: J. Wright & Sons Ldt., 1983, pp 82–112.
19. Bruce BG, Habliston DL, Hedberg KK, Hayes GO: J Histochem Cytochem 34:339, 1986.
20. Baird A, Esch F, Gospodarowicz D, Guillemin R: J Bio Chem 24:7855–7859, 1985.
21. Sensenbrenner M, Pettman B, Labourdette G, Weibel M: Horm Cell Reg 9:345, 1985.
22. Morrison RS, De Vellis J, Lee YL, Bradshaw RA, Eng LF: J Neurosci Res 14:167, 1985.
23. Huang SS, Tsai CC, Adams SP, Huang JS: Biochem Biophys Res Commun 144:81, 1987.
24. Schubert D, Ling N, Baird A: J Cell Biol 104:635, 1987.
25. Eccleston PA, Silberberg DH: Dev Brain Res 21:315, 1985.
26. Plouet J, Mascarelli F, Lagente O, Lorans G, Faure JP, Courtois Y: In Agarth E, Ehinger B (eds): New York: Elsevier, 1986, pp 311–320.
27. Campochiaro PA, Blaydes SH: Doc Ophtalmol Proc Ser 50:401, 1987.
28. Schweigerer L, Neufeld G, Gospodarowicz D: Invest Ophtalmol Vis Sci 28:1838, 1987.
29. Schweigere L, Malerstein B, Neufeld G, Gospodarowicz D: Biochem Biophys Res Commun 143:934, 1987.
30. Jeanny JC, Fayein N, Moenner M, Chevallier B, Barritault D, Courtois Y: Exp Cell Res 171:63, 1987.
31. Chamberlain CG, McAvoy JW: Curr Eye Res 6:1165, 1987.
32. Petroutsos G, Courty J, Guimares R, Pouliquen Y, Barritault D, Plouet J, Courtois Y: Curr Eye Res 3:593, 1984.
33. Muller G, Courty J, Courtois Y, Clere B, Barritault D: J Français Ophtalmol 8:187, 1985.
34. Fredj-Reygrobellet D, Plouet J, Delayre T, Baudoin C, Bouvet F, Lapalus P: Curr Eye Res 6:1205, 1987.

Journal of Cellular Biochemistry 39:129–137 (1989)
Growth Factors and Their Receptors: Genetic Control and Rational Application 107–115

Identification of Tyrosine-Phosphorylated Colony-Stimulating Factor 1 (CSF-1) Receptor and a 56-Kilodalton Protein Phosphorylated in Intact Human Cells in Response to CSF-1

Richard D. Huhn, Mary Elizabeth Cicione, and A. Raymond Frackelton, Jr.

Department of Medicine, Roger Williams General Hospital, Brown University, Providence, Rhode Island 02908

Colony-stimulating factor 1 (CSF-1) selectively supports the survival, proliferation, and maturation of hemopoietic cells of the monocyte/macrophage lineage. Although the cellular receptor for CSF-1, (the c-*fms* protein) is a protein-tyrosine kinase activated by the binding of CFS-1, the role of phosphorylation of cellular proteins in CSF-1 signal transduction is poorly understood. Therefore, we examined the CSF-1-stimulated phosphorylation of cellular proteins in human BeWo choriocarcinoma cell line (known to express the c-*fms* protein). BeWo cells were metabolically labeled with $^{32}P_i$, stimulated with recombinant human CSF-1, and extracted with detergent. Phosphotyrosyl proteins were isolated from detergent extracts by affinity chromatography on a highly specific antibody to phosphotyrosine. Rapid phosphorylation of 170-kd protein, followed closely by the phosphorylation of a 56-kd protein, was observed in response to CSF-1. The 170-kd phosphotyrosyl protein bound to wheat germ agglutinin and was secondarily immunoprecipitated with a specific anti-*fms* serum, consistent with its identity as the CSF-1 receptor. Although purified human macrophages that proliferate in culture in response to CSF-1 are not generally accessible, CSF-1 did stimulate the phosphorylation of a 56-kd protein in intact mononuclear leukocytes from human peripheral blood. Thus, the BeWo cell line may represent a good model for the study of CSF-1-stimulated cellular protein phosphorylation.

Key words: **colony-stimulating factor 1, protein phosphorylation, hematopoietic growth factors, c-*fms*, phosphotyrosyl proteins, tyrosine kinases**

The survival, proliferation, and maturation of hemopoietic cells of the monocyte/macrophage lineage are selectively supported by the hematopoietic growth fac-

Richard D. Huhn's present address is Department of Biochemistry, St. Jude Children's Research Hospital, Memphis, TN 38101.

Received February 1, 1988; accepted August 9, 1988.

tor, colony stimulating factor 1 (CSF-1, M-CSF) [1,2]. Interestingly, roles for CSF-1 in pregnancy and embryogenesis have also been proposed since levels of CSF-1 in blood and other tissues are elevated during pregnancy and since receptors for CSF-1 are expressed both in normal placental tissues and in the human choriocarcinoma cell lines BeWo and JEG-3 [3–5].

The receptor for CSF-1 is identical to the product of c-*fms*, the cellular homologue of the v-*fms* oncogene (the transforming gene of the McDonough strain of feline sarcoma virus) [6,7]. The CSF-1 receptor is a ligand-activated protein-tyrosine kinase, analogous to the receptors for epidermal growth factor, platelet-derived growth factor, insulin, and other peptide growth factors [8–15]. The protein-tyrosine kinase activity of the activated receptors for these growth factors has been strongly implicated in the stimulation of cellular growth, although the specific substrates of the activated receptors, as well as their physiological roles, are generally undefined.

Human CSF-1 is a dimeric glycoprotein highly homologous to the murine factor and is thought to function in similar physiologic role in human hematopoiesis [16–19]. While methods for the culture of murine bone marrow-derived CSF-1-dependent macrophages have been developed for study of the action of CSF-1 in vitro, purified human bone marrow-derived macrophages are not generally accessible [20]. Thus, examination of the role of CSF-1 in human hemopoiesis and the mechanisms of CSF-1 signal transduction in human cells has been difficult. In this study, we have utilized a highly specific monoclonal antibody to phosphotyrosine to affinity purify phosphotyrosyl proteins from CSF-1-stimulated human BeWo cells and normal human peripheral blood monocytes to study the phosphorylations of cellular proteins that may be involved in transducing the CSF-1 growth signal. Utilizing this technique, we have isolated the CSF-1 receptor and 56-kd proteins from BeWo cells and peripheral blood monocytes that become phosphorylated on tyrosine in response to CSF-1.

MATERIALS AND METHODS

Cells and Materials

The human BeWo choriocarcinoma cell line [21] was obtained from the American Type Culture Collection (Rockville, MD) and was grown in RPMI-1640 medium supplemented with non-essential amino acids (Flow Laboratories, McLean, VA), and 1 mM sodium L-pyruvate, 2 mM L-glutamine, 500 U/ml penicillin, 500 μg/ml streptomycin, and 10% fetal bovine serum (GIBCO, Grand Island, NY).

Platelet-free human peripheral blood monocytes were isolated by sedimentation on a cushion of Ficoll/metrizoate, sequential washes in Ca^{++}-free, Mg^{++}-free Puck's saline G containing 0.3 mM EDTA followed by serum containing 5 mM EDTA, and plastic adherence in tissue culture dishes, essentially as described by Pawlowski et al. [22].

Metabolic labeling, growth factor stimulation, and immunoprecipitations. Confluent cultures of BeWo cells or the adherent monocytes derived from 2×10^7 washed mononuclear cells in 35-mm culture dishes were incubated with $^{32}P_i$ at 1 mCi/ml for 3 h at 37°C, then transferred to ice and treated with 10^3 U/ml recombinant human CSF-1 (rhCSF-1, Cetus lot #A422, a gift of P. Ralph) or control medium (without growth factor) for 30 min. Cells were then extracted on ice with 1% Triton X-100 in buffer containing protease, phosphatase, and kinase inhibitors, as described [11]. Tyrosine- phosphorylated proteins were isolated from the extracts by microbatch immunoaffinity chromatography on monoclonal 1G2 antiphosphotyrosine antibody linked to Sepharose beads, competitively eluted with the hapten phenyl phosphate, and resolved by reducing SDS-PAGE, as described [11,23]. The 1G2 monoclonal antiphosphotyrosine antibody is highly specific for the phenyl phosphate moiety of phosphotyrosine: phosphoserine, phosphothreonine, phosphohistidine, and mono- and tri-phosphonucleotides are not recognized [13,24] (Frackelton AR Jr, Posner MR, Mermelstein F, manuscript in preparation).

^{32}P-labeled *fms*-related proteins were isolated from cell extracts or antiphosphotyrosine-purified phosphotyrosyl proteins by reacting them with specific rabbit antisera v-*fms* (a gift of C. Rettenmier) followed by adsorption on protein A-Sepharose (Pharmacia) as described [24]. Isolated proteins were solubilized by heating SDS sample buffer and were resolved by SDS-PAGE and autoradiography.

Isolation of glycosylated phosphotyrosyl proteins. Hapten-eluted phosphotyrosyl proteins from 1G2-immunoaffinity chromatography were applied to 25 μl of wheat germ agglutinin-Sepharose (WGA-Sepharose) (Pharmacia) and incubated 60 min at 2°C. After unbound proteins were collected, the beads were washed and the glycosylated proteins were eluted with 40 μl of 400 mM N-acetylglucosamine in 1% Triton X-100 extraction buffer. The unbound and eluted proteins were analyzed by reducing SDS-PAGE.

Staphylococcal V8 protease peptide mapping. Proteins were extracted from individual gel bands, precipitated with trichloroacetic acid, and digested with staphylococcal V8 protease type XVIII (Sigma) as described [25]. The resulting peptide fragments were resolved by SDS-PAGE (12.5% acrylamide) and visualized by autoradiography.

Phosphoamino acid analysis. Phosphoproteins in individual bands were extracted from polyacrylamide gels and then subjected to partial acid hydrolysis and two-dimensional thin-layer electrophoresis, as described [24,26,27]. Labeled phosphoamino acids were detected by autoradiography.

RESULTS

Cellular proteins phosphorylated on tyrosine in response to CSF-1 would be candidates for signal-transducing proteins, possibly substrates of CSF-1 receptor kinase itself. As an initial approach to studying CSF-1 signal transduction, we isolated phosphotyrosyl proteins from ^{32}P-labeled, CSF-1-stimulated BeWo cells by affinity chromatography on antiphosphotyrosine antibody-linked Sepharose. Phosphorylations of 170-kd and 56-kd phosphotyrosyl proteins were dramatically stimulated by CSF-1 (Fig. 1A). Phosphorylation of a less prominent 50-kd phosphotyrosyl protein in response to CSF-1 was also seen. Phosphopeptide mapping with staphylococcal V8 protease indicated that the 56-kd and 170-kd proteins of BeWo cells were unrelated (data not shown). Phosphoamino acid analysis of the 56-kd gel band of BeWo cells revealed that it contained substantial amounts of phosphotyrosine as well as phosphoserine and a trace of phosphothreonine (Fig. 1B).

The BeWo cell line was chosen for study of CSF-1-stimulated protein phosphorylations because of its simple maintenance in culture, but since no proliferative or maturational responses of this cell line to CSF-1 have been observed, we also examined adherent peripheral blood mononuclear cells, which are known to express the c-*fms* product and to bind ^{125}I-CSF-1 [4,5]. Mononuclear leukocytes were isolated

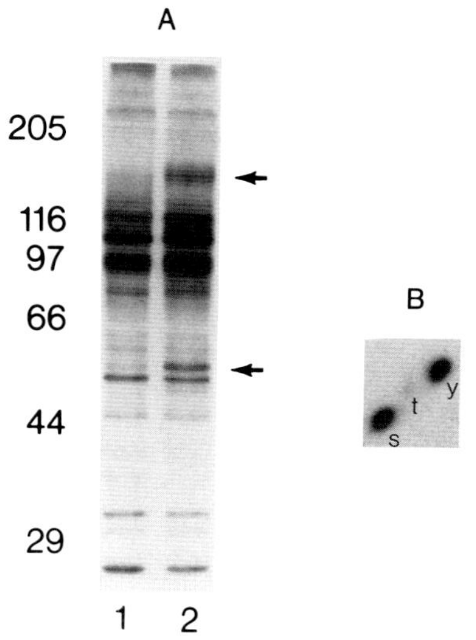

Fig. 1. **A:** Phosphorylation of 170-kd and 56-kd proteins in response to CSF-1. BeWo cells were metabolically labeled for 3 h with a mCi/ml $^{32}P_i$ and treated with 10^4U/ml rhCSF-1 **(Lane 2)** or control medium **(Lane 1)** for 30 min at 0–4°C prior to extraction with 1% Trition X-100 buffer containing inhibitors of proteases, phosphatases, and kinases. Phosphotyrosyl proteins were isolated by antiphosphotyrosine affinity chromatography and analyzed by SDS-PAGE. Autoradiogram exposure was 60 h. **B:** Phosphotyrosine content of 56-kd protein of CSF-1 stimulated BeWo cells. The 56-kd SDS-PAGE gel band of figure 1 was excised and the eluted protein was subjected to partial acid hydrolysis and 2-dimensional thin-layer electrophoresis. Phosphoamino acids were visualized by autoradiography. Exposure was 13 days with intensifying screen at −70°C.

from peripheral blood of a healthy human donor by Ficoll/metrizoate centrifugation and enriched for mature monocytes by plastic adherence. Brief metabolic labeling of these cells with $^{32}P_i$ followed by stimulation with rhCSF-1 also resulted in the phosphorylation of 170-kd and 56-kd proteins isolated by antiphosphotyrosine affinity chromatography (Fig. 2). Phosphorylations of 95-kd and 78-kd antiphosphotyrosine-isolated proteins were also enhanced in peripheral blood monocytes by CSF-1 (not observed in BeWo). Thus the phosphorylation of a 56-kd protein in response to CSF-1 appears not to be a peculiarity of the BeWo cell line, but rather to be a phenomenon common to human cells responding to CSF-1.

To explore further the relationship of the 56-kd and 170-kd proteins, we examined the kinetics of their phosphorylations in response to CSF-1. BeWo cells metabolically labeled with $^{32}P_i$ were exposed to rhCSF-1 at 0–4°C (to inhibit membrane protein internalization/degradation and protein-tyrosine phosphatases) for various times up to 30 min prior to extraction and isolation of phosphotyrosyl proteins (Fig. 3). ^{32}P-labeling of the 170-kd protein was first detected at one minute of rhCSF-1 stimulation and reached peak intensity by 5 min. Phosphorylation of the 56-kd protein could not be detected until 5 min and was still increasing in intensity at 30 min. This result is consistent with a mechanistic relationship between the phosphorylations of the 170-kd and the 56-kd proteins, possibly as kinase and substrate, respectively.

We suspected the 170-kd phosphoprotein to be the c-*fms* (CSF-1 receptor) since its molecular weight was in the appropriate range and its phosphorylation occurred specifically in response to CSF-1. To further characterize the 170-kd protein, we

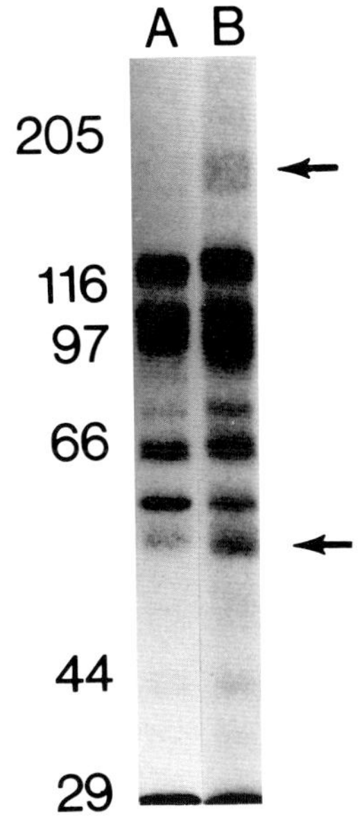

Fig. 2. Phosphorylation of 170-kd and 56-kd proteins in monocytes in response to CSF-1. Peripheral blood monocytes were isolated by Ficoll/metrizoate centrifugation and plastic adherence, then metabolically labeled with $^{32}P_i$ for 3 h and treated with 10^4 U/ml rhCSF-1 (**Lane B**) or control medium (**Lane A**). The cells were extracted with 1% Triton X-100 in buffer containing inhibitors of proteases, phosphatases, and kinases, then subjected to antiphosphotyrosine affinity chromatography and SDS-PAGE as described in the text. Autoradiogram was exposed for 16 h with intensifying screen at −70°C.

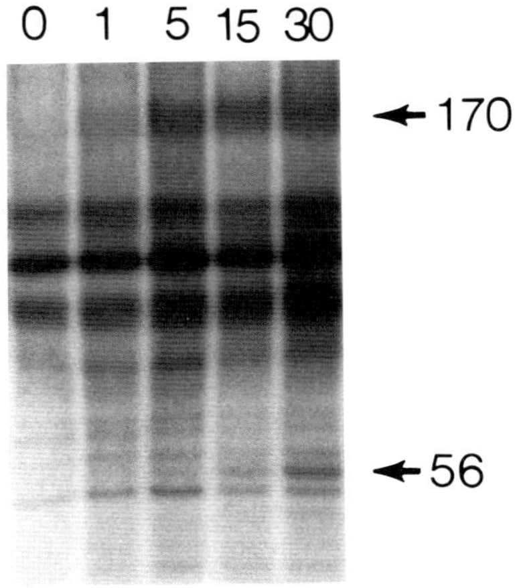

Fig. 3. Time dependence of 170-kd and 56-kd protein phosphorylations in response to CSF-1. BeWo cells were incubated with $^{32}P_i$ for 3 h, then treated with 10^4 U/ml rhCSF-1 on ice for various times from 0 to 30 min (as shown) prior to extraction for antiphosphotyrosine affinity chromatography and SDS-PAGE. Autoradiogram was exposed for 72 hr.

examined whether it was adsorbed by wheat germ agglutinin, since the CSF-1 receptor is known to be a glycosylated membrane protein [4,5,28]. Phosphotyrosyl proteins were isolated from BeWo cells that were metabolically labeled with $^{32}P_i$ and stimulated with rhCSF-1. The eluted phosphotyrosyl proteins were partitioned on WGA-Sepharose beads, then eluted with N-acetylglucosamine (Fig. 4A). A 170-kd phosphotyrosyl protein with affinity for wheat germ agglutinin was isolated only from the CSF-1-stimulated cells, consistent with its identity as the autophosphorylated c-*fms* product. Direct immunoprecipitation of detergent extracts of ^{32}P-metabolic-labeled BeWo cells with anti-*fms* serum demonstrated a phosphoprotein of identical molecular weight and similar appearances on SDS-PAGE to the 170-kd phosphoprotein isolated by antiphosphotyrosine affinity chromatography (data not shown). Final determination of the identity of the CSF-1-stimulated 170-kd protein was achieved by its ability to be secondarily immunoprecipitated from hapten eluates by the anti-*fms* serum (Fig. 4B).

DISCUSSION

In this study, we have isolated proteins phosphorylated on tyrosine from cells expressing the CSF-1 receptor to examine cellular protein phosphorylations in re-

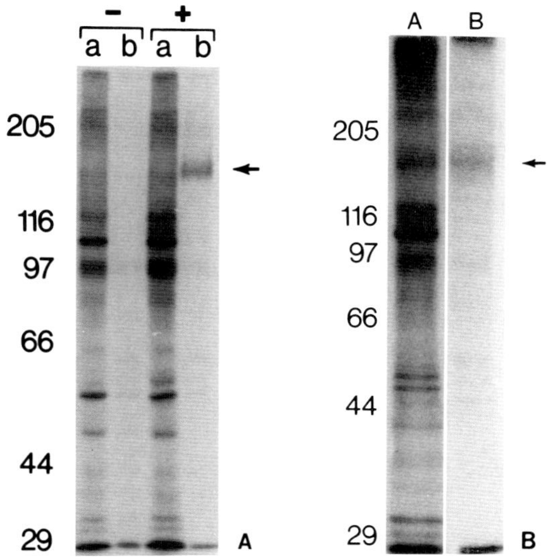

Fig. 4. **A:** Identification of glycosylated 170-kd CSF-1-stimulated phosphotyrosyl protein from BeWo cells. BeWo cells were incubated for 3 h with $^{32}P_i$, and then either stimulated (**+**) or not stimulated with rhCSF-1 (—) (for 30 min at 0–4°C). Phosphotyrosyl proteins were isolated from detergent extracts, as described in Materials and Methods. Hapten-eluted phosphotyrosyl proteins were partitioned on wheat germ agglutinin-Sepharose (WGA-Sepharose); bound proteins were eluted with N-acetylglucosamine. **a lanes:** Phosphotyrosyl proteins that did not bind to WGA-Sepharoses (supernatants of incubation step). **b lanes:** Phosphotyrosyl proteins eluted from WGA-Sepharose by N-acetylglucosamine. Autoradiogram exposure was 24 h with intensifying screen at −70°C. **B:** Identification of tyrosine-phosphorylated c-*fms* protein from CSF-1 stimulated BeWo cells. BeWo cells were incubated for 3 h with $^{32}P_i$ and stimulated with rhCSF-1. Phosphotyrosyl proteins isolated from detergent extracts by antiphosphotyrosine affinity chromatography (**Lane A**) were reacted with 5 μl anti-*fms* serum and immunoprecipitated on protein A-Sepharose (**Lane B**). Phosphoproteins were analyzed by SDS-PAGE and visualized by autoradiography. Exposure was 8 h with intensifying screen at −70°C.

sponse to CSF-1 stimulation. We observed phosphorylations of a 170-kd glycoprotein and 56-kd protein in response to rhCSF-1 in cells of the BeWo choriocarcinoma cell line. We believe the 170-kd phosphotyrosyl protein is the c-*fms* product (CSF-1 receptor) for several reasons: its size is in the appropriate range by SDS-PAGE; its phosphorylation occurs rapidly in response to rhCSF-1; it binds to wheat agglutinin, consistent with its being glycosylated (as the CSF-1 receptor is known to be); its electrophoretic appearance is similar to a 170-kd phosphoprotein that is immunoprecipitated by *fms* antiserum from BeWo cells and whose phospholabeling is enhanced in vivo by rhCSF-1; and it can be secondarily immunoprecipitated from hapten eluates by *fms* antiserum. A similarly rapid phosphorylation on tyrosine of the CSF-1 receptor in vivo in response to CSF-1 has recently been demonstrated in the murine BAC1.2F5 cell line by Downing et al. [29].

The previously published size of the human c-*fms* phosphoprotein detected by immune complex kinase reactions is approximately 150 kd [4,5]. The discrepancy in molecular weights between this protein and the protein we see in CSF-1-stimulated cells could be due to the effects of glycosylation or phosphorylation differences on electrophoretic behavior, or could reflect limited proteolysis. This small discrepancy is presently unresolved.

The time course of appearance of the phosphorylated 56-kd protein following the CSF-1-stimulated phosphorylation of the 170-kd protein is consistent with the activation of a 170-kd CSF-1 receptor kinase by autophosphorylation, much like the insulin receptor [30]. Only after autophosphorylation would the kinase be able to phosphorylate a cellular substrate. However, this time course may reflect the activation of secondary kinases or inhibition of phosphotyrosine phosphatases. The 56-kd protein does not appear to be a component of a macromolecular complex with the 170-kd protein, since WGA-Sepharose isolated the 170-kd protein independent of the 56-kd protein. Further studies of noncovalent associations between the 170-kd and 56-kd proteins and assays in vitro of phosphorylations of cellular proteins by purified c-*fms* kinase may help answer the question of whether the 56-kd protein is a substrate of the c-*fms* protein.

The identity of the 56-kd protein phosphorylated in response to CSF-1 has yet to be determined. It is intriguing to consider the possibility that it may be related to the *src* family of protein-tyrosine kinases. However, attempts to isolate it by immunoprecipitation either directly from detergent extracts of ^{32}P-labeled CSF-1-stimulated cells or from the mixture of hapten-eluted phosphotyrosyl proteins with monoclonal 327 *src* antibody [31] were unsuccessful (data not shown). Thus, it does not appear to be the c-*src* protein itself. Whether it may be one of the other *src*-family proteins, such as *yes, hck,* or *syn/slk,* remains to be determined [32–34]. It is especially interesting that proteins of similar sizes, as yet unidentified, or phosphorylated on tyrosine (along with other cellular proteins) in response to other peptide growth factors (epidermal growth factor in A431 cells, platelet-derived growth factor in 3T3 cells, and interleukin 3 in murine hematopoietic progenitor cells) and constitutively in v-*abl*-transformed cells and in chronic myelogenous leukemia cells [11,24,26,35]. The possible relationships between these proteins are currently under investigation.

From the study of BeWo cells, the 56-kd protein appeared to be the major candidate for a molecule that could transduce the CSF-1 signal from the CSF-1 receptor. However, no proliferative or morphologic effects of CSF-1 on BeWo cells have been observed. Therefore, since we were primarily interested in the regulatory actions of CSF-1 on growth and differentiation of hemopoietic cells, it was important that a similar 56-kd protein phosphorylation was found in response to CSF-1 in normal human monocytes. The phosphorylations of two additional proteins appears to be enhanced in peripheral blood monocytes by CSF-1. The observation of different spectra of phosphotyrosyl proteins between the peripheral blood monocytes and BeWo cells is not especially surprising, though, since these cells are of such disparate lineages. Unfortunately, peptide mapping of the 56-kd proteins from the BeWo cells and peripheral blood monocytes has been precluded by technical difficulties such as very limited digestion by staphylococcal V8 protease or chymotrysin and low signal strength in the 56-kd band of peripheral blood monocytes. Thus, while the relationship between the two 56-kd proteins remains to be firmly established, the BeWo cell line in culture may nevertheless represent a good model for some biochemical aspects of the CSF-1 response.

The results of this present study do not necessarily implicate the phosphorylation of the 56-kd protein in cell proliferation per se. As mentioned, the growth of BeWo cells is not affected by CSF-1 (although the fetal bovine serum in which they are cultured would be expected to contain a small concentration of CSF-1). Furthermore, peripheral blood mononuclear leukocytes isolated by plastic adherence are enriched for mature monocytes that are well along the pathway of differentiation to macrophages [36]. These cells may have very limited potential for proliferation in response to CSF-1. Maturing monocyte/macrophage cells do require CSF-1 for survival, however. Thus, the phosphorylation of the 56-kd protein could conceivably be involved in transducing a "viability" signal, or perhaps signals for differentiation-related processes. The development of culture methods for human monocyte/macrophage precursors that proliferate in response to CSF-1 will enable investigations to help answer this question. The correlation of CSF-1-stimulated proliferation with the phosphorylation of an homologous protein in murine systems would help to evaluate its potential role in growth signal transduction.

ACKNOWLEDGMENTS

We wish to thank Peter Ralph (Cetus Corporation) for the generous gift of rhCSF-1, Carl Rettenmier (St. Jude Children's Research Hospital) for the gift of anti-*fms* serum, Joan Brugge (SUNY Stonybrook) for the monoclonal 327 *src* antibody, and E. Richard Stanley (Albert Einstein College of Medicine) for many helpful discussions. This study was supported by grants from the Rhode Island Foundation and the Ocean State Charities Trust to R.D.H. and from the NIH (R01-CA39235 to A.R.F. and CA13943 to the Cancer Research Center at Roger Williams General Hospital). R.D.H. is a fellow of the Pharmaceutical Manufacturer's Association Foundation.

REFERENCES

1. Stanley ER, Guilbert LJ, Tushinski RJ, Bartelmez SH: J Cell Biochem 21:151–159, 1983.
2. Stanley ER, Heard PM: J Biol Chem 252:4305–4312, 1977.
3. Bartocci A, Pollard JW, Stanley ER: J Exp Med 164:956–961, 1986.
4. Woolford J, Rothwell V, Rohrschneider L: Mol Cell Biol 5:3458–3466, 1985.
5. Rettenmier CW, Sacca R, Furman WL, Roussel MF, Holt JT, Nienhuis AW, Stanley ER, Sherr CJ: J Clin Invest 77:1740–1746, 1986.

6. Sherr CJ, Rettenmier CW, Sacca R, Roussel MF, Look AT, Stanley ER: Cell 41:665–676, 1985.
7. Sacca R, Stanley ER, Sherr CJ, Rettenmier CW: Proc Natl Acad Sci USA 83:3331–3335, 1986.
8. Hunter T, Cooper JA: Cell 24:741–752, 1981.
9. Bertics PJ, Gill GN: J Biol Chem 260:14642–14647, 1985.
10. Bishayee S, Ross AH, Womer R, Scher CD: Proc Natl Acad Sci USA 83:6756–6760, 1986.
11. Frackelton AR Jr, Tremble PM, Williams LT: J Biol Chem 259:7909–7915, 1984.
12. Pang DT, Sharma BR, Shafer JA, White MF, Kahn CR: J Biol Chem 260:7131–7136, 1985.
13. Tornqvist HE, Gunsalus JR, Nemenoff RA, Frackelton AR Jr, Pierce MW, Avruch J: J Biol Chem 263:350–359, 1988.
14. Goustin AS, Leof EB, Shipley GD, Moses HL: Cancer Res 46:1015–1029, 1986.
15. Yeung YG, Jubinsky PT, Sengupta A, Yeung DCY, Stanley ER: Proc Natl Acad Sci USA 84:1268–1271, 1987.
16. Das SK, Stanley ER, Guilbert LJ, Forman LW: Blood 58:630–641, 1981.
17. Das SK, Stanley ER: J Biol Chem 257:13679–13684, 1982.
18. Kawasaki ES, Ladner MB, Wang AM, Van Arsdell J, Warren MK, Coyne MY, Schweickart VL, Lee M, Wilson KJ, Boosman A, Stanley ER, Ralph P, Mark DF: Science 230:291–296, 1985.
19. Wong GG, Temple PA, Leary AC, Witek-Giannotti JS, Yang Y, Cialetta AB, Chung M, Murtha P, Kriz R, Kaufman RJ, Ferenz CR, Sibley BS, Turner KJ, Hewick RM, Clark SC, Yanai N, Yokota H, Yamada M, Saito M, Motoyoshi K, Takaku F: Science 235:1504–1508, 1987.
20. Guilbert LJ, Stanley ER: J Biol Chem 261:4024–4032, 1986.
21. Patillo R, Gey GO: Cancer Res 28:1231–1236, 1968.
22. Pawlowski NA, Kaplan G, Hamill AL, Cohn ZA, Scott WA: J Exp Med 158:393–412, 1983.
23. Laemmli UK: Nature 227:680–685, 1970.
24. Huhn RD, Posner MR, Rayter SI, Foulkes JG, Frackelton AR Jr: Proc Natl Acad Sci USA 84:4408–4412, 1987.
25. Cleveland DW, Fischer SG, Kirschner MW, Laemmli UK: J Biol Chem 252:1102–1106, 1977.
26. Frackelton AR Jr, Ross AH, Eisen HN: Mol Cell Biol 3:1343–1352, 1983.
27. Cooper JA, Sefton BM, Hunter T: Methods Enzymol 99:387–402, 1983.
28. Rettenmier CW, Chen JH, Roussel MF, Sherr CJ: Science 228:320–322, 1985.
29. Downing JR, Rettenmier CW, Sherr CJ: Mol Cell Biol 8:1795–1799, 1988.
30. Petruzelli LM, Ganguly S, Smith CR, Cobb MH, Rubin CS, Rosen OM: Proc Natl Acad Sci USA 79:6792–6796, 1982.
31. Lipsich LA, Lewis AJ, Brugge JS: J Virol 48:352–360, 1983.
32. Sudol M, Hanafusa H: Mol Cell Biol 6:2839–2846, 1986.
33. Quintrell N, Lebo R, Varmus H, Bishop JM, Pettenati MJ, LeBeau MM, Diaz MO, Rowley JD: Mol Cell Biol 7:2267–2275, 1987.
34. Kawakami T, Pennington CY, Robbins KC: Mol Cell Biol 6:4195–4201, 1986.
35. Morla AO, Schreurs J, Miyajima A, Wang JYJ: Mol Cell Biol 8:2214–2218, 1988.
36. Tushinski RJ, Oliver IT, Guilbert LJ, Tynan PW, Warner JR, Stanley ER: Cell 28:71–81, 1982.

Journal of Cellular Biochemistry 39:139–151 (1989)
Growth Factors and Their Receptors: Genetic Control and Rational Application 117–129

Effects of Platelet-Derived Growth Factor and Fibroblast Growth Factor on Free Intracellular Calcium and Mitogenesis

Robert W. Tucker, David T. Chang, and Kimberly Meade-Cobun

Johns Hopkins Oncology Center, Baltimore, Maryland 21205

Although increased free intracellular calcium (Ca_i) may be one of the main regulators of cell growth and differentiation, studies in cell populations have implied that not all growth factors produce Ca_i increases. In order to examine in more detail whether Ca_i increases were related to mitogenesis, we used digital image analysis of intracellular Fura-2 fluorescence to measure Ca_i in individual BALB/c 3T3 cells stimulated with either platelet-derived growth factor (PDGF) or fibroblast growth factor (FGF). We found that PDGF induced larger and more prolonged Ca_i increases than FGF did, but that both growth factors induced an initial rapid increase in Ca_i (<2 min) followed by a later sustained increase (>20 min). Only the prolonged Ca_i increase required extracellular calcium. Following PDGF treatment (1–8 units/ml), the percentage of cells with a large peak Ca_i increase (>twofold) correlated with the percentage of cells made competent (subsequent growth in 1% platelet-poor-plasma). In contrast, purified bovine basic FGF (200–800 pg/ml) and recombinant human acidic FGF (10–300 ng/ml) produced peak Ca_i increases that were not directly correlated with mitogenesis. In addition, concentrations of intracellular Quin 2 that inhibited Ca_i transients also inhibited PDGF stimulation but not FGF stimulation of mitogenesis. Thus, Ca_i increases are necessary for mitogenesis in BALB/c 3T3 cells stimulated by PDGF, but not that stimulated by FGF.

Key words: calcium, Fura-2, growth factors, competence, PDGF, autoradiography, digital image analysis, FGF

Increases in free intracellular calcium (Ca_i) are among the earliest events stimulated by growth factors [1–3]. However, no evidence has been presented to indicate whether these early ionic signals are produced by all growth factors or whether Ca_i increases are required for mitogenic events. Platelet-derived growth factor (PDGF) definitely increases Ca_i [1–3], but Ca_i increases induced by fibroblast growth factor (FGF) are controversial. In one study of cell populations, FGF produced small increases in Ca_i [4] measured by Quin-2 fluorescence, but in another study no Ca_i change was detected using aequorin luminescence [5]. Moreover, only

Received February 1, 1988; accepted August 17, 1988.

small and inconsistent increases in ^{45}Ca efflux were measured in cell populations stimulated with FGF, in contrast to large increases of ^{45}Ca efflux induced by PDGF [Tucker RW, Meade-Cobun K, Phair R; unpublished observation]. Yet, short treatment (20 min) of BALB/c 3T3 cells with FGF produced a larger mitogenic effect (more cells competent to initiate DNA synthesis) than that produced by the same short exposure to PDGF. These studies suggested that Ca_i increases that were detectable in cell populations (e.g., large synchronous increases in most cells) correlated with mitogenesis induced by PDGF but were not necessary for mitogenesis induced by FGF. However, significant questions remain about whether small, asynchronous, or localized increases in Ca_i, which are difficult to measure in cell populations, can contribute to mitogenesis induced by FGF, and whether Ca_i increases are actually required for mitogenesis induced by PDGF.

Determining the relationship between Ca_i increases and DNA synthesis in cell populations has been difficult. Both Ca_i increases and DNA synthesis can be heterogeneous and variable from time to time and from cell to cell. Thus, measurements of Ca_i in cell populations represent spatial and temporal averages only, and will underestimate asynchronous Ca_i increases or will fail to detect localized Ca_i changes. In contrast, analysis of single cells can be used to compare Ca_i changes and DNA synthesis in the same cell or population and to define spatial changes in Ca_i. In the present study we used digital image analysis of intracellular Fura-2 to measure Ca_i changes in single quiescent BALB/c 3T3 cells stimulated with either PDGF or FGF. Both PDGF and FGF induced similar spatial and temporal increases in Ca_i, but only the Ca_i increases produced by PDGF correlated with mitogenesis; moreover, buffering of Ca_i changes by intracellular Quin 2 inhibited PDGF-stimulated but not FGF-stimulated mitogenesis. Thus, Ca_i increases appear to be necessary for mitogenesis stimulated by PDGF, but not for that stimulated by FGF.

MATERIALS AND METHODS

Cell Culture and Materials

An original stock of BALB/c 3T3 cells (Clone A-31) was obtained from G. Todaro (Frederick, MD), and cells were grown in Dulbecco modified Eagle medium (DME; Gibco) supplemented with 5% calf serum (CS; Gibco). BALB/c 3T3 cells (8×10^4/dish) were plated and allowed to become quiescent during 6 days of growth in DME + 0.5% CS at 37°C and 7% CO_2.

Purified PDGF (AB heterodimer) was obtained from outdated human platelets (Dr. W.J. Pledger, Vanderbilt University) and stored as 4×10^4 units/ml in 1 M acetic acid at −30°C. Immediately before use, PDGF stock was diluted 1:400 with 0.1% bovine serum albumin (BSA) in calcium-free Krebs/Hepes buffer (in mM; NaCl 120, KCl 4, KH_2PO_4 1, $MgSO_4$ 1, Hepes 1.2) supplemented with 5 mM glucose and amino acids (minimal essential medium, Gibco). Purified basic FGF was obtained from Dr. G.D. Shipley (Oregon Health Sciences University) and was prepared as previously described from bovine brain [8]. In addition, recombinant human acidic FGF was obtained from Chiron (Emeryville, CA). Both FGF preparations were stored in 0.1% gelatin in chelexed water and were diluted in 0.1% gelatin in Krebs/Hepes buffer for experiments. Platelet-poor plasma (PPP) was prepared from human serum as previously described [7].

Autoradiography

All cultures were assayed for DNA synthesis by autoradiography of cells incubated with growth factors in tritiated thymidine (1 μCi/ml; 60 Ci/mg, New England Nuclear) as previously described [8]. Briefly, after treatment with growth factor for 1 h and subsequent incubation with 1% PPP and tritiated thymidine for 48 h, cell monolayers were fixed in 95% ethanol, dipped in Kodak emulsion (NTB-2), and incubated in the dark for 48 h. Emulsion-covered slides were then developed and fixed as previously described [8], and the percentage of cells with labeled nuclei was determined. The percentage of cells made competent for DNA synthesis was measured as the percentage of labeled cells stimulated by growth factor followed by 1% PPP, minus the percentage of labeled cells stimulated by 1% PPP alone. In all experiments the percentage of cells able to initiate DNA synthesis in the presence of 1% PPP alone was less than 10%.

Fura-2 Loading

Fura-2/AM loading of BALB/c 3T3 cells was performed as a modification of previously described procedures [9]. Briefly, BALB/c 3T3 cells were cooled for 15 min on ice, then incubated for 2 h at room temperature with 25 μM Fura-2/AM, 0.03% Pluronic (Pluronic F127, BASF Wyandotte Corp.), 0.01% dimethyl sulfoxide (DMSO), and 0.75% PPP in Krebs/Hepes (in mM; NaCl 120, KCl 4, KH_2PO_4 1, $MgSO_4$ 1, $CaCl_2$ 1.3, Hepes 1.2). Fura-2 loading under these conditions did not enhance or inhibit DNA synthesis stimulated by 20% CS (data not shown).

Intracellular Fura-2 concentration (750 uM) was estimated from the fluorescence of material released from cells permeabilized with 0.001% digitonin in Krebs/Hepes buffer for 20 min. The excitation spectrum of permeabilized cell material was measured using a Perkin-Elmer fluorescence spectrophotometer (4-nm slits) with a Xenon lamp (in collaboration with Dr. Reynafarje, Johns Hopkins Medical School) and emission filters set at 500 nm.

Digital Imaging of Intracellular Fura-2

Ca_i was calculated from the fluorescence of intracellular Fura-2 as previously described [10]. Briefly, the microscope system consisted of an inverted epifluorescence microscope (Leitz Diavert), 100× or 40× Nikon UV-Fluor (N.A. 1.3) objectives, and a DC-stabilized 50-W mercury arc lamp. All optical components in the excitation pathway were either quartz or glass, with high transmittance for ultraviolet (UV) wavelengths. Fluorescent images of cells, buffer, and Fura-2 solutions were obtained at 340 and 380 nm excitation using narrow-band (± 8 nm) interference filters (Ditric Optics, Hudson, MA), and emitted light was collected through a 500-nm broad-band filter (± 40 nm) (Ditric Optics). All calculations and digitizations were performed using a microcomputer image-analysis system (BIAS, Loats Associates, Inc., Westminster, MD) based on an IBM/AT. An Epyx silicon video board was used to digitize a central region containing 128 horizontal × 240 vertical pixels. Thirty-two frames were averaged per wavelength, and images were acquired every 20 s, with 256 grey-level precision (8 bit). Ca_i images were photographed from the monitor using a Minolta 35-mm camera with Plus-X film.

Calculations of Ca_i

Ca_i was calculated from the relation $Ca_i = Kd\,B(R - R_{min})/(R_{max} - R)$ where K_d is the dissociation constant for Fura-2, B = Fura-2 (380 nm)/Ca Fura-2 (380 nm),

R = cell (340 nm)/cell (380 nm), R_{min} = Fura-2 (340 nm)/Fura-2 (380 nm), and R_{max} = Ca Fura-2 (340 nm)/Ca Fura-2 (380 nm) (with excitation wavelength in parentheses). Fura-2 was a solution of Fura-2 (10–100 uM) in MOPS buffer (in mM: KCl 100; KMOPS 10) containing 10 mM EGTA, and Ca Fura-2 was a Ca^{2+}-saturated solution of Fura-2 in MOPS buffer containing 1 mM $CaCl_2$. K_d = 220 nM, in agreement with previous reports [11,12]. Background images of buffer solutions in the dish were subtracted from cell images and Fura-2 solution images to correct for background noise (dark current) of the video camera and inherent fluorescence of the optical system. The spatial heterogeneity of illumination and camera sensitivity were compensated for by the inclusion of ratios of the uniform solutions of Fura-2 in the calculations of Ca_i.

RESULTS

PDGF and FGF Both Increase Ca_i

Average basal Ca_i was 57 ± 22 nM (S.D.) for 151 cells. For any one cell, the basal Ca_i varied by 4.6% (S.D./mean), so that a 10% increase (1.1-fold) in Ca_i was significantly different from basal Ca_i ($P < .05$). Both PDGF and FGF induced transient increases in average Ca_i in quiescent BALB/c 3T3 cells (Figs. 1, 2). The Ca_i increase appeared to be generalized, occurring in both the nucleus and cytoplasm. The Ca_i increase was also biphasic, consisting of an initial Ca_i increase lasting 2 min, followed by a second Ca_i increase persisting for at least 20 min (the period of observation) (Fig. 3). The prolonged (>20 min) elevation of Ca_i (>1.5-fold) oc-

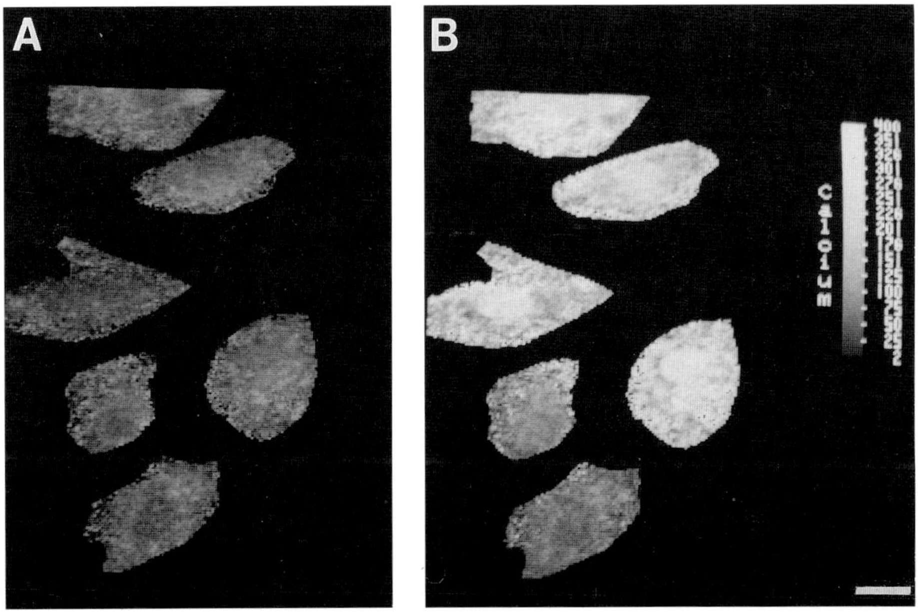

Fig. 1. PDGF increased Ca_i in both nucleus and cytoplasm. Images of Ca_i depict spatial variation of Ca_i in the cell, with Ca_i quantitated in nM and displayed as different intensities in the grey scale. **A:** Quiescent BALB/c 3T3 cells. **B:** 3 min after PDGF (8 units/ml) treatment of same cells as in A. Bar = 25 μm.

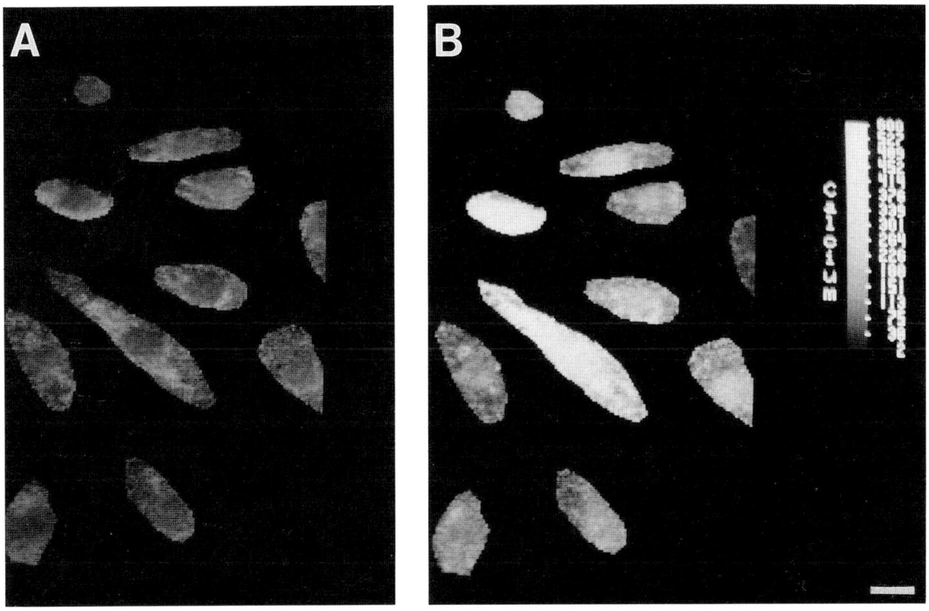

Fig. 2. FGF (human recombinant) increased Ca_i throughout the cell. **A:** Quiescent BALB/c 3T3 cells. **B:** 2.3 Min after FGF (30 ng/ml) treatment of same cells as in A. Bar = 25 μm.

curred in more than 50% of PDGF-treated cells (n = 54) and in only 10% of FGF-treated cells (n = 46). Only the initial Ca_i peak was independent of extracellular calcium (Fig. 3B,D). At maximal stimulation, PDGF produced tenfold increases in Ca_i, while FGF produced only sixfold increases (Fig. 4–6). Detailed examination of the localization of Ca_i changes did not reveal any striking spatial pattern of Ca_i increase that was consistently correlated with either growth factor stimulation. Occasionally, Ca_i increases started around the nucleus or in the periphery of the cell, but these initial changes quickly led to the more generalized Ca_i increases shown in Figures 1, 2, and at present we have not determined any pattern to these localized transient increases. Thus, detailed examination of changes in Ca_i distribution have revealed many similarities in Ca_i increases induced by PDGF and FGF. Only the increased magnitude of the initial peak Ca_i change and the presence of the prolonged Ca_i increase distinguished PDGF-treated cells from FGF-treated cells.

Ca_i Increase Induced by PDGF Correlated With Mitogenesis

Increasing concentrations of PDGF stimulated a larger percentage of cells to respond with Ca_i increases, so that the percentage of cells with Ca_i increases in each of three categories (<twofold, two- to fourfold, >fourfold) rose progressively as the dose of PDGF increased (Fig. 4). Larger concentrations of PDGF also increased the magnitude of the peak Ca_i increase in each cell. In fact, cells treated with maximal doses (8 units/ml) of PDGF responded with particularly large (>fourfold) increases in peak Ca_i . Since increasing PDGF concentration (2-h exposure) also produced a progressive increase in the percentage of cells made competent for DNA synthesis (Table I) we compared Ca_i increase and DNA synthesis induction in the same cell population. For example, a high PDGF dose (8 units/ml) induced competence in the

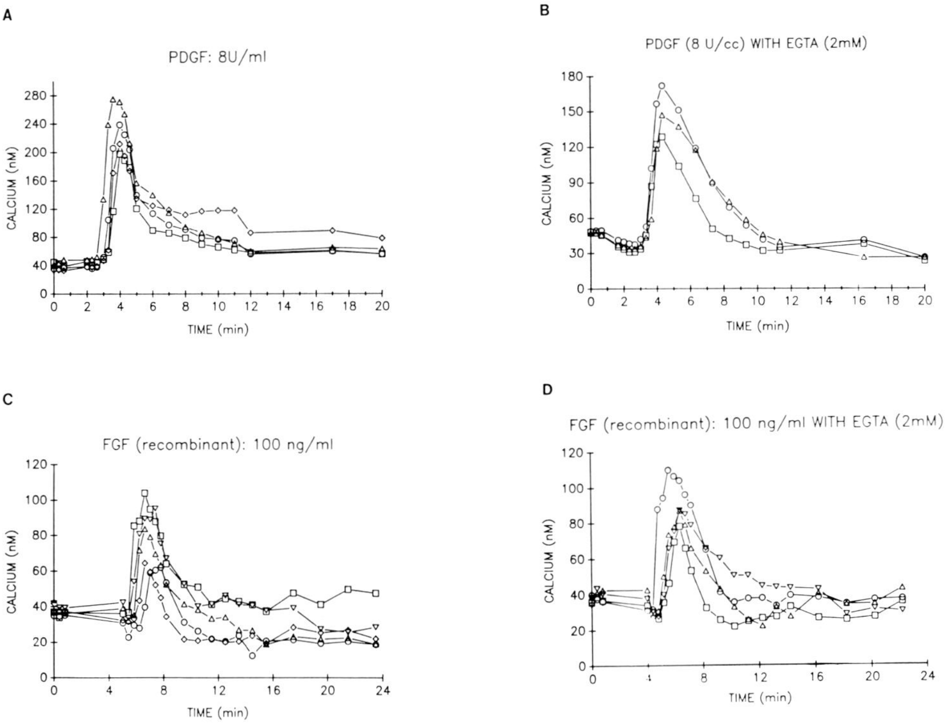

Fig. 3. PDGF and FGF produce Ca_i increases, with initial Ca_i change independent of extracellular Ca^{2+}. Time course of average cellular Ca_i in individual cells following stimulation of quiescent BALB/c 3T3 cells with treatment indicated. **A:** PDGF (8 units/ml). **B:** PDGF (8 units/ml) + EGTA (2 mM). **C:** Human recombinant FGF (100 ng/ml). **D:** Human recombinant FGF (100 ng/ml) + EGTA (2 mM). Growth factors were added 1 min after Ca_i measurement began.

majority of the quiescent cells and stimulated an increase in Ca_i to more than four times basal levels in most cells (54%) and more than two times basal levels in almost all cells (>92%). Thus, as the percentage of cells with an increase in Ca_i rose in response to increased concentrations of PDGF, so also did the percentage of cells made competent to initiate DNA synthesis.

Ca_i Increases Induced by FGF Did Not Directly Correlate With Mitogenesis

In contrast to PDGF, FGF stimulated Ca_i increases that did not correlate with the induction of competence for DNA synthesis (Table II). For human recombinant acidic FGF, there was very little difference in the Ca_i increase stimulated by low (10 ng/ml) and high (300 ng/ml) doses (Fig. 5). However, the high dose of FGF (300 ng/ml) made 75% of the cells competent, while the low dose (10 ng/ml) induced competence in only 1% of cells (Table II). Therefore, in cells stimulated by recombinant FGF there was no correlation between the percentage of cells with Ca_i increases and the percentage of cells made competent. In addition, low concentrations of purified bovine basic FGF (200 pg/ml) produced barely detectable increases (>1.2-fold) in Ca_i (Fig. 6) and yet 45% of the same cells were made competent by a 1-hr exposure to this concentration of FGF (Table II). Thus, neither purified bovine basic FGF nor recombinant human acidic FGF produced Ca_i increases in a percentage

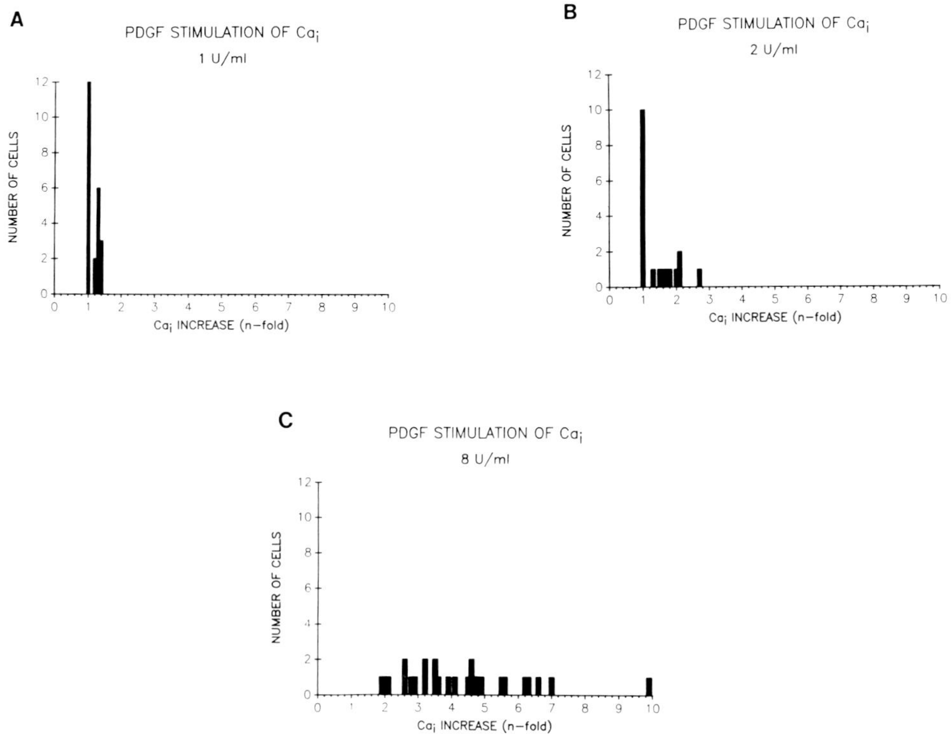

Fig. 4. PDGF produced dose-related increases in Ca_i . Frequency histograms of the magnitude of peak Ca_i after PDGF treatment. Increased Ca_i expressed as n-fold increases (stimulated average Ca_i/basal average Ca_i). Three different concentrations of PDGF were studied. **A:** 1 units/ml. **B:** 2 units/ml. **C:** 8 units/ml.

of cells that correlated with the percentage of cells made competent for DNA synthesis.

Quin 2 Inhibited Ca_i Increases Induced by PDGF and FGF

As Figure 7 illustrates, pretreatment of quiescent cells with 25 to 35 μM Quin 2/AM for 30 min resulted in obliteration of the initial transient Ca_i increase induced by both PDGF and FGF. In some cells, PDGF still increased Ca_i to a low plateau level that was not maintained. BAPTA-loaded cells showed even more profound inhibition of Ca_i increases but were not viable enough for later mitogenic studies. Permeabilization of Quin 2-loaded cells with digitonin (0.001%) showed that the intracellular concentration of Quin 2 was approximately 2 mM. Thus, this intracellular concentration of Quin 2 was sufficient to buffer Ca_i changes, thereby preventing growth factor-induced transient increases and plateau Ca_i increases.

Quin 2 Inhibited Mitogenesis Induced by PDGF but Not That Induced by FGF

Quin 2 (25 μM) loading prevented up to 90% of PDGF-stimulated competence, but less than 1% of the FGF-stimulated competence in the same cell population (Table III). In order to identify whether the inhibition of DNA synthesis was specifically

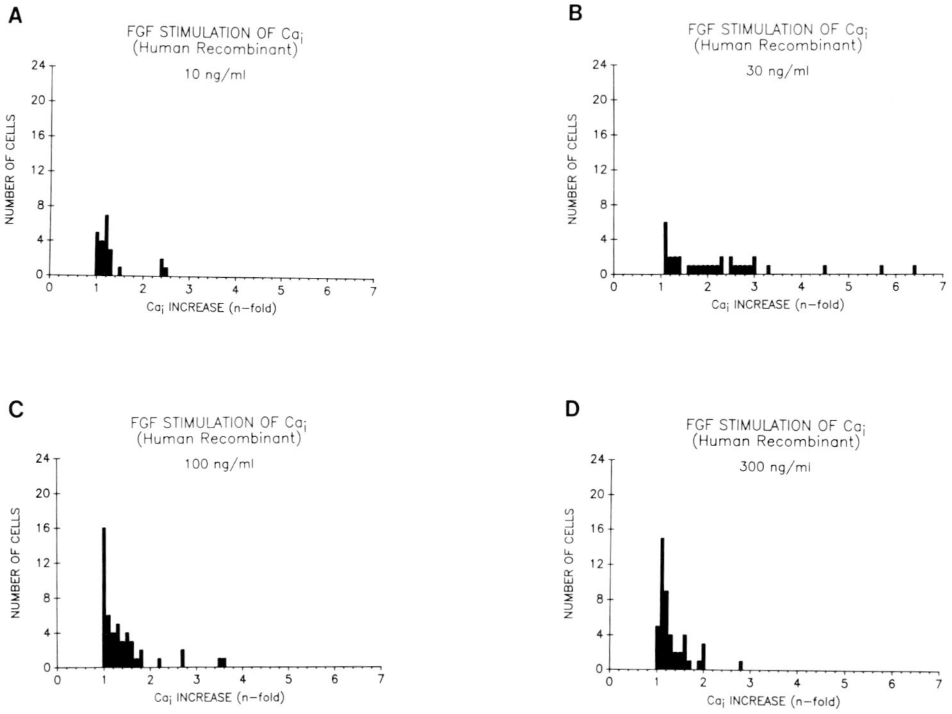

Fig. 5. Human recombinant FGF stimulated increases in Ca_i that were not dose-related. Frequency histograms of magnitude of peak Ca_i after treatment with FGF. **A:**10 ng/ml. **B:**30 ng/ml. **C:**100 ng/ml. **D:**300 ng/ml. Ca_i changes defined as n-fold increases (stimulated average Ca_i /basal average Ca_i).

related to the prevention of Ca_i increases, we performed three controls. *First*, we showed that inhibition of DNA synthesis was totally reversible, by using large concentrations (20%) of serum to stimulate DNA synthesis in Quin 2-loaded cells (data not shown). In addition, the continued stimulation of Quin 2-loaded cells by FGF also indicated the lack of nonspecific inhibition (Table III). *Second*, we used two pH indicators (calcein blue and carboxy-calcein blue) to load cells under the same conditions as those used for Quin 2 loading. These compounds contain three ester groups and following deesterification would produce amounts of intracellular formaldehyde similar to that produced by Quin 2, which contains four ester groups. Unlike Quin 2, however, calcein blue and carboxy-calcein blue do not bind calcium. We found that calcein blue and carboxy-calcein blue did not inhibit calcium transients and did not prevent competence induced by PDGF (Table III). Therefore, the production of intracellular formaldehyde by deesterification of Quin 2/AM cannot account for inhibition of DNA synthesis by Quin 2/AM. *Third*, TPEN (10–35 μM), a chelator of intracellular heavy metals, also did not inhibit Ca_i transients or DNA synthesis (Table III). Thus, Quin 2 inhibits competence for DNA synthesis induced by PDGF but not that induced by FGF. The mechanism of inhibition is independent of heavy metal chelation and intracellular formaldehyde production and most probably depends on the buffering of Ca_i increases.

DISCUSSION

By using digital image analysis of intracellular Fura-2 fluorescence in single cells, we have been able to examine the relationship between Ca_i increases and DNA

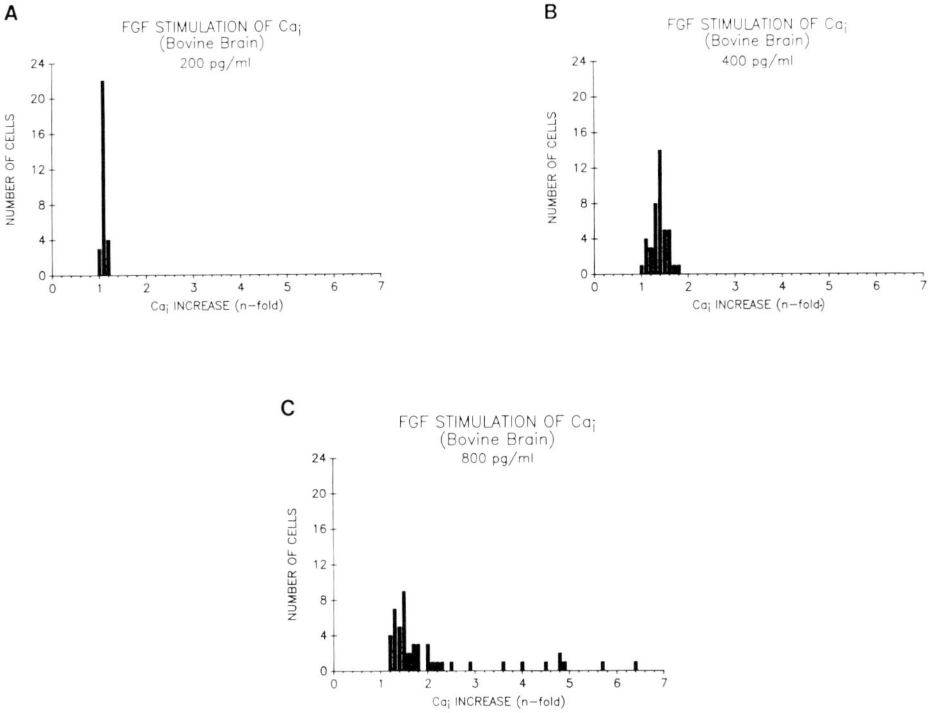

Fig. 6. Bovine brain FGF stimulated increases in Ca_i .Frequency histograms of the magnitude of peak Ca_i after treatment with different concentrations of FGF. **A:** 200 pg/ml. **B:** 400 pg/ml. **C:** 800 pg/ml. Ca_i responses expressed as n-fold increases (stimulated average Ca_i /basal average Ca_i).

TABLE I. Correlation Between Peak Ca_i Increase and DNA Synthesis Stimulated by PDGF

PDGF (units/ml)	N (No. cells)	Ca_i (% with n-fold increase)[a]				DNA synthesis (% competent)[b]
		1.0	<2	2–4	>4	
.5						0 (1)
1	23	52	48	0	0	27 ± 27 (2)
2	19	52	27	21	0	36 ± 30 (2)
8	26	0	8	38	54	77 ± 7 (4)

[a]Ca_i increase calculated as described for N (No. of cells). The n-fold increase in stimulated Ca_i was determined by stimulated peak Ca_i (cell average) divided by basal Ca_i (cell average).
[b]Percent of cells in DNA synthesis determined by autoradiography of cells stimulated with the indicated concentrations of PDGF for 2 h, followed by incubation in 1% PPP for 47 h. Results expressed as mean ± S.E.M. (No. of experiments).

synthesis in considerable detail. We found that both PDGF and FGF produced Ca_i increases in quiescent BALB/c 3T3 cells, but only PDGF treatment induced Ca_i increases that were directly dose-related and that correlated to subsequent DNA synthesis. Low concentrations of bovine brain FGF produced no cells with Ca_i increases (>1.2-fold) and a substantial percentage (>30%) of competent cells, while large concentrations of recombinant FGF produced a large percentage of competent cells but only a few cells with large (>twofold) increases in Ca_i. Moreover, buffering of Ca_i transients by intracellular Quin 2 prevented PDGF-stimulated but not FGF-

TABLE II. Correlation Between Peak Ca_i Increase and DNA Synthesis Stimulated by FGF

Stimulus (conc)	N (No. cells)	Ca_i (% with n-fold increase)[a] 1.0	<2	2–4	>4	DNA synthesis (% competent)[b]
Bovine brain FGF (pg/ml)						
50						11 (1)
100						17 ± 7 (2)
200	29	3	97	0	0	45 ± 14 (2)
400	42	2	98	0	0	34 ± 15 (3)
800	49	0	67	18	15	52 ± 15 (5)
1,600						38 ± 16 (2)
3,200						66 (1)
Human recombinant FGF (ng/ml)						
10	23	23	65	12	0	1 ± 0 (2)
30	33	0	48	42	10	19 ± 8 (3)
100	49	34	57	9	0	46 ± 1 (3)
300	49	11	81	8	0	75 ± 9 (2)

[a]Ca_i increase calculated as described for N (No. of cells). The n-fold increase in stimulated Ca_i was determined by stimulated peak Ca_i (cell average) divided by basal Ca_i (cell average).
[b]Percent of cells in DNA synthesis determined by autoradiography of cells stimulated with the indicated concentrations of FGF for 1 h, followed by incubation in 1% PPP for 48 h. Results expressed as mean ± S.E.M. (No. of experiments).

stimulated mitogenesis. Thus, both PDGF and FGF produced Ca_i increases, but only PDGF stimulation required Ca_i increases for the induction of competence for DNA synthesis.

Our documentation of the role of Ca_i in the induction of mitogenesis in PDGF-stimulated cells confirms previous work showing that PDGF stimulates increases in Ca_i [1–3] and that such changes correlate with mitogenesis in aequorin-loaded cell populations [5]. We were able to extend this analysis and use the inhibition of Ca_i increases with Quin 2 to obtain the first evidence that PDGF stimulation of mitogenesis depends on Ca_i increases. However, such clear dependence of PDGF-stimulated mitogenesis on Ca_i increases appears inconsistent with a study showing that phorbol ester (TPA) could inhibit PDGF stimulation of Ca_i increases without preventing DNA synthesis [2]. However, TPA may stimulate mitogenesis via activation of protein kinase C independently of Ca_i increases, thereby bypassing any inhibitory effect of TPA on Ca_i increase. Thus the Ca_i pathway stimulated by PDGF may not be absolutely obligatory for the stimulation of DNA synthesis when PDGF is combined with other factors. In addition, the Ca_i pathway contributing to mitogenesis can also be induced by other factors. For example, calcium ionophore A23187 induces both competence and transient Ca_i increases very similar to those stimulated by PDGF [R.W. Tucker, unpublished observations]. These results indicate that situations in which Ca_i increases are required for mitogenesis can now be documented by digital image analysis of Fura-2 fluorescence in individual cells.

This same analysis also showed that Ca_i increases are not required for FGF-stimulated mitogenesis. A low concentration of purified FGF (200 pg/ml) from bovine brain is capable of inducing competence in >45% of cells with only barely detectable (>1.2-fold) increases in Ca_i . Moreover, both low (10 ng/ml) and high (100 ng/ml)

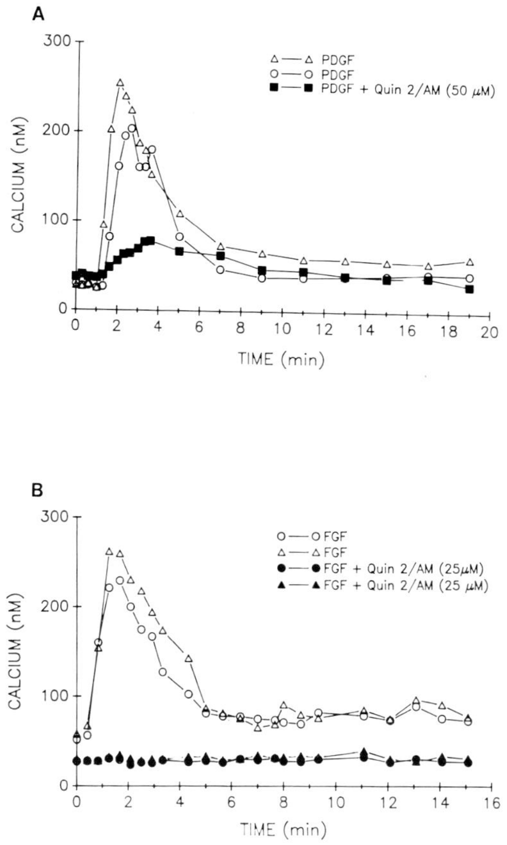

Fig. 7. Quin 2 inhibited Ca_i increases induced by PDGF and FGF. Pretreatment of quiescent BALB/c 3T3 cells with Quin 2/AM (25–50 μM) inhibited (**A**) PDGF stimulation and (**B**) FGF stimulation of Ca_i increases. Quin 2 itself was weakly fluorescent, bleached rapidly, and did not contribute significantly to the Ca^{2+}-sensitive intracellular fluorescence.

concentrations of recombinant FGF induced similar increases in Ca_i (Figs. 4, 5), but stimulated a vastly different percentage of cells to initiate DNA synthesis (1% and 46%, respectively). Finally, the concentration of recombinant FGF that induces a maximal Ca_i increase is lower than that required to stimulate maximal DNA synthesis. These results are compatible with our previous aequorin studies [5] in which large concentrations of unpurified FGF (Collaborative Research) stimulated competence without producing detectable Ca_i increases. In the present study we have extended this analysis to show that certain concentrations of purified FGF (both human recombinant and bovine brain) can cause Ca_i changes, but that the Ca_i changes are not directly related to stimulation of competence for DNA synthesis. In fact, intracellular concentrations of Quin 2 that inhibit PDGF-stimulated mitogenesis do not inhibit FGF-stimulated mitogenesis (Table III). Thus, Ca_i increases must be induced by a nonmitogenic component of purified FGF preparations. Previous workers have indeed suggested that a contaminant of purified FGF preparations produces small increases in Ca_i, as measured in hamster fibroblast populations using Quin-2 fluorescence [4]. However these Ca_i increases measured by Quin 2 were abolished by EGTA, whereas

TABLE III. Effect of Inhibitors on DNA Synthesis

Pretreatment[c] (μM)	Percent inhibition of competence stimulated by[a]	
	PDGF[b]	FGF (bovine)
Quin 2/AM	90	0
10	82	0
25	91	—
35		
Calcein blue/AM	7	—
25	1	—
35		
Carboxy-calcein blue/AM	1	—
25	0	—
35		
TPEN	0	—
10	0	—
35		

[a]Competent cells determined by percent of labeled cells as indicated in text. Percent inhibition of competence = [1 − % competence (PDGF + inhibitor)/% competence (PDGF)].
[b]Cell cultures pretreated with inhibitor for 30 min in Krebs/HCO_3- solution, at 37°C, in CO_2 incubator.
[c]PDGF (AB heterodimer) (8 units/ml) was used for Quin 2/AM experiments comparing PDGF and bovine brain FGF (800 pg/ml); PDGF (AB heterodimer) (16 units/ml) was used for Quin 2, calcein, and carboxy-calcein blue comparison, and PDGF (AA homodimer) (30 ng/ml) was used for TPEN experiments.

the FGF-stimulated Ca_i increases described in this report were not. In addition, since purified FGF from two different sources (with presumably different contaminants) both produced Ca_i increases, a contaminant of FGF preparations was probably not responsible for the Ca_i increases reported here. These discrepancies between our results and previous reports [4] may result from the study of different cell lines or from a more sensitive detection of Ca_i changes using single cell analysis of Fura-2 fluorescence in the present study. In fact, our results were recently confirmed by experiments using anti-PIP2 antibodies to prevent the generation of diacylglycerol and inositol trisphosphates (IP3) (and subsequent release of intracellular calcium stores); only PDGF-stimulated, and not FGF-stimulated, mitogenesis depended on constant PIP turnover and, by inference, on Ca_i increases [15]. Other changes in Ca^{2+} metabolism, such as depletion of Ca^{2+} stores, could still contribute to the activation of genes involved in mitogenesis. Indeed, Ca^{2+} depletion, rather than Ca_i increases, appears to be important in stimulating heat shock genes in *Drosophila* [13] and in stimulating glucose-deprivation genes in hamster fibroblasts [14]. However, appreciable Ca^{2+} depletion in BALB/c 3T3 cells can only occur if Ca^{2+} is released from an intracellular store, a finding incompatible with the negligible increase in ^{45}Ca efflux measured in FGF-stimulated 3T3 cells (R.W. Tucker, K. Meade-Cobun, R. Phair, unpublished observations). Thus, Ca^{2+} changes in both the cytosol and intracellular compartments do not appear to be important in FGF-stimulated DNA synthesis.

There are undoubtedly multiple pathways for inducing proliferation of cells. Ca_i measurements using Fura-2 in this study have shown that PDGF and FGF, two polypeptide factors that can induce competence in BALB/c 3T3 cells, use stimulatory pathways that differ in their dependence on Ca_i increases. Inhibition studies using

Quin 2 have further documented that the calcium transients are required for proliferation stimulated by PDGF, but not by FGF. Thus, Ca_i increases are not part of a common pathway used by all competence factors. In the future, digital image analysis of second messengers (e.g., Ca_i transients) in individual cells will be important in defining the relative contributions of multiple overlapping pathways to mitogenic stimulation.

ACKNOWLEDGMENTS

This work was supported by NIH grant Ca-34472. We gratefully acknowledge the superb secretarial services of Carmen Cardona and the technical assistance of Angela Broadt. The generous gifts of purified PDGF from Dr. W.J. Pledger, basic FGF from Dr. G.D. Shipley, and acidic FGF from Dr. L. Cousens (Chiron Corp.) are especially appreciated.

REFERENCES

1. Mix LL, Dinerstein RJ, Villereal ML: Biochem Biophys Res Commun 119:69–75, 1984.
2. McNeil PL, McKenna MP, Taylor DL: J Cell Biol 101:372–379, 1985.
3. Moolenaar WH, Tertoolen LGJ, de Laat SW: J Biol Chem 259:8066–8089, 1984.
4. Magnaldo I, L'Allemain G, Chambard JC, Moenner M, Barritault D, Pouyssegur J: J Biol Chem 261:16916–16922, 1986.
5. Tucker RW, Snowdowne KW, Borle AB: Eur J Cell Biol 41:347–351, 1986.
6. Shipley GD, Childs CB, Volkenant ME, Moses HL: Cancer Res 44:710–716, 1984.
7. Pledger JW, Stiles CD, Antoniades HN, Scher CD: Proc Natl Acad Sci USA 74:4481–4484, 1977.
8. Tucker RW, Pardee AB, Fujiwara K: Cell 17:527–535, 1979.
9. Poenie M, Alderton J, Steinhardt R, Tsien RY: Science 233:886–889, 1986.
10. Tucker RW, Meade-Cobun K, Loats H: In Fiskum G (ed): "Cell Calcium Metabolism." New York: Plenum Press, 1988 (in press).
11. Grynkiewicz G, Poenie M, Tsien RY: J Biol Chem 260:3440–3450, 1985.
12. Williams DA, Fogarty KE, Tsien RY, Fay FS: Nature 318:558–561, 1985.
13. Drummond IAS, McClure SA, Poenie M, Tsien RY, Steinhardt RA: Mol Cell Biol 6:1767–1775, 1986.
14. Drummond IAS, Lee AS, Resendez E, Steinhardt RA: J Biol Chem 282:12801–12805, 1987.
15. Matuoka K, Fukami K, Nakanishi O, Kawai S, Takenawa T: Science 239:640–643, 1988.

Journal of Cellular Biochemistry 39:153–166 (1989)
Growth Factors and Their Receptors: Genetic
Control and Rational Application 131–144

Retroviruses Expressing Different Levels of the Normal Epidermal Growth Factor Receptor: Biological Properties and New Bioassay

Thierry J. Velu, Laura Beguinot, William C. Vass, Ke Zhang, Ira Pastan, and Douglas R. Lowy

Laboratories of Cellular Oncology (T.J.V., W.C.V., K.Z., D.R.L.) and Molecular Biology (L.B., I.P.), National Cancer Institute, National Institutes of Health, Bethesda, Maryland 20892; University Institute of Microbiology, 1353 Copenhagen K, Denmark (L.B.)

Two retroviral DNAs that encode the normal human epidermal growth factor (EGF) receptor hEGFR have been generated by inserting a h*EGFR* cDNA into two different retroviral vectors. One DNA (pCO11-EGFR-neo) also contained a linked selectable marker gene (*neo*^{R}). The other (pCO12-EGFR) only expresses h*EGFR*. When introduced into NIH3T3 cells, the two DNAs and the viruses derived from them induced a fully transformed phenotype, including focal transformation and growth in agar or low serum, but transformation depended entirely upon EGF being present in the growth medium. Compared with pCO11-EGFR-neo, pCO12-EGFR induced EGF-dependent transformation 2–5 times more efficiently and expressed higher numbers of receptors (4×10^5 vs. 1×10^5 EGF receptors per cell). The results indicate that transforming potential is directly related to the number of EGF receptors. In defined, serum-free medium that contained only very low concentrations of insulin (0.6 μg/ml) and transferrin (0.6 μg/ml), hEGFR-virus infected cells were able to grow with EGF as the only growth factor. Moreover, daily incubation of the cells with EGF for only 30 min was sufficient to induce growth. NR6 cells, which lack endogenous EGF receptors, were transformed as efficiently as NIH3T3 cells by the hEGFR virus. The dose-dependent growth response to EGF of infected NR6 cells grown in serum-free medium can be used as a highly sensitive bioassay for the quantitative assessment of EGF and transforming growth factor type α (TGFα). This bioassay is at least as sensitive as previously reported radioimmunoassays and can measure a much wider concentration range (10 pg–100 ng/ml). Uninfected NR6 cells or NR6 cells infected by helper virus alone can be used as controls for the EGF specificity of growth stimulation.

Key words: **cell transformation, neoplastic, receptor, epidermal growth factor, transforming growth factor, oncogenes, genetic vectors, retrovirus, bioassay**

Received March 10, 1988; accepted July 26, 1988.

INTRODUCTION

Growth factors and their receptors have been hypothesized to serve an important physiologic role in normal growth and differentiation, as well as to contribute to the pathogenesis of some tumors [1]. The possible relationship between this class of proteins and oncogenesis has been underlined by the finding that several viral oncogenes (v-*onc*) of acute transforming retroviruses have been derived from genes that encode growth factors or growth factor receptors. These include the v-*sis* oncogene, which originated from the gene encoding the B chain of platelet-derived growth factor (PDGF) [2], v-*erbB* from the epidermal growth factor (EGF) receptor gene [3], and v-*fms* from the colony stimulating factor-1 (CSF-1) receptor gene [4].

Compared with their normal cellular homologs (c-*onc*), many oncogenes, including v-*erbB* and v-*fms*, have been found to possess deletions and/or mutations that significantly increase their transforming activity. By contrast, the most striking alteration of c-*erbB* in human tumors seems to be its overexpression. Elevated levels of apparently normal EGF receptors, secondary to amplification or overexpression of c-*erbB*, have been observed in several types of cancer, most of which are of epidermoid origin [5,6]. Amplification and overexpression have also been found in several tumor-derived human cell lines, including the A431 epidermoid carcinoma cell line [7–11]. These observations have fostered the belief that increased numbers of EGF receptors might contribute to tumor formation.

To test this hypothesis experimentally, we inserted the human c-*erbB* proto-oncogene (h*EGFR*) cloned from A431 cells into a retroviral vector [12]. NIH3T3 cells transfected with the h*EGFR* retroviral DNA or infected with the corresponding rescued h*EGFR* virus expressed high levels of normal human EGF receptors (4×10^5 receptors per cell) and developed a fully transformed phenotype in the presence of EGF. The cell transformation depended completely upon EGF in that no biological alterations were detected in its absence. Cells expressing high levels of EGF receptors formed tumors in male nude mice (which produce significant amount of endogenous EGF) [13], while control cells did not. Tumor development was much faster when exogenous EGF was administered to the mice [12]. Using a cDNA from A431 cells, Di Fiore et al. [14] confirmed the ability of an overexpressed normal EGF receptor to induce EGF-dependent NIH3T3 cell transformation.

To extend our EGF receptor studies, we have now placed the A431 h*EGFR* gene in a second retroviral vector and compared this second DNA and its recovered virus to the previously described DNA and virus. We have found that the transforming potential of the two h*EGFR*-expressing viruses correlated with their EGF receptor expression and that short daily incubations with EGF are sufficient to induce growth of cells expressing high EGF receptor levels. Using serum-free conditions, we have also demonstrated that the growth of h*EGFR*-expressing cells is sensitive to very low levels of EGF (10 pg/ml); this growth response can be used as a highly sensitive bioassay for the quantitative assessment of EGF and transforming growth factor type α (TGFα).

MATERIALS AND METHODS

Retroviral Vector Construction

The 12-kb plasmid pCO11-EGFR-neo was derived from pGV16, which is a murine retroviral based shuttle vector that contains a selectable genetic marker (*neo*^{R})

(Fig. 1) [15]. Initially, the v-ras^H 2.1-kb BamHI-EcoRI fragment from the Harvey murine sarcoma virus (Ha-MuSV) genome was inserted between the 5′ long terminal repeat (LTR) and the neo^R gene, giving rise to plasmid pBW1160 [16]. The PstI site present downstream from the *ras* coding sequences had been replaced with XhoI. The original h*EGFR* clone was a 4.2- kb XbaI-NcoI cDNA fragment, derived from pMMTV-EGFR [17], that contained the complete EGF-receptor coding sequences; the fragment was made SacII-XhoI compatible by linker ligation. To generate pCO11-EGFR-neo, the 0.8-kb SacII-XhoI fragment of pBW1160, containing the entire coding region of v-ras^H, was removed and replaced by the 4.2-kb SacII-XhoI h*EGFR* fragment. In pCO11-EGFR-neo, h*EGFR* is inserted within 1.3-kb of noncoding sequences (BamHI-EcoRI) derived from Ha-MuSV DNA, and its expression is under the control of the pGV16 LTR, which was derived from the Moloney murine sarcoma virus (Mo-MuSV).

Plasmids pCO12-EGFR and pCO13-st have been described previously [12]. The 15 kb pCO12-EGFR was derived from a full-length Ha-MuSV DNA cloned in pBR322 (pCO6-HX) in which h*EGFR* has been inserted in place of the *ras* gene (Fig. 1). Plasmid pCO13-st is identical with pCO12-EGFR, except that a universal translation terminator linker was inserted at the SmaI site at codon 220 in *hEGFR* so that no functional hEGF receptor is transcribed.

Cell Culture and Transfection Assays

NIH3T3 cells [18], NR6 [19], and KB cells (American Type Culture Collection, Rockville, Maryland) were maintained in Dulbecco's modified Eagle (DME) medium

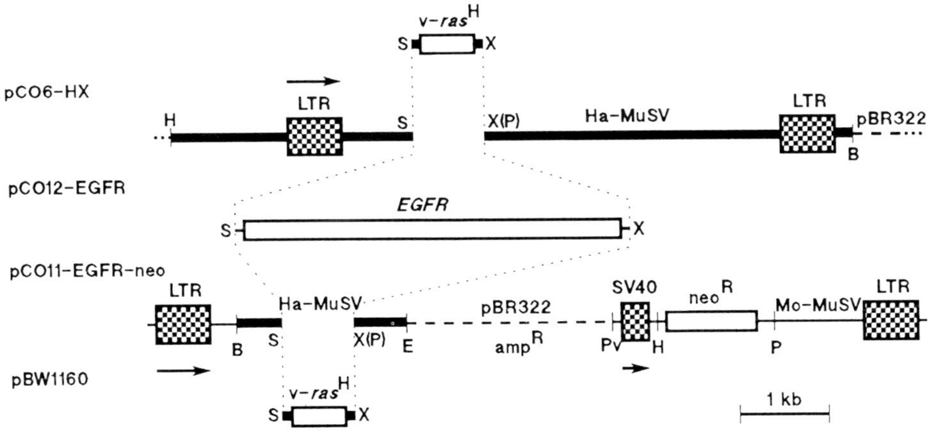

Fig. 1. Construction of the normal hEGF receptor retroviral genomes. Plasmid pCO12-EGFR was derived from the Harvey murine sarcoma virus (Ha-MuSV) cloned in pBR322 (pCO6-HX). pCO11-EGFR-neo was constructed from pGVl6 [15] in which p21 v-ras^H with flanking sequences had been cloned into the BamHI and EcoRI sites (pBW1160) [16]. In both pCO6-HX and pBW1160, the PstI site of Ha-MuSV at nucleotide 1759 [46] has been replaced by an XhoI linker [16]. In both cases, the entire coding region of v-ras^H (SacII-XhoI) was removed and replaced by the 4.2-kb fragment coding for the full-length h*EGFR* cDNA (XbaI-NcoI) [17], which was made compatible by linker ligation (SacII-XhoI). Ha-MuSV sequences are shown in bold lines; pBR322, in broken lines. Restriction endonucleases sites: H, HindIII; S, SacII; P, PstI; X, XhoI; B, BamHI; and Pv, PvuII. Open boxes: coding sequences of v-ras^H, *EGFR*, and neo^R genes; cross-hatched boxes, Ha-MuSV and Mo-MuSV LTRs, and SV40 promotor; arrows, direction of transcription.

supplemented with 50 1U/ml penicillin, 50 μg/ml streptomycin, and 10% heat inactivated fetal calf serum (FCS).

For transfection experiments, 3 × 10^5 NIH3T3 cells were plated in 35-mm dishes. Twenty-four hours later, cells were transfected with 1 μg of the h*EGFR* containing plasmid and 5 μg of carrier NIH3T3 DNA using the calcium phosphate method [20]. The cells were treated as previously described, except that dimethyl sulfoxide was not used [18]. As a selectable marker, 100 ng of a 6-kb plasmid carrying the neo^R gene under the control of an SV40 promoter (pSV2neo) [21] was cotransfected with pCO12-EGFR and pCO13-st. Cells that were not treated with G418 received EGF (Collaborative Research, Inc., Bedford, Massachusetts) 2 days after the transfection (20 ng/ml final concentration). Foci were counted after 3 weeks. When G418 (Geneticin, GIBCO, Grand Island, New York) was used for selection (500 μg/ml final concentration), it was added 2 days after the transfection. EGF was added 5 days later to some G418 treated dishes. G418-resistant colonies were counted after 10 days in G418. When cultivated without EGF, the colonies were grown to confluency for studies on mass cultures. If grown with EGF, individual colonies were selected on the basis of their transformed phenotype to obtain a suitable group of cells from which a high-titer virus that expressed h*EGFR* could be rescued by infection with a helper virus.

Growth in semisolid media was studied by plating a layer of 0.4% soft agar (DIFCO, Detroit, Michigan) containing 10^5 cells in suspension on the top of a firm basal layer of 0.6% agar. A feeder layer of 0.4% agar was applied weekly. The agar was prepared either in DME medium containing 10% FCS or in the serum-free medium which is defined below. In some plates, EGF was added at a final concentration of 100 ng/ml.

Generation of hEGER Expressing Viruses

G418-resistant colonies which acquired the transformed phenotype upon the addition of EGF were previously found to be the ones expressing high levels of EGF receptors [12]. To obtain high-titer viruses that expressed h*EGFR*, these colonies were superinfected with a helper virus, amphotropic murine leukemia virus (MuLV) [22], or Moloney MuLV, at a multiplicity of infection of about 0.1. Adsorption was performed for 1 h at 37°C in the presence of 4 μg/ml polybrene (Sigma Chemical Co., St. Louis, Missouri). Culture fluid from infected cells was harvested 10 days later, stored at −70°C, and used as virus stocks. Virus titrations were performed, after filtering stocks through a cellulose acetate filter (0.45 μm), by testing serial dilutions for their ability to induce EGF dependent focal tranformation of NIH3T3 cells.

Experiments in Serum-Free Medium

A modification of previously described conditions [23] was used. Sixty-millimeter dishes were precoated for 2 h at 37°C with 150 μg poly-D-lysine/ml water (molecular weight: 540,000; 1 ml/plate), rinsed with phosphate-buffered saline (PBS), and then coated overnight at 37°C with 40 μg human fibronectin/ml of DME medium (2 ml/plate). After washing the plates with PBS, 10^5 cells were plated in DME medium containing 10% FCS to allow attachment. Three hours later, the cells were washed and refed with defined, serum-free medium consisting of a 1:1 mixture of DME and Ham F12 medium (GIBCO) supplemented with 20 mM HEPES (pH 7.3),

42 μg/ml histidine, and 10 ml/liter ITS+ (Collaborative Research, Inc., Bedford, Massachusetts), resulting in a final concentration of 6.25 μg/ml insulin, 6.25 μg/ml transferrin, 6.25 ng/ml selenium, 1.25 mg/ml bovine serum albumin (BSA), and 5.35 μg/ml linoleic acid. In some experiments, only 1 ml/liter ITS+ was added so that the final concentration of the ITS+ components was decreased by a factor of 10. These conditions are called "low" defined medium. In some experiments, platelet derived growth factor (PDGF, 5 U/ml) and basic fibroblast growth factor (bFGF, 2 ng/ml) were used at saturating concentrations. HEPES and histidine were from Sigma Chemical Co. (St. Louis, Missouri). All other reagents were obtained from Collaborative Research, Inc. (Bedford, Massachusetts).

Thymidine Incorporation

DNA synthesis was assessed by ^{3}H-thymidine (25 Ci/mmol, Amersham, Arlington Heights, Illinois) incorporation. Cells were incubated overnight with ^{3}H-thymidine (1 μCi/ml) in defined medium containing different concentrations of EGF. The cells were then washed three times with PBS containing 0.1% cold thymidine, frozen at −20°C, washed again three times with ice-cold 0.2 N perchloric acid (PCA), and solubilized in 1 M NaOH. PCA-insoluble material was frozen in dry ice, thawed in ice water, washed with 0.2 N PCA, resuspended in 0.5 N PCA, and hydrolyzed at 90°C for 20 min. Incorporated ^{3}H-thymidine was assayed by counting the supernatant in a scintillation counter.

Immunoprecipitation

Cells were seeded at 2×10^6/100 mm dish in complete medium. Eight hours later, cells were radiolabeled for 12 h with 250 μCi/ml ^{35}S-methionine (1,000 Ci/mmol, Amersham, Arlington Heights, Illinois), in 4 ml of methionine-free Eagle's medium containing 2% FCS. Cells were harvested by scraping into 3 ml of ice-cold STE buffer (20 mM TRIS HCl pH 7.4, 0.15 M NaCl, 1 mM EDTA, 1% aprotinin), centrifuged at 500*g* for 5 min, and solubilized in 1 ml of lysis buffer (20 mM HEPES pH 7.4, 1% Triton X-100, 10% glycerol, 1% aprotinin). Cell lysates were clarified at 100,000*g* for 60 min. Immunoprecipitation was performed as previously described [24], using a rabbit polyclonal anti-EGF receptor antibody, Ab 2913 [25], or an anti-hEGF receptor monoclonal antibody (EGFR1, Amersham, Arlington Heights, Illinois), and proteins were analyzed by SDS-PAGE on 8.5% gels [26]. Gels were fluorographed [27] and subjected to autoradiography. The amount of EGF receptor was quantitated by excising the immunoprecipitated band. Counts in the gel slices were extracted with 2 ml of 30% H_2O_2, 1% NH_4OH for 24 h and then counted by liquid scintillation after addition of 10 ml Aquasol.

RESULTS

We have recently reported development of a high-titer retrovirus that expressed hEGFR under the control of the Ha-MuSV LTR; the viral DNA was called pCO12-EGFR and the resulting virus hEGFR (Fig. 1) [12]. To obtain a second virus that expressed lower levels of h*EGFR* and contained a selectable marker unrelated to cell transformation, we introduced the h*EGFR* cDNA into another retroviral vector. This latter construction, called pCO11-EGFR-neo, has the coding sequences of the EGF receptor under the control of the Mo-MuSV LTR and contains the *neo*^{R} marker gene under the control of a SV40 promoter (Fig. 1). As had been true of pCO12-EGFR, transfection of pCO11-EGFR-neo onto NIH3T3 cells induced multiple foci of transformed cells only when EGF was added to the medium (Table I). The transforming efficiency was somewhat lower and the foci smaller for pCO11-EGFR-neo than for pCO12-EGFR, suggesting that the latter expressed higher levels of hEGF receptor. No transforming activity was found in the absence of EGF, nor for the premature termination mutant pCO13-st with or without EGF.

Previous experiments had shown that high-titer virus could be obtained from cells cotransfected with pCO12-EGFR and pSV2neo by growing them with EGF and G418 and selecting individual morphologically transformed colonies. We therefore selected cells for G418 resistance, after transfection of pCO11-EGFR-neo by itself and cotransfection of pCO12-EGFR with pSV_2neo (Table I). When the cells transfected with these two constructions were grown in the presence of EGF, many of the G418-resistant colonies that developed were composed of morphologically transformed cells that were refractile and spindle shaped and grew over each other. Approximatively 20% of the pCO11-EGFR-neo transfected colonies were morphologically transformed, while almost 50% of the colonies that arose from cotransfection of pCO12-EGFR and pSV2neo were transformed (Table I). These results again suggested that pCO11-EGFR-neo might express lower levels of hEGF receptors than pCO12-EGFR. As expected, no transformed colonies were seen in the absence of EGF, nor in pCO13-st transfectants even with EGF. Immunofluorescence studies, using an hEGFR-specific antibody (EGFR1), indicated that foci and transformed G418-resistant colonies expressed easily detectable levels of hEGF receptors, with an excellent correlation between the high expression of EGF receptors and the transformed phenotype (data not shown).

Viruses able to transmit an EGF-dependent transformed phenotype were obtained using a helper virus to superinfect G418-resistant, morphologically transformed colonies expressing hEGF receptors. The viruses derived from pCO11-EGFR-neo

TABLE I. h*EGFR* Transforming Activity on NIH3T3 Cells

	pCO11-EGFR-neo		pCO12-EGFR		pCO13-st	
	+EGF	−EGF	+EGF	−EGF	+EGF	−EGF
Foci per μg DNA[a]	250	0	800	0	0	0
Transformed/total G418 colonies per μg DNA (%)	260/1,420 (18%)	0/1,380 (0%)	550/1,140 (48%)	0/1,220 (0%)	0/1,310 (0%)	0/1,120 (0%)
Growth in agar[c]	2.6%	0	8.8%	0	0	0
Titer of h*EGFR* virus/ml[d]	10^4		10^7		N.D.[b]	
EGF receptors/cell (type)[e]	1×10^5 (human)		4×10^5 (human)		10^4 (mouse)	

[a]Results are expressed per μg of h*EGFR* plasmid DNA.
[b]N.D., not determined.
[c]To be able to compare the EGF-induced growth in soft agar for each construction, cells for this experiment came from G418-selected mass cultures that had not been treated with EGF before being placed in agar. 10^5 cells were seeded in suspension. Results are expressed as the % of suspended cells that formed colonies in agar.
[d]h*EGFR* viruses were derived from G418 colonies expressing high levels of hEGF receptors.
[e]Receptor determinations were made on morphologically transformed cells that had been selected in EGF and G418 after transfection of the pCO11-EGFR-neo and pCO12-EGFR DNAs. Similar results were obtained for cells infected by hEGFR or by hEGFR-neo viruses that were derived from these transfected cells.

and pCO12-EGFR have been designated hEGFR-neo and hEGFR viruses, respectively. When titrated on NIH3T3 cells, similar results were obtained when the viruses were assayed by indirect immunofluorescence or by EGF dependent foci. The titer of the hEGFR-neo virus was found to be about three orders of magnitude lower (10^4 focus-forming units [ffu]/ml) than for the hEGFR virus (10^7 ffu/ml) (Table I).

To rule out the possibility that the viruses obtained from the morphologically transformed cells displayed the EGF-dependent transformed phenotype as a result of the selective pressure imposed by incubation with EGF prior to superinfection, we also tested culture fluid from a mass culture of NIH3T3 cells that had been cotransfected with pCO12-EGFR and pSV$_2$neo, selected with G418 in the absence of EGF, and superinfected with a helper virus. Although the titer (10^3 ffu/ml) was significantly lower than the hEGFR virus derived from pCO12-EGFR transfected cells selected with EGF, this virus readily induced EGF dependent transformation of NIH 3T3 cells.

The ability of pCO11-EGFR-neo-transfected and hEGFR-neo-virus-infected cells to confer anchorage-independent growth in semisolid medium was also studied. As was true of pCO12-EGFR cells [12], pCO11-EGFR-neo cells grew in agar in an EGF-dependent manner when the growth medium was 10% FCS or a defined, serum-free formulation. The number and the average size of the colonies appeared to be larger with pCO12-EGFR cells than with pC011-EGFR-neo cells. No growth was seen in the absence of EGF, nor in control cells in the presence or absence of EGF (Table I).

To study the integrity of the hEGF receptor, its function, and its relative abundance in pC011-EGFR-neo and pCO12-EGFR cells, we performed EGF binding, immunoprecipitation, immunofluorescence, and in vitro phosphorylation studies. Quantitative binding with ^{125}I-labeled EGF of pCO11-EGFR-neo- and pCO12-EGFR cells produced results that correlated with the biological data; pCO11-EGFR-neo cells contained about 1–2 $\times$ 10^5 human EGF receptors per cell, while pCOl2-EGFR contained about 3–5 $\times$ 10^5 (Table I). Parental NIH3T3 and pCO13-st cells expressed fewer than 1 $\times$ 10^4 mouse EGF receptors per cell. In immunoprecipiation assays, we found that the EGF receptors from both constructions migrated as an approximatively 175-kd band (Fig. 2). These bands comigrated with hEGF receptors from KB cells (data not shown), which express about 1.5 $\times$ 10^5 normal receptors per cell [24,28,29]. The content of ^{35}S-methionine in the EGF receptor bands confirmed the quantitative binding studies; pCO12-EGFR cells contained about four times more receptors than pCO11-EGFR-neo cells, and 20–50-fold more than control NIH3T3 cells (data not shown, obtained using a rabbit polyclonal antibody, Ab 2913, that recognized mouse and human EGF receptors). Furthermore, the estimated number of EGF receptors found in KB cells resembled that of pCO11-EGFR-neo cells, confirming again the level of EGF receptor expression in these cells. Immunofluorescent studies indicated that the hEGF receptors of pCO11-EGFR-neo and pCO12-EGFR cells were normally internalized upon the addition of EGF (Beguinot and Pastan, unpublished observations). Finally, the kinase activity of the hEGF receptors, as measured by autophosphorylation and phosphorylation of exogenous substrates, correlated with the amount of receptors expressed by the various clones and was EGF stimulated (Beguinot and Pastan, unpublished observations). These results suggest the hEGF receptors encoded by the cDNA are structurally and functionally normal.

We also tested the ability of pCO12-EGFR-transfected NIH3T3 cells to grow in low serum (1%), in serum-free defined medium, and in "low" defined medium, as

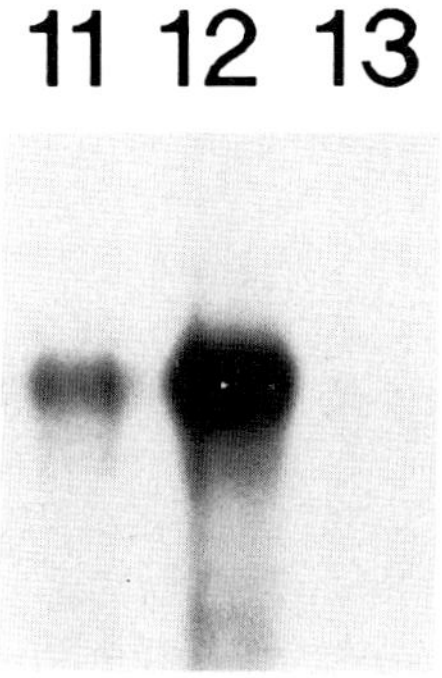

Fig. 2. Immunoprecipitation of the hEGF receptor. ^{35}S-labeled cell extracts were prepared and immunoprecipitated with an anti-hEGFR monoclonal antibody (EGFR1, Amersham) from a mass culture of NIH3T3 cells that had been selected with G418 after transfection of the DNAs (**11**, pCO11-EGFR-neo; **12**, pCO12-EGFR; **13**, pCO13-st). No EGF was added to the medium to avoid the selection of cells carrying high levels of EGF receptors and to be able to compare the relative EGF receptor expression. pCO12-EGFR cells contained about four times more EGF receptors than pCO11-EGFR-neo cells. The hEGF-receptors migrated as a ~175-kd band.

TABLE II. EGF-Dependent Growth

	No. cell doublings in 8-day assay[a]					
	pCO12-EGFR NIH3T3 cells			Control NIH3T3 cells		
	1% FCS[b]	"Low" DM[b]	DM[b]	1% FCS[b]	"Low" DM[b]	DM[b]
No EGF	1.2	2.0	3.3	1.3	1.9	3.1
+EGF (30 min/day)[c]	N.D.[d]	N.D.[d]	5.1	N.D.[d]	N.D.[d]	3.2
+EGF (continuous)[c]	3.9	5.8	6.4	1.5	2.1	3.5

[a] 10^5 cells, previously grown in the absence of EGF, were plated in 60-mm dishes. Results are expressed as No. of cell doublings after 8 days of culture, and are the mean of two independent experiments. Cell counts were performed on duplicate plates for each experiment.
[b] Growth conditions in 1% FCS, in defined medium (DM), and in "low" DM are described in Materials and Methods.
[c] EGF was added to the medium at 20 ng/ml, either continuously, or 30 min per day. All plates were washed twice (at the end of the 30 min incubation) and media were changed daily.
[d] N.D., not determined.

described in the Materials and Methods. Table II summarizes the results, which are expressed as the number of cell doublings after 8 days of growth. In the absence of EGF, pCO12-EGFR cells grew similarly to control NIH3T3 cells. When grown with EGF, pCO12-EGFR cells grew 2–3 times faster than control cells and achieved a higher final cell density. The EGF dependent difference was greater in "low" defined medium than in regular defined medium, showing again that the growth advantage of pCO12-EGFR cells was highly dependent upon the presence of EGF and that it persisted even in very poor growth conditions. Moreover, the EGF-dependent growth of pCO12-EGFR cells did not require the continuous presence of EGF. Daily incubation of the cells with EGF for 30 min only, although not as efficient as the continuous presence of EGF, was sufficient to induce EGF-dependent growth. Similar results were obtained with hEGFR virus (Fig. 3). After 10 days of continuous EGF treatment, hEGFR-infected cells were still growing (1.2×10^7/60-mm dish), while

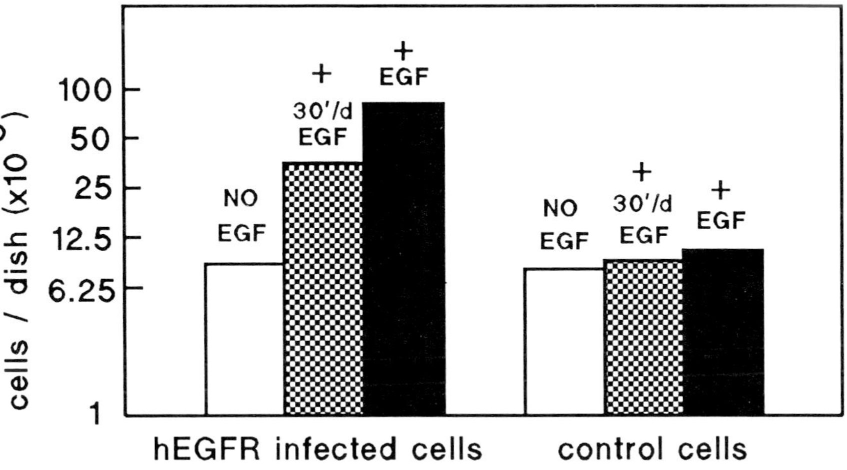

Fig. 3. Cell growth dependence on hEGFR and EGF. 1×10^5 NIH3T3 cells infected by pCO12-EGFR virus or by the amphotropic helper virus as control were seeded in 60-mm dishes coated with poly-D-lysine and fibronectin and kept in DME medium supplemented with 10% FCS to allow cell attachment. After 4 h, cells were washed and refed with serum-free defined medium. Black columns: cells in defined medium containing 20 ng/ml EGF; checkered columns: cells incubated daily for 30 min with medium containing 20 ng/ml EGF, then washed twice and refed with defined medium. The other plates were also washed twice every day, and refed with their respective medium. Results are given as number of cells/60-mm dish, counted after 8 days of culture, and are the mean of duplicate samples in two different experiments.

infected cells cultured without EGF were dying, as were control cells grown in the presence or absence of EGF.

NR6 cells, a variant of the Swiss 3T3 cells which do not express endogenous EGF receptors [19], were also infected with hEGFR virus. The virus conferred EGF-dependent focal transformation on NR6, with a titer similar to that obtained on NIH3T3 cells. When the hEGFR-virus-infected NR6 cells (NR6ER cells) were cultured in serum-free medium, EGF was found to induce dose-dependent cell growth, as assessed by ^{3}H-thymidine incorporation into DNA (Fig. 4). In this assay, the NR6ER cells displayed a detectable growth response to extremely low concentrations of EGF (10 pg/ml). Maximal growth response to EGF was seen at 100–1,000 ng/ml; no attenuation was noted even at the highest concentration tested (1,000 ng/ml). Thymidine incorporation measured in NIH3T3 cells infected by hEGFR virus yielded similar results (data not shown). As expected, parental NR6 cells or NR6 cells infected with the helper virus alone (NR6c cells) did not show a growth response to EGF. In contrast to NR6 cells, control NIH3T3 cells did respond somewhat to EGF, since they express about 10^4 endogenous mouse EGF receptors (data not shown). The sensitivity to EGF of NIH3T3 cells infected with the helper virus alone (or parental NIH3T3 cells) was equivalent to that of hEGFR-infected NIH3T3 cells, but saturation occurred at 1 ng/ml.

The EGF dose-response curve of NR6ER cells, as measured by ^{3}H-thymidine incorporation, was used as a highly sensitive bioassay for the quantitative measurement of the ligands that stimulate the EGF receptor: EGF and TGFα. The specificity of the bioassay can be assessed by testing the sample on NR6c cells. When the culture fluid of normal rat kidney (NRK) cells that expressed a synthetic TGFα gene (1/28

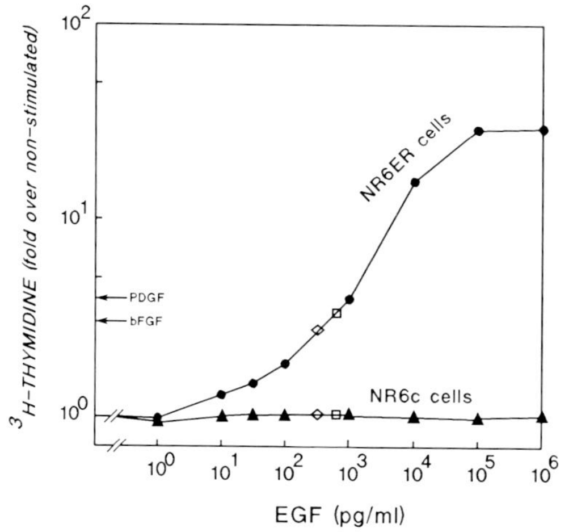

Fig. 4. Dose-dependent growth response to EGF of NR6 cells expressing (or not) high levels EGF receptors. NR6 cells (1×10^5) infected by pCO 12-EGFR virus (NR6ER cells, **solid circles**) or by helper virus alone (NR6c cells, **solid triangles**) were plated on coated dishes. Cells were grown for 48 h in DME medium with 10% FCS. Then they were placed in "low" defined medium (see Materials and Methods) without or with different concentrations of EGF (range: $1–1 \times 10^6$ pg/ml). The medium was changed daily. Forty-eight hours later, ^{3}H-thymidine was added at a final concentration of 1 μCi/ml, in the presence of the different concentrations of EGF. After overnight incubation, ^{3}H-thymidine incorporation was measured as described in Materials and Methods. The supernatant of a cell line producing TGFα (1/28 Cl.3 NRK cells) [30] and two growth factors, PDGF and bFGF, were also tested. The TGFα producing cells were grown in DME medium with 10% FCS to subconfluence, washed twice, and refed with "low" defined medium. After an overnight incubation and then filtration, the supernatant was tested directly (**open squares**) or diluted 1:1 with fresh "low" defined medium (**open diamonds**), on NR6ER and NR6c cells. The levels of growth stimulation induced by PDGF (5 U/ml) and bFGF (2 ng/ml) are indicated by an arrow and were quite identical for either NR6ER or NR6c cells. Results are from a single experiment. Two other experiments gave similar results.

C1.3 NRK cells; see Fig. 4, legend) [30] was tested by the NR6ER assay, it was found to contain an activity that was equal to 0.7 ng/ml EGF. This result is similar to the amount of TGFα determined by a different assay [30]. The specificity of the activity was demonstrated by the failure of the culture fluid of this cell line to stimulate NR6c cells. Culture fluid from parental NRK cells also failed to stimulate NR6ER and NR6c cells (data not shown). As further specificity controls, two other growth factors (PDGF and bFGF), used at saturating conditions, were found to stimulate NR6ER and NR6c cells equally (Fig. 4).

DISCUSSION

The use of retroviruses to study the biological and biochemical properties of a protein encoded by a cloned gene has some advantages over DNA transfection. The efficiency of viral infection is several orders of magnitude higher than DNA transfection, which means that infection with high titer virus can, in a single step, induce high expression of the protein in the majority of cells. In contrast, the inefficiency of DNA-mediated gene transfer means that no more than 1% of the recipient cells will stably express the transfected gene. Consequently, infection can preclude any theoretical selective bias that might be imposed by DNA transfection and the isolation of cells that express high level of the gene. Infection can also be used in some cell types that poorly support stable transfection of DNA.

These advantages led us to study the normal human EGF receptor by placing it in retroviral vectors so that expression of the gene could be studied via infection, as well as by DNA transfection [12]. In an initial study, an h*EGFR* cDNA was inserted in a vector derived from Ha-MuSV, yielding pCO12-EGFR. To obtain a high-titer EGFR virus, we found it useful to select individual colonies of morphologically transformed cells by cotransfecting pCO12-EGFR with a selectable marker (neo^R) and growing the transfectants in G418 and EGF. In the current study, clone pCO11-EGFR-neo was constructed by placing the h*EGFR* cDNA in a retroviral vector that contained the selectable marker neo^R. h*EGFR*-neo virus was obtained by growing pCO11-EGFR-neo-transfected cells in G418 and EGF, selecting morphologically transformed colonies, and superinfecting the cells with a helper virus. The EGF receptors expressed by the two constructions were found to be structurally and functionally normal by several criteria, including their size, their ability to bind to EGF and be internalized in response to this ligand, their kinase activity, and their EGF-dependent biologic activity.

When the biologic properties of the two DNAs and the two viruses recovered by helper virus infection were tested, all were able to induce the fully transformed phenotype on NIH3T3 cells, including focal transformation, growth in low serum, and growth in agar. These alterations in the growth properties of the cells occurred only in the presence of EGF. The transfected or infected cells did not display any detectable biologic changes in the absence of EGF. Moreover the continuous presence of the growth factor was not required to stimulate growth. Incubation of the cells with EGF for only 30 min per day (the shortest time tested) was sufficient to induce a significant increase in growth. This result parallels previous in vivo data obtained with pCO12-EGFR transfected NIH3T3 cells [12]. When injected into male mice, these cells formed tumors more rapidly in animals given daily subcutaneous injections of EGF, which would have increased circulating EGF levels only for a short time. We also noted that EGF growth stimulation of NIH3T3 or NR6 cells infected with EGFR virus was not inhibited by continuous exposure to EGF concentrations as high as 1 μg/ml. Other cells carrying higher numbers of EGF receptors, such as human epidermoid carcinoma A431 cells (2×10^6 receptors per cell), have been reported to display growth inhibition at high concentrations of EGF [31,32].

The titer of EGFR virus was about three orders of magnitude higher than that of hEGFR-neo virus. We believe that two important differences between the viruses account, at least in part, for the much higher titer found in hEGFR virus. First, the very large predicted size (12 kb) of the hEGFR-neo RNA implies that it is probably packaged into virions much less efficiently than the 9-kb hEGFR viral RNA [33]. Second, our quantitative assessment of the average number of hEGF receptors per cell indicated that pCO12-EGFR-transfected cells and hEGFR-virus-infected cells expressed about four times as many receptors per cell as pCO11-h*EGFR*-neo cells or cells infected with hEGFR-neo virus.

An important biologic difference between the two constructions was that pCO11-EGFR-neo induced EGF-dependent cell transformation less efficiently than pCO12-EGFR. This result correlated with the latter DNA expressing more hEGF receptors than the former, which strongly suggests that the transforming potential of the EGF receptor depends upon the level of its expression, as well as on ligand concentration.

Similar, but indirect, observations have been reported previously by Santon et al. [34], who observed that the degree of gene amplification and of EGF receptor expression in different A431 clonal variants correlated with the growth of these tumor cells in vivo and in vitro.

The lack of detectable biologic activity from overexpression of h*EGFR* in the absence of ligand suggests that the receptor is stringently regulated so that it will be virtually inactive without ligand. These observations are qualitatively similar to those reported on c-*fms*, which encodes the colony stimulating factor-1 (CSF-1) receptor [35]. Overexpression of the c-*fms* gene induced transformation of NIH3T3 cells that depended upon CSF-1.

Our studies provide direct experimental support for the hypothesis that increased numbers of EGF receptors can contribute to oncogenicity. Several clinical studies of malignancies, especially cancers of the breast and lung, have suggested that EGF receptor overexpression is associated with increased malignancy, as defined by tumor spread, recurrence, and poor prognosis [36–39]. Similar data have recently been reported with the *HER-2/neu* oncogene, which displays sequence homology with *EGFR*, although a putative ligand has not yet been identified for its encoded protein. Amplification and/or overexpression of this gene in human tumors, especially in breast cancers, has been found to correlate with short survival and time to relapse [40]. In constrast to *EGFR*, the ability of overexpressed *HER-2/neu* to induce morphologic transformation of NIH3T3 cells does not appear to depend upon the exogenous addition of a hypothetical ligand [41].

The observation that biologic activity of EGF receptors, even when expressed at more than 10^5 receptors per cell, is entirely ligand dependent, suggests that this receptor contributes to the growth of tumor cells principally via an autocrine or paracrine mechanism. Indeed, increased production of TGFα by different types of cancers and tumor cell lines has been found to be frequently associated with increased levels of EGF receptor [42,43]. Elevated levels of TGFα have also been found in the urine of patients with disseminated breast cancer [44]. The accessibility of the EGF receptor on the cell membrane and the dependence of activity upon the binding of its ligand make it a potentially attractive target for therapy.

The dose-response relationship between EGF concentration and cell proliferation of NR6 cells infected with EGFR virus, as assessed by ^{3}H-thymidine incorporation, can be an extremely sensitive bioassay for the quantitative determination of EGF and TGFα. When the culture fluid of a cell line producing TGFα was tested in this assay, the observed EGF-like activity correlated with the concentration obtained previously with other assays [30]. The detection limit of the bioassay is 10 pg/ml, which is at least as sensitive as that reported for radioimmunoassays (RIA) [45]. The bioassay can measure EGF and TGFα concentrations that vary by four orders of magnitude (10 pg–100 ng/ml = 10^{-12}–10^{-8} M), which is much wider than by RIA. NR6 cells infected by helper virus alone (or parental NR6 cells) can be used to determine the specificity of growth stimulation. NR6 cells are superior to NIH3T3 cells for this assay. NIH3T3 cells contain low numbers of endogenous EGF receptors so that EGF and TGFα will induce a growth response, while the lack of endogenous EGF receptors on parental NR6 cells means that they will not respond to EGF and TGFα.

ACKNOWLEDGMENTS

We thank P. Tambourin, R. Feldman, and M. Willingham for very helpful discussions; B. Willumsen for providing clone pBW1160; T. Robins for the vector pGV16; S. Watanabe for 1/28 CL.3 NRK cells; A. Papageorge for EGF receptor immunoprecipitations; and B. Hosier for excellent technical assistance.

REFERENCES

1. Weinberg RA: Science 230:770, 1985.
2. Waterfield MD, Scrace GT, Whittle N, Stroobant P, Johnsson A, Wasteson A, Westermark B, Heldin C-H, Huang JS, Deuel TF: Nature 304:35, 1983.
3. Downward J, Yarden Y, Mayes E, Scrace G, Totty N, Stockwell P, Ullrich A, Schlessinger J, Waterfield MD: Nature 307:521, 1984.
4. Sherr CJ, Rettenmier CW, Sacca R, Roussel MF, Look AT, Stanley ER: Cell 41:665, 1985.
5. Libermann TA, Nusbaum HR, Razon N, Kris R, Lax I, Soreq H, Whittle N, Waterfield MD, Ullrich A, Schlessinger J: Nature 313:144, 1985.
6. Hendler FJ, Ozanne BW: J Clin Invest 74:647, 1984.
7. Lin CR, Chen WS, Kruiger W, Stolarsky LS, Weber W, Evans RM, Verma IM, Gill GN, Rosenfeld MG: Science 224:843, 1984.
8. Merlino GT, Xu Y-H, Ishii S, Clark AJL, Semba K, Toyoshima K, Yamamoto T, Pastan I: Science 224:417, 1984.
9. Ullrich A, Coussens L, Hayflick JS, Dull TJ, Gray A, Tam AW, Lee J, Yarden Y, Libermann TA, Schlessinger J, Downward J, Mayes ELV, Whittle N, Waterfield MD, Seeburg PH: Nature 309:418, 1984.
10. Xu Y-H, Richert N, Ito S, Merlino GT, Pastan I: Proc Natl Acad Sci USA 81:7308, 1984.
11. King CR, Kraus MH, Williams LT, Merlino GT, Pastan IH, Aaronson SA: Nucleic Acids Res 13:8477, 1985.
12. Velu TJ, Beguinot L, Vass WC, Willingham MC, Merlino GT, Pastan I, Lowy DR: Science 238:1408, 1987.
13. Tsutsumi O, Kurachi H, Oka T: Science 233:975, 1986.
14. Di Fiore PP, Pierce JH, Fleming TP, Hazan R, Ullrich A, King CR, Schlessinger J, Aaronson SA: Cell 51:1063, 1987.
15. Jhappan C, Vande Woude GF, Robins TS: J Virol 60:750, 1986.
16. Willumsen BM, Papageorge AG, Kung H-F, Bekesi E, Robins T, Johnsen M, Vass WC, Lowy DR: Mol Cell Biol 6:2646, 1986.
17. Clark AJL, Beguinot L, Ishii S, Ma DP, Roe BA, Merlino GT, Pastan I: Biochem Biophys Acta 867:244, 1986.
18. Lowy DR, Rands E, Scolnick, EM: J Virol 26:291, 1978.
19. Pruss RM, Herschman HR: Proc Natl Acad Sci USA 74:3918, 1977.
20. Graham FL, van der Eb AJ: Virology 52:456, 1973.
21. Southern PJ, Berg P: J Mol Appl Genet 1:327, 1982.
22. Chattopadhyay SK, Oliff AI, Linemeyer DL, Lander MR, Lowy DR: J Virol 39:777, 1981.
23. Zhan X, Goldfarb M: Mol Cell Biol 6:3541, 1986.
24. Beguinot L, Lyall RM, Willingham MC, Pastan I: Proc Natl Acad Sci USA 81:2384, 1984.
25. Beguinot L, Werth D, Ito S, Richert N, Willingham MC, Pastan I: J Biol Chem 261:1801, 1986.
26. Laemmli UK: Nature 227:680, 1970.
27. Bonner WM, Laskey RA: Eur J Biochem 46:83, 1974.
28. King AC, Cuatrecasas P: J Biol Chem 257:3053, 1982.
29. Lyall RM, Pastan I, Willingham MC: J Cell Physiol 122:166, 1985.
30. Watanabe S, Lazar E, Sporn MB: Proc Natl Acad Sci USA 84:1258, 1987.
31. Gill GN, Lazar CS: Nature 293:305, 1981.
32. Barnes DW: J Cell Biol 93:1; 1982.
33. Cepko CL, Roberts BE, Mulligan RC: Cell 37:1053, 1984.
34. Santon JB, Cronin MT, MacLeod CL, Mendelsohn J, Masui H, Gill GN: Cancer Res 46:4701, 1986.

35. Roussel MF, Dull TJ, Rettenmier CW, Ralph P, Ullrich A, Sherr CJ: Nature 325:549, 1987.
36. Sainsbury JRC, Farndon JR, Needham GK, Malcolm AJ, Harris AL: Lancet 1:1398, 1987.
37. Macias A, Azavedo E, Hägerström T, Klintenberg C, Pérez R, Skoog L: Anticancer Res 7:459, 1987.
38. Veale D, Ashcroft T, Marsh C, Gibson GJ, Harris AL: Br J Cancer 55:513, 1987.
39. Tahara E, Sumiyoshi H, Hata J, Yasui W, Taniyama K, Hayashi T, Nagae S, Sakamoto S: Jpn J Cancer Res 77:145, 1986.
40. Slamon DJ, Clark GM, Wong SG, Levin WJ, Ullrich A, McGuire WL: Science 235:177, 1987.
41. Di Fiore PP, Pierce JH, Kraus MH, Segatto O, King CR, Aaronson SA: Science 237:178, 1987b.
42. Derynck R, Goeddel DV, Ullrich A, Gutterman JU, Williams RD, Bringman TS, Berger WH: Cancer Res 47:707, 1987.
43. Dickson RB, Huff KK, Spencer EM, Lippman ME: Endocrinology 118:138, 1986.
44. Stromberg K, Hudgins WR, Orth DN: Biochem Biophys Res Commun 144:1059, 1987.
45. Oka Y, Orth DN: J Clin Invest 72:249, 1983.
46. Soeda E, Yasuda S: In Weiss R, Teich N, Varmus H, Coffin J (ed): "RNA Tumor Viruses," Ed. 2 Suppl. Cold Spring Harbor, NY: Cold Spring Harbor Laboratory, 1985, pp 928–939.

Journal of Cellular Biochemistry 39:167–173 (1989)
Growth Factors and Their Receptors: Genetic
Control and Rational Application 145–151

Overexpression of the c-erbB-2 Protein in Human Breast Tumor Cell Lines

Nancy E. Hynes, Heinz A. Gerber, Suzanne Saurer, and Bernd Groner

Ludwig Institute for Cancer Research, Bern Branch, Inselspital (N.E.H., S.S., B.G.), and Department of Pathology, University of Bern, Freiburgstr. 30 (H.A.G.), 3010 Bern, Switzerland

The c-erbB-2 proto-oncogene is amplified in a high percentage of primary human breast tumors, suggesting that the overexpression of this gene may be involved in the development of human breast cancer. We have investigated five human breast tumor cell lines and have detected amplified c-erbB-2 gene copies in two of them. This amplification leads to overexpression of the c-erbB-2 protein. In addition, two other cell lines have elevated protein levels without gene amplification, suggesting that other mechanisms can lead to overexpression of the c-erbB-2 protein. These results are similar to those that we obtained during a study of primary breast tumors (Berger et al.: Cancer Res 48:1238–1243, 1988). These breast tumor cell lines should be useful for an analysis of c-erbB-2 expression and of the mechanisms that in some cases lead to overexpression.

Key words: gene amplification, mammary cancer, tyrosine kinase, proto-oncogene

INTRODUCTION

Altered proto-oncogene expression has been implicated in the development of certain human and animal tumors [1]. Proto-oncogenes can be activated by point mutations or genomic DNA rearrangements, leading to the production of altered protein products. Another mechanism of activation is the overproduction of an apparently normal protein product that in some cases is due to DNA amplification. Various proto-oncogenes have been shown to be amplified in different tumor cells [2], and studies of the N-myc gene were the first to show that there is a relationship between N-myc amplification and survival of neuroblastoma patients [3].

The c-erbB-2 gene was first isolated because of its sequence similarity with v-erbB and human epidermal growth factor (EGF) receptor probes [4–6]. Based upon the deduced amino acid sequence, it has been postulated that c-erbB-2 encodes a growth factor receptor similar to the EGF receptor. There are two mechanisms that

Received March 29, 1988; accepted August 25, 1988.

lead to the oncogenic activation of c-erbB-2. First, a point mutation in the *trans*-membrane domain of the *neu* gene, the rat homologue of c-erbB-2, has been found in rat neuroblastoma DNA [7]. Second, overexpression of the normal human c-erbB-2 protein leads to transformation of NIH/3T3 fibroblasts [8,9]. Recent studies have shown that the c-erbB-2 proto-oncogene is amplified in up to 33% of primary human breast tumors [10–13] and breast tumor cell lines [14]. The detection of this consistent alteration suggests that overexpression of the c-erbB-2 proto-oncogene is involved in the development of human breast cancer. We have shown that there is a good correlation between c-erbB-2 gene amplification and high c-erbB-2 protein expression [13]. In addition, we have observed that some primary breast tumors with single-copy c-erbB-2 DNA sequences stained strongly with the c-erbB-2 specific antiserum [13], suggesting that mechanisms other than gene amplification may lead to elevated protein levels. This phenomenon was more closely studied using human breast tumor cell lines. In this paper we present the results of an analysis of c-erbB-2 DNA, RNA, and protein levels. As seen for the primary tumors, some of the tumor cell lines overexpress the c-erbB-2 protein in the absence of gene amplification.

MATERIALS AND METHODS

Nucleic Acid Isolation and Hybridization Analyses

High molecular weight DNA was isolated from the different cell lines and analyzed for the c-erbB-2 gene copy number as described [13]. Total RNA was isolated by the guanidinium-cesium chloride method [15]. The RNA dot-blot analysis was done by applying twofold dilutions of glyoxal denatured total RNA in 20 × SSC to replicate nitrocellulose filters mounted in a dot-blot apparatus. Filters were hybridized with the c-erbB-2 probe [16] or a control human cDNA probe as described [17]. Hybridization signals were quantitated using an LKB laser densitometer. Northern blot analysis on total SK-BR-3 cell line RNA was performed as described [18]

Immunoblotting

Cells were lysed in 50 mM Tris, pH 7.4, 5 mM EGTA, 1% Triton X-100, 150 mM NaCl, and 3 mM paramethylsulfonyl fluoride, and the supernatant was tested for protein concentration using the method of Bradford. Different amounts of extract were electrophoresed through a 6% polyacrylamide gel and electroblotted onto a nitrocellulose filter. The filter was exposed to the c-erbB-2-specific antiserum 21N [19], and the immunoglobulin reactive band was visualized by ^{125}I-labeled protein A.

Immunohistochemistry

Cells were fixed overnight in 4% formalin, pH 7.0, washed in PBS, suspended in 0.9% agar, and then dehydrated and embedded in paraffin. The cells were stained for c-erbB-2 protein by an immunohistochemical technique on 7-μ paraffin sections using the avidin-biotin complex immunoperoxidase assay [20]. The c-erbB-2-specific antipeptide antisera, 20N and 21N, were used in the assay [13,19]. The dilutions of the antisera and the control staining were as described previously [13]. The scoring is described in the legend to Table I.

RESULTS

In our previous work [13], 51 primary human breast tumor DNAs were analyzed for c-erbB-2 gene copy number. Thirteen (25%) of the tumor DNAs contained

TABLE I. Relative Level of c-erbB-2

Cell line	DNA	RNA	Protein: Western	Protein: Immunohistochemical
BT20	1	1	4	++/+
HBL-100	1	1	1	+/−
MDA-MB231	1	1	1	+/−
MDA-MB361	6–8	23	20	+++
SKBR3	10	23	20	+++
T47D	1	1	4	++

Quantitation of the DNA, RNA, and Western blots was done using a laser densitometer. The immunohistochemical staining for c-erbB-2 on formalin-fixed, paraffin-embedded cells was scored as follows: strong cytoplasmic with a predominant plasma membrane staining (+++), strong cytoplasmic staining (++), weak cytoplasmic staining (+), and little or no staining (+/−).

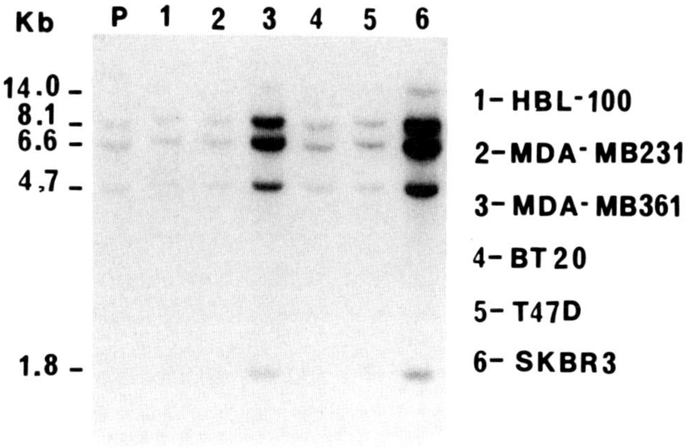

Fig. 1. Analysis of the c-erbB-2 gene in genomic DNA. Five micrograms of DNA isolated from placenta (**P**), a human breast epithelial cell line, HBL-100 (lane **1**), and five different human breast tumor cell lines (lanes **2–6**) were digested with the restriction enzyme EcoR1and analyzed by the Southern blotting technique [21]. The filter was hybridized with a 3.0-kb HindIII-KpnI fragment containing c-erbB-2 cDNA sequences [16]. The sizes of the hybridizing fragments are indicated on the left.

multiple gene copies. Using an immunohistochemical technique, we determined that the majority of these overexpressed the protein, but additional tumors with no apparent gene amplification also stained strongly with the c-erbB-2-specific antiserum [13]. These results suggested that mechanisms other than gene amplification may lead to elevated levels of c-erbB-2 protein. To study this result in more detail, we analyzed five human breast tumor cell lines quantitatively for their c-erbB-2-specific DNA, RNA, and protein content. The cell lines were also stained for c-erbB-2 protein using the same immunohistochemical technique that we have previously employed [13].

The c-erbB-2 gene copy number in six different cell lines was determined by the blotting technique of Southern [21]. The results of the DNA analysis are shown in Figure 1. The SK-BR-3 and MDA-MB361 breast tumor cells have, respectively, ten- and seven-fold c-erbB-2 gene amplification compared with placenta (P), a human breast epithelial cell line, HBL-100, and three other breast tumor cell lines, MDA-MB231, BT-20, and T47D.

To test for c-erbB-2 overexpression at the RNA level, serial two-fold dilutions of total cellular RNA were subjected to a dot- blot analysis. The filter was hybridized with a c-erbB-2 cDNA probe. The specificity of this cDNA probe is shown in the Northern analysis of Figure 2C. RNA isolated from SK-BR-3 cells contains the approximately 5-kb c-erbB-2 transcript that has previously been described [11, 14]. A duplicate control filter was hybridized with a control human cDNA probe that we have recently cloned. The cDNA is specific for a relatively abundant 1.2-kb mRNA that is transcribed to the same extent in all human cell lines that have been tested [17]. The results of the RNA analyses are shown in Figure 2A,B. As expected, the SK-BR-3 and MDA-MB361 cells that have amplified c-erbB-2 gene copies have the highest amounts of c-erbB-2 mRNA. These cells express approximately 20 times more c-erbB-2 mRNA than do the other four cell lines.

The level of c-erbB-2 protein was analyzed in each of the cell lines using the Western protein blotting technique and the 21N antipeptide antiserum. This serum has been well characterized by ourselves and others [12,13,19]. In Western blotting analyses, the 21N antiserum detects a protein of approximately 185-kd in all the human and rodent cell extracts that we have tested. The antiserum rarely reacts with other proteins on the filter (unpublished observations). Different amounts of total cellular protein were separated by electrophoresis, transferred to nitrocellulose, and incubated with the 21N antiserum. The bound antibodies were revealed by incubation with [^{125}I]Protein A. The results are shown in Figure 3. The MDA-MB231 and HBL-100 cells have low levels of the p185, c-erbB-2 protein. A similar low level of c-erbB-2 protein is also present in the human epidermoid carcinoma cell line, A431, a cell line that overexpresses the structurally related EGF receptor protein [22]. The c-erbB-2 protein in 75 and 120 μg of cellular protein from, respectively, the T47D and BT-20 cell lines appears as a strong band that is approximately equivalent to the

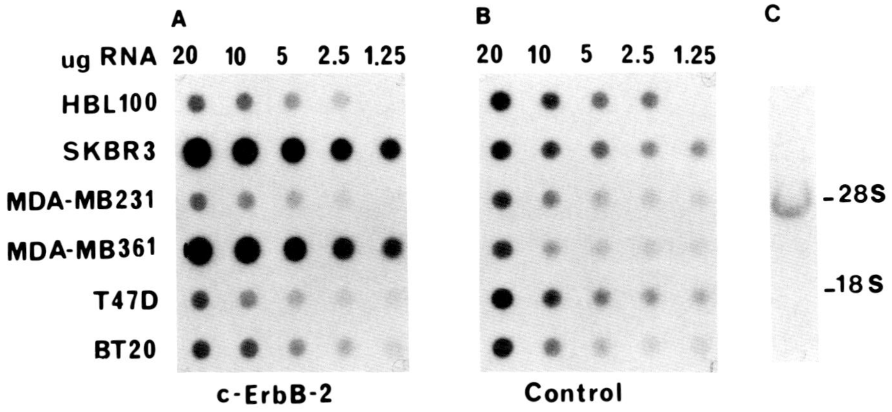

Fig. 2. Quantitation of the c-erbB-2 RNA levels in different cell lines. Twofold dilutions of total cellular RNA from the indicated cell lines were applied to duplicate nitrocellulose filters. One filter **(A)** was hybridized with a c-erbB-2 cDNA probe [16], while the other **(B)** was hybridized with a human cDNA that is equally expressed in all human cells that have been tested [17]. This serves to control for the amounts of RNA applied to the filters. Quantitation of the results is shown in Table I. **(C)** Fifteen micrograms of total RNA isolated from the SK-BR-3 cell line was electrophoresed, blotted, and hybridized with a c-erbB-2 cDNA probe [16].

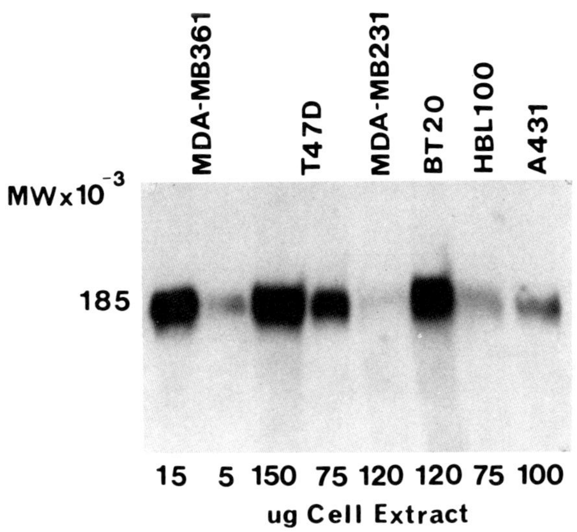

Fig. 3. Detection of c-erbB-2 protein by Western blotting. The indicated amounts of cellular protein from each of the human cell lines were separated on a 6% polyacrylamide gel, transferred to nitrocellulose exposed to the 21N c-erbB-2 specific antiserum [19], and visualized by ^{125}I-labeled protein A. The molecular weight of the c-erbB-2 protein is indicated on the left.

amount in 15 μg of MDA-MB361 cell extract. The latter cells and SK-BR-3 cells have equivalent amounts of c-erbB-2 protein (data not shown).

A quantitation of the results shown in Figures 1–3 are presented in Table I. In addition, the results of an immunohistochemical analysis for c-erbB-2 protein in each of the cell lines is also presented. The cells were formalin-fixed and paraffin-embedded before staining. This approximately mimics the state of the primary breast tumor material that has been stained for c-erbB-2 protein [13]. The results show that the two cell lines with amplified c-erbB-2 gene copies, SK-BR-3 and MDA-MB361, have the highest amounts of c-erbB-2 RNA and protein. They also display the strongest staining with the c-erbB-2 specific antisera. The BT-20 and T47D cell lines also contain elevated levels of c-erbB-2 protein when compared with the HBL-100 and MDA-MB231 cell lines. The former two cell lines do not contain amplified gene copies, and we have not seen an obvious increase in c-erbB-2 RNA when compared to the level in the HBL-100 or MDA-MB231 cells. Slightly elevated levels, for example, two-fold, of c-erbB-2 RNA would be difficult to quantitate. In the immunohistochemical analysis, these cells are also clearly positive for c-erbB-2 staining. The HBL-100 and MDA-MB231 cells contain low levels of c-erbB-2 RNA and protein; they were barely, if at all, stained in the immunohistochemical analysis. In conclusion, the results show that the c-erbB-2 protein level in human breast tumor cell lines can be elevated either in the presence or absence of gene amplification. In addition, the results from the immunohistochemical and the Western analyses for c-erbB-2 protein are in good agreement. Thus, the ABC immunoperoxidase staining technique can be used for the detection of c-erbB-2 protein in primary breast tumors.

DISCUSSION

It has been shown that c-erbB-2 gene amplification [10–14] and protein over-expression [12,13] occur in a high percentage of breast tumors. Our own work has shown that there is a statistically significant correlation between c-erbB-2 protein expression and parameters, such as lymph node involvement, which are used in breast cancer prognosis [13]. This suggests that the detection of c-erbB-2 protein in breast tumor sections may have value in predicting the course of the disease. The role that the c-erbB-2 protein plays in the development of breast cancer is unclear. Its normal function in the growth and differentiation of breast cells is also unknown. To understand more about this protein it will be important to study the mechanisms that control its expression. The results presented here show that c-erbB-2 RNA and protein are expressed in all the breast tumor cell lines examined. This agrees with other published work in which breast [14] as well as other tissues and cell lines were found to be positive for c-erbB-2 expression [6,19]. The promoter region of the c-erbB-2 gene has been sequenced, and it is different from the promoter of the closely related EGF receptor gene. It possesses a TATA box and a CAAT box as well as two putative SP1 binding sequences. In addition, its transcription appears to initiate at multiple sites [23,24]. Therefore, its expression may be controlled by multiple protein factors of which one or more may be present in most cell types.

The results also show that the SK-BR-3 and MDA-MB361 cells that contain amplified c-erbB-2 gene copies and elevated RNA levels also contain the most c-erbB-2 protein. The other four cell lines examined have similar amounts of c-erbB-2 RNA, yet the T47D and BT-20 cells contain approximately fourfold more c-erbB-2 protein than MDA-MB231 and HBL-100 cells. Some of these cell lines have been examined previously, and it was also found that the HBL-100 and BT-20 cells expressed the same low level of c-erbB-2 RNA [14]. In this previous study, the c-erbB-2 protein level was not determined. One mechanism that could explain the difference in c-erbB-2 protein levels in cells with similar c-erbB-2 mRNA levels is differences in protein stability. It has been observed for a number of peptide growth factor receptors that they are down-regulated by their ligand. Fibroblasts grown in the presence of EGF have only 20% of the EGF receptors present on the untreated control cells [25]. The ligand for c-erbB-2 (*neu*) has not yet been described, but addition of *neu* specific monoclonal antibodies to cells expressing the rat *neu* protein causes p185 *neu* down-regulation. This down-modulation is controlled by changes in protein stability [26]. Cells expressing low levels of c-erbB-2 protein could also be synthesizing its ligand. Such a situation would lead to continuous down-modulation of the c-erbB-2 protein. A comparison of the c-erbB-2 protein stability in the different breast tumor cell lines could help clarify this point.

This study shows that high levels of c-erbB-2 protein are present in four out of five of the breast tumor cell lines that have been studied. The elevated protein level is seen either in the presence or absence of gene amplification. The phenomenon of protein overexpression without gene amplification has also been reported for the EGF receptor [2]. One can speculate that those transformed cells that express high levels of growth factor receptors have a proliferative advantage in a tumor. The deregulation of c-erbB-2 expression at different levels of control of gene expression might indicate different mechanisms of transformation. Whereas gene amplification might be restricted to the c-erbB-2 locus, transcriptional deregulation or posttranscriptional effects are most likely mediated by *trans*-acting factors. Insights into the components involved in c-erbB-2 deregulation might provide targets for directed intervention with the growth of tumor cells.

ACKNOWLEDGMENTS

We thank Drs. Mark Berger (Hospital of the University of Pennsylvania, Philadelphia) and William Gullick (ICRF, Hammersmith Hospital, London) for helpful discussions and for supplying us with the c-erbB-2 antisera. The c-erbB-2 cDNA was a gift from Dr. T. Yamamoto (University of Tokyo, Japan).

REFERENCES

1. Bishop JM: Science 235:305–311, 1987.
2. Alitalo K, Schwab M: Adv Cancer Res 47:235–281, 1986.
3. Brodeur GM, Seeger RC, Schwab M, Varmus HE, Bishop JM: Science 224:1121–1124, 1984.
4. Semba K, Kamata N, Toyoshima K, Yamamoto T: Proc Natl Acad Sci USA 82:6497–6501, 1985.
5. King CR, Kraus MH, Aaronson SA: Science 229:974–976, 1985.
6. Coussens L, Yang-Feng TL, Chen Y-CLE, Gray A, McGrath J, Seeburg PH, Libermann TA, Schlessinger J, Francke U, Levinson A, Ullrich A: Science 230:1132–1139, 1985.
7. Schechter AL, Stern DF, Vaidyanathan L, Decker SJ, Drebin JA, Green MI, Weinberg RA: Nature 312:513–516, 1984.
8. Di Fiore PP, Pierce JH, Kraus MH, Segatto O, King CR, Aaronson SA: Science 237:178–182, 1987.
9. Hudziak RM, Schlessinger J, Ullrich A: Proc Natl Acad Sci USA 84:7159–7163, 1987.
10. Slamon DJ, Clark GM, Wong SG, Levin WJ, Ullrich A, McGuire WL: Science 235:177–182, 1987.
11. Van de Vijver M, van de Bersselaar R, Devilee P, Cornelisse C, Peterse J, Nusse R: Mol Cell Biol 7:2019–2023, 1987.
12. Venter DJ, Tuzi NL, Kumar S, Gullick WJ: Lancet ii:69–72, 1987.
13. Berger MS, Locher GW, Saurer S, Gullick WJ, Waterfield MD, Groner B, Hynes NE: Cancer Res 48:1238–1243, 1988.
14. Kraus MH, Popescu NC, Amsbaugh SC, King CR: EMBO J 6:605–610, 1987.
15. Glisin V, Crkvenjakov R, Byus C: Biochemistry 13:2633–2637, 1974.
16. Yamamoto T, Shantaro I, Akiyama T, Semba K, Nomura N, Miyajima N, Saito T, Toyoshima K: Nature 319:230–234, 1986.
17. Kozma SC, Redmond SMS, Xiao-Chang F, Saurer SM, Groner B, Hynes NE: EMBO J 7:147–154, 1988.
18. Redmond SMS, Reichmann E, Müller RG, Friis RR, Groner B, Hynes NE: Oncogene 2:259–265, 1988.
19. Gullick WJ, Berger MS, Bennett PLP, Rothbard JB, Waterfield MD: Int J Cancer 40:246–254, 1987.
20. Hsu SM, Raine M, Fanger H: J Histochem Cytochem 29:577–580, 1981.
21. Southern EM: J Mol Biol 98:503–517, 1975.
22. Downward J, Yarden Y, Mayes E, Scrace G, Totty N, Stockwell P, Ullrich A, Schlessinger J, Waterfield MD: Nature 307:521–527, 1984.
23. Ishii S, Imamoto F, Yamanashi Y, Toyoshima K, Yamamoto T: Proc Natl Acad Sci USA 84:4374–4378, 1987.
24. Tal M, King CR, Kraus MH, Ullrich A, Schlessinger J, Givol D: Mol Cell Biol 7:2597–2601, 1987.
25. Carpenter G, Cohen S: Annu Rev Biochem 48:193–216, 1979.
26. Drebin JA, Link VC, Stern DF, Weinberg RA, Greene MI: Cell 41:695–706, 1985.

Journal of Cellular Biochemistry 39:175–184 (1989)
Growth Factors and Their Receptors: Genetic Control and Rational Application 153–162

Transforming Growth Factor β: Possible Roles in the Regulation of Normal and Leukemic Hematopoietic Cell Growth

Jonathan R. Keller, Garwin K. Sing, Larry R. Ellingsworth, and Francis W. Ruscetti

Biological Carcinogenesis Development Program, Program Resources, Inc. (J.R.K.) and Laboratory of Molecular Immunoregulation, Biological Response Modifiers Program (G.K.S., F.W.R.), NCI-Frederick Cancer Research Facility, Frederick, Maryland, 21701; Collagen Corporation, Palo Alto, California 94303 (L.R.E.)

We have recently demonstrated that transforming growth factor (TGF)-β1 and TGF-β2 are potent inhibitors of the growth and differentiation of murine and human hematopoietic cells. The proliferation of primary unfractionated murine bone marrow by interleukin-3 (IL-3) and human bone marrow by IL-3 or granulocyte/macrophage colony-stimulating factor (GM-CSF) was inhibited by TGF-β1 and TGF-β2, while the proliferation of murine bone marrow by GM-CSF or murine and human marrow with G-CSF was not inhibited. Mouse and human hematopoietic colony formation was differentially affected by TGF-β1. In particular, CFU-GM, CFU-GEMM, BFU-E, and HPP-CFC, the most immature colonies, were inhibited by TGF-β1, whereas the more differentiated unipotent CFU-G, CFU-M, and CFU-E were not affected. TGF-β1 inhibited IL-3-induced growth of murine leukemic cell lines within 24 h, after which the cells were still viable. Subsequent removal of the TGF-β1 results in the resumption of normal growth. TGF-β1 inhibited the growth of factor-dependent NFS-60 cells in a dose-dependent manner in response to IL-3, GM-CSF, G-CSF, CSF-1, IL-4, or IL-6. TGF-β1 inhibited the growth of a variety of murine and human myeloid leukemias, while erythroid and macrophage leukemias were insensitive. Lymphoid leukemias, whose normal cellular counterparts were markedly inhibited by TGF-β, were also resistant to TGF-β1 inhibition. These leukemic cells have no detectable TGF-β1 receptors on their cell surface. Last, TGF-β1 directly inhibited the growth of isolated Thy-1-positive progenitor cells. Thus, TGF-β may be an important modulator of normal and leukemic hematopoietic cell growth.

Key words: **hematopoiesis, transforming growth factor β, colony-stimulating factors, leukemic cells, interleukin-3**

A family of polypeptide growth factors that regulates cell growth has recently been identified and includes 1) two distinct forms of transforming growth factor beta

Received June 2, 1988; accepted August 3, 1988.

(TGF-β), TGF-β1 and TGF-β2 [1–3], which have been shown to be multifunctional regulators of cell growth; 2) inhibin, which suppresses follicle-stimulating hormone secretion; 3) Mullerian inhibitory substance, which causes regression of the Mullerian duct in the male embryo; and 4) the predicted product of the Drosophilian decapentaplegic gene complex [2]. The cDNA sequences for this family of growth factors indicate that the mature proteins are synthesized from the carboxyl-terminal half of larger precursors, and share homologies to each other within this region and, in particular, show conservation in seven out of nine cysteines [1]. TGF-β1 has been purified to homogeneity from platelets, placenta, kidney, and bone and is a 25-kilodalton (Kd) disulfide-linked homodimeric protein whose sequence is remarkably conserved with a single amino acid substitution between mouse and man [1].

TGF-β1 was originally identified in the supernatants of transformed fibroblasts by its ability to induce the anchorage-independent growth of NRK 49F fibroblasts in soft agar. More recently, it has been shown to enhance the proliferation of two mesenchymal cell types, osteoblasts, responsible for new bone formation, and Schwann cells, involved in the repair of peripheral nerves and synthesis of myelin [3]. TGF-β1 has also been described as a potent inhibitor of proliferation of a variety of cell types in vitro including epithelial cells, embryonic fibroblasts, endothelial cells, and T and B lymphocytes [3]. Thus, it has been postulated that TGF-β1 may act as an autocrine or paracrine negative regulator of cell growth.

TGF-β is also known to exert biological effects unrelated to cell proliferation, including the induction of extracellular matrix proteins such as collagen and fibronectin in human, rat, mouse, and chicken fibroblasts [4], and the inhibition of the degradation of extracellular matrix proteins. Local injection of TGF-β1 in vivo results in a strong fibrotic response including activation of fibroblasts to produce collagen. Also, TGF-β1 accelerates the healing of incisional wounds in rats. Therefore, TGF-β1 has been proposed to play a role in inflammation and wound healing [4].

Immunohistochemical studies using antibodies to the N-terminus of TGF-β1 to localize the production of TGF-β1 showed that TGF-β1 is produced by cells in centers of active hematopoiesis including fetal liver and bone marrow, and active lymphopoiesis in the Hassall's corpuscles [5]. Therefore, experiments were designed to determine whether TGF-β1 might be involved in regulating normal hematopoietic cell growth and differentiation.

RESULTS

For these studies, we employed the bone marrow soft agar colony assay, which measures the clonal growth of hematopoietic progenitors and has historically been used to determine the factors and conditions required for normal hematopoietic cell growth and differentiation. This assay has facilitated the identification and cloning of the colony-stimulating factors (CSFs) [6], which are polypeptide growth factors that enhance the proliferation and differentiation of hematopoietic cells in vitro. Using this assay and purified CSFs, we designed experiments to determine whether TGF-β1 could enhance or inhibit the growth of hematopoietic cells in vitro.

Effect of TGF-β on Normal Bone Marrow Progenitor Cell Growth

The initial experiments to examine the effect of TGF-β1 on the growth of hematopoietic cells used ^{3}H-thymidine proliferation assays with freshly aspirated murine bone marrow cells cultured in microtiter wells for 72 h in the presence of the various CSFs. For these studies, homogeneous TGF-β1 purified from bovine bone was used. In contrast to two recent reports [7,8] stating TGF-β2 had no effect on hematopoiesis, our comparative study of the effects of TGF-β1 and TGF-β2, each purified from two different sources, showed little or no difference between the two. Thus, all the results presented here will have used TGF-β1. The CSFs employed were interleukin 3 (IL-3), which promotes the colony formation (CFU) of mixed lineage colonies that include granulocytes (G), macrophages (m), megakaryocytes (M), and erythroid cells (E) (CFU-GEMM); granulocyte/macrophage-CSF (GM-CSF), which promotes the growth of granulocytes and macrophages (CFU-GM); granulocyte-CSF (G-CSF), which promotes granulocyte growth; and erythropoietin (EPO), which promotes the terminal differentiation of erythroid cells. The results demonstrate that TGF-β1 alone has no effect on the proliferation of hematopoietic cells but can inhibit between 50% and 70% of the IL-3-driven bone marrow proliferation in a dose-dependent manner with an ED_{50} of 5–10 pM. In contrast, G-CSF-, GM-CSF–, and EPO-driven proliferation is unaffected by TGF-β1. To determine what progenitors and which lineages may have been affected by TGF-β1, bone marrow cells were plated in soft agar in the presence of the CSFs. IL-3-induced colony formation was inhibited in a dose-dependent manner with an ED_{50} of 5–10 pM, while G-CSF- and GM-CSF-driven colony formation were unaffected (Table I). Finally, even though the colony formation induced by IL-3 was significantly inhibited by TGF-β1, clusters of differentiated myeloid cells were consistently observed.

Since the inhibition of IL-3-driven proliferation was incomplete and ranged from 50% to 70% and since clusters of differentiated cells were observed in soft agar assays containing IL-3 and TGF-β1, we hypothesized that TGF-β1 might selectively regulate early hematopoietic progenitor cell growth. To test this hypothesis, we studied the effect of TGF-β1 on the growth of the most primitive CFU found in normal marrow populations, the multipotential CFU, CFU-GEMM. The CFU-GEMM develops in response to two signals—namely, IL-3, which provides the early signal, and EPO, which promotes the terminal differentiation of erythroid cells. In addition, we examined the primitive erythroid lineage colony, the burst-forming unit (BFU-E), which also uses the same two growth signals, IL-3 and EPO. As shown in Table I, TGF-β1 is a potent inhibitor of CFU-GEMM and BFU-E formation; however,

TABLE I. Effect of TGF-β1 on Murine Hematopoietic Colony Formation in Soft Agar*

		No. of colonies		
CSF	TGF-β1	CFU-GM	CFU-GEMM	BFU-E
None	+ or −	0	0	0
IL-3	−	52 ± 5	0	0
IL-3	+	6 ± 3	0	0
IL-3 + EPO	−	58 ± 3	5 ± 2	11 ± 3
IL-3 + EPO	+	4 ± 2	0	0
GM-CSF	−	113 ± 12	0	0
GM-CSF	+	387 ± 11	0	0
G-CSF	−	20 ± 4	0	0
G-CSF	+	25 ± 4	0	0

*Colonies were scored after 6–8 days of incubation and are expressed as the mean of triplicate experiments. The growth factors were supplemented to the cultures at the following concentrations: TGF-81 (10 ng/ml), IL-3 (50 U/ml), EPO (2 U/ml), GM-CSF (1 ng/ml), and G-CSF (100 U/ml).

TGF-β1 had no effect on erythropoietin-driven proliferation of splenocytes from phenylhydrazine-treated mice (spleen cells are highly enriched for erythroblasts [9]) or erythroid colony formation, CFU-E, from normal bone marrow. Taken together, the results suggest that TGF-β1 selectively inhibits the growth of early progenitor cells, while the proliferation and differentiation of the more committed or differentiated progenitors are unaffected.

To determine if TGF-β1 could also regulate human hematopoietic cells, comparative studies were performed on freshly aspirated human bone marrow. In proliferation assays, TGF-β1 inhibited 50–70% of the IL-3- and GM-CSF-driven proliferation in a dose-dependent manner with an ED_{50} of 5–15 pM. In contrast, G-CSF driven proliferation was unaffected. Similar results were observed in the colony assays, in that TGF-β1 inhibited GM-CSF and IL-3 induced CFU-GM, CFU-GEMM, and BFU-E progenitor cell growth, while G-CSF-induced colony formation was unaffected (Table II). In addition, similar to results seen with murine hematopoietic cells, clusters of terminally differentiated progeny of a single lineage were consistently seen in cultures containing TGF-β1 and GM-CSF or IL-3. However, in contrast to results with murine hematopoietic cells, human GM-CSF-induced colony formation was inhibited by TGF-β1. This may reflect differences in the two species, since human GM-CSF, unlike murine GM-CSF, is a potent inducer of the more primitive progenitors including CFU-GEMM [10]. Therefore, these results are consistent with the concept that TGF-β1 selectively regulates early hematopoietic progenitor cells while the more differentiated and committed progenitors are unaffected.

Effect of TGF-β1 on Primitive Murine Progenitor Populations

To determine whether the effect of TGF-β1 on hematopoietic cells was direct or indirect, pure populations of hematopoietic progenitors were isolated from heterogeneous populations of murine bone marrow cells. This was accomplished using a previously described system which demonstrated that IL-3 was required for the induction of Thy-1 antigen expression on Thy-1-negative (Thy-1−) bone marrow cells. Single-cell analysis of fluorescence-activated cell sorted (FACS) Thy-1+ cells has shown that these cells have the potential to differentiate to the various hematopoietic cell types, and contain progenitors with multipotent, bipotent, and unipotent differentiation potential [11]. Therefore, two aspects of this system were studied—first, what are the effects of TGF-β1 on IL-3-induced Thy-1 antigen expression; and second, does TGF-β1 inhibit the growth of isolated Thy-1+ progenitor cells in soft

TABLE II. Effect of TGF-β1 on Human Hematopoietic Colony Formation in Soft Agar*

		No. of colonies		
CSF	TGF-β1	CFU-GM	CFU-GEMM	BFU-E
None	+ or −	0	0	0
GM-CSF + EPO	−	62 ± 23	9 ± 3	43 ± 8
GM-CSF + EPO	+	3 ± 2	0	0
IL-3 + EPO	−	15 ± 2	8 ± 2	37 ± 7
IL-3 + EPO	+	11 ± 1	0	0
G-CSF	−	48 ± 5	0	0
G-CSF	+	49 ± 4	0	0

*Colonies were scored after 14 days and the results are expressed as the mean of triplicate experiments. The growth factors were supplemented to the medium at the following concentrations: TGF-β1 (10 ng/ml), GM-CSF, IL-3, and G-CSF (10 ng/ml).

agar colony assays? As shown in Table III, TGF-β1 inhibited IL-3-induced Thy-1 antigen expression in a dose-dependent manner with an ED_{50} of 5–10 pM. When these cultures were examined morphologically using a Jenner's stain, there was an absence of myeloid blast cells normally present in cultures supplemented with IL-3. Next, to demonstrate the direct action of TGF-β1 on hematopoietic progenitor cell growth, Thy-1+ progenitors were isolated by fluorescence-activated cell sorting from bone marrow cultures grown for seven days in medium supplemented with IL-3, and then plated into soft agar colony assays. As previously demonstrated [12], IL-3 induces the formation of mixed-lineage colonies that contain granulocytes, macrophages, and mast cells as well as pure monolineage colonies of each cell type. However, in addition to the previous results, Thy-1+ progenitors also give rise to CFU-GEMM in cultures supplemented with IL-3 and EPO. The results show that TGF-β1 inhibits greater than 50% of the total colony formation, which includes all of the multipotential progenitors, CFU-GEMM and CFU-GM, while colonies and clusters containing single lineages of granulocytes and macrophages are unaffected. As summarized in Figure 1, TGF-β1 inhibits the induction of Thy-1 antigen on

TABLE III. Effect of TGF-β1 on IL-3-Induced Thy-1 Expression*

Stimulator	TGF-β1	% Thy-1 expressed	% Inhibition
None	+ or −	0	0
IL-3	−	28	−
IL-3	20 ng/ml	6	79
IL-3	2 ng/ml	7	75
IL-3	0.2 ng/ml	21	25
IL-3	0.02 ng/ml	27	4

*Bone marrow cells were depleted of Thy-1+ cells by complement-mediated cytotoxicity and then plated in medium containing 100 U/ml IL-3 with the indicated concentrations of TGF-β1. Thy-1 expression was determined by FACS analysis after a 7-day incubation period.

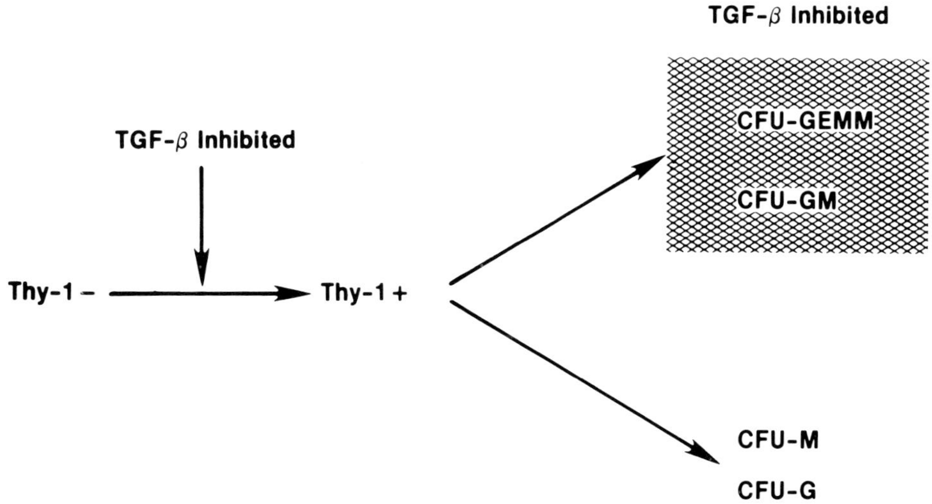

Fig 1. Effect of TGF-β on Thy-1 expression and on the proliferation of IL-3-induced Thy-1-positive cells.

hematopoietic progenitors and can directly act on isolated Thy-1 progenitors selectively to inhibit multipotential cell growth and differentiation.

To determine at what stage in hematopoietic development progenitor cells acquire the ability to respond to TGF-β, bone marrow cells were obtained from the femurs of mice injected with 5-fluorouracil (5-FU), a compound that enriches for noncycling primitive hematopoietic progenitors [13]. One class of hematopoietic progenitor cell in post 5-FU bone marrow is the highly proliferative potential colony-forming cell (HPP-CFC) that proliferates in vitro to generate large macroscopic colonies (greater than 0.5 mm). Therefore, we examined the effect of TGF-β1 on the growth of the three classes of HPP-CFC which have been defined. The first class of HPP, HPP-CFC-1, which can be induced to proliferate in response to interleukin 1 [IL-1 (previously known as hematopoietin-1)] and CSF-1, represents the most primitive HPP progenitor capable of reconstituting the hematopoietic system [14]. It can be inhibited by TGF-β1 in a dose-dependent manner, with an ED_{50} of 10–30 pM. The second class of HPP, HPP-CFC-2, which has been shown to be derived from HPP-CFC-1 [15] and is induced to proliferate by IL-3 and CSF-1, was also inhibited by TGF-β1 with similar dose-dependent kinetics. Finally, a third class of HPP, HPP-CFC-HLGF-1, induced to proliferate from normal bone marrow cells in response to hemato-lymphopoietic growth factor (HLGF-1) and CSF-1 [16], was also inhibited. Thus, TGF-β1 can inhibit the growth of the most primitive hematopoietic progenitors detectable in vitro. Since the HPP progenitors are quiescent cells, the mechanism of negative regulation by TGF-β probably involves maintaining the progenitor cells in a noncycling state. A summary of those progenitors whose growth can be inhibited by TGF-β is presented in Figure 2.

Effect of TGF-β1 on Leukemic Myeloid Cell Growth

Since the results obtained with normal hematopoietic progenitors suggest that TGF-β1 plays a role in the regulation of hematopoietic cell growth, experiments were designed to determine whether TGF-β1 also inhibits the growth of leukemic cell lines. Thus, we examined a variety of murine cell lines, including growth factor-dependent "preleukemic" progenitor cell lines blocked in early stages of differentiation as shown by phenotype, function and morphology and factor independent myeloid leukemic cells [17–19] (Table IV). TGF-β1 inhibits the proliferation of all the IL-3-dependent murine progenitor cell lines tested to date, regardless of their derivation. In addition, the growth of NFS-60 cells, which proliferate in response to IL-3, GM-CSF, G-CSF, CSF-1, IL-6, and IL-4, was inhibited by TGF-β1 in a dose-dependent manner with an ED_{50} of 5–10 pM regardless of the growth factor employed. TGF-β1 did not induce the differentiation of those lines that were inhibited. The inhibitory effects of TGF-β1 on the proliferation of factor dependent cell lines could be specifically neutralized with a polyclonal antiserum to TGF-β1 [5].

Examination of factor-independent murine leukemic cell lines revealed that cell lines with an early blastic phenotype were inhibited by TGF-β, but those with a more mature phenotype were not (Table IV). WEHI-3B (myelomonocytic) and GG2EE (promonocytic) were inhibited while the macrophage cell lines P388D1 and J774, the erythroid cell lines TP-3 and DS-19 and the mastocytoma P815 were insensitive to the effects of TGF-β1.

We next explored the question of whether factor dependence or independence plays a role in TGF-β sensitivity. To address this question two IL-3-dependent cell

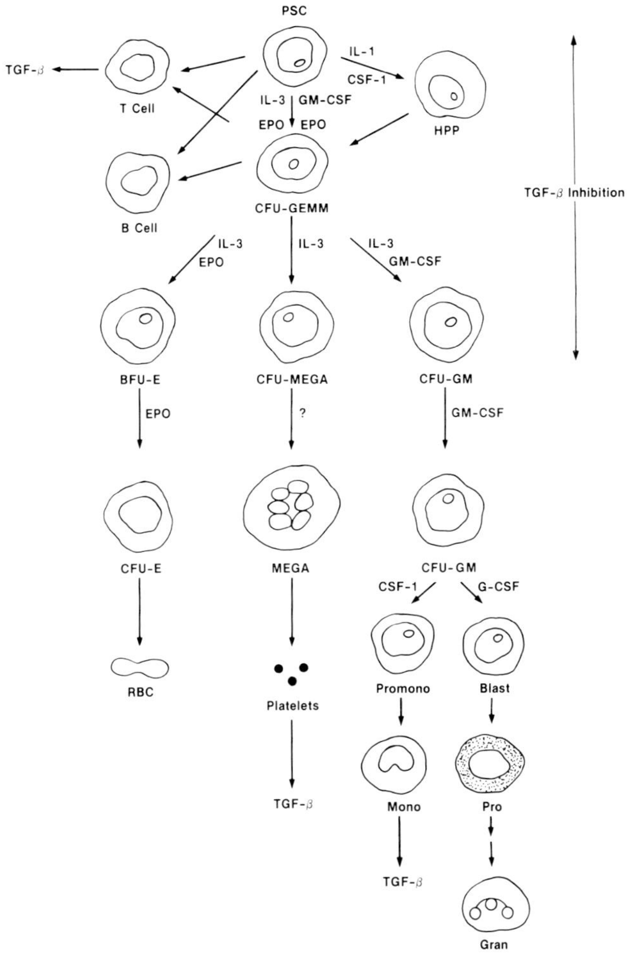

Fig 2. Scheme of TGF-β inhibition.

lines, NFS-60 and 32D-cl23, were infected with oncogene-containing retroviruses to abrogate their requirement for factor dependence [previously demonstated, 19]. In separate experiments, retroviruses containing v-abl, v-src, and v-fms were used to make factor-indepenent variants NFS-60 and 32D-cl23. In all cases, the factor-independent cell lines obtained retained the ability to be inhibited by TGF-β, suggesting that the response to TGF-β was not related to factor dependence.

TGF-β Receptor Differences on Human Leukemic Cells

A study of the effect of TGF-β on the growth of human myeloid and lymphoid cell lines revealed that all myelomonocytic cell lines except the promyelocytic HL-60 cell line were inhibited by TGF-β (Table V). In contrast to our data with myeloid leukemic cells, four out of four mature B-lymphoid leukemic cell lines and four out of four mature T-lymphoid leukemic cell lines were not inhibited by TGF-β (Table V), while their normal counterparts were sensitive to TGF-β [20,21]. To determine whether the differential response of these cell lines to TGF-β could be at the level of

TABLE IV. Effect of TGF-β1 On Leukemic Cell Lines*

Cell line	Lineage	Factor	% TGF-β1 inhibition
FDC-P1	Myeloid	IL-3	80–90
32D-c123	Myeloid	IL-3	70–80
NFS-60	Myeloid	IL-3	60–70
NFS-60	Myeloid	GM-CSF	60–70
NFS-60	Myeloid	G-CSF	60–70
NFS-60	Myeloid	CSF-1	60–70
NFS-60	Myeloid	IL-4	60–70
WEHI-3	Myelomonocytic	—	80–90
P388D1	Macrophage	—	0
J774A.1	Macrophage	—	0
TP-3	Erythroid	—	0
DS-19	Erythroid	—	0
P815	Mastocytoma	—	0

*TGF-β1 was diluted into cell proliferation assays to generate a dose-response curve starting at a concentration of 40 ng/ml. The range represents the maximum percent inhibition obtained from the dose-response curves from three separate experiments.

TABLE V. Effects of TGF-β1 in the Proliferation of Human Leukemia Cells*

Cell line	Cell type	TGF-β inhibition		TGF-β1 receptors
		ED-50 (pM)	Maximal (%)	
U937	Promonocyte	24	60	+++
THP-1	Monocyte	24	45	+++
KG-1	Myeloblast	14	70	+++
HL-60	Promyelocyte	–	0	0
RL	B-cell lymphoma	–	0	0
DB	B-cell lymphoma	–	0	0
HT	B-cell lymphoma	–	0	0
SR	Lymphoma	–	0	0
Hut 78	T-cell lymphoma	–	0	±
Hut 102	T-cell lymphoma	–	0	±
MT-2	T-cell lymphoma	–	0	±
CM-235	T-cell lymphoma	–	0	0

*TGF-β1 was diluted into cell proliferation assays to generate a dose-response curve starting at a concentration of 40 ng/ml. The range represents the maximum percent inhibition obtained from the dose-response curves from three separate experiments.

receptor expression, ligand-binding and affinity-labeling studies were performed. Three structurally distinct cell surface glycoproteins—a 280-Kd, an 85-Kd and a 65-Kd species—have been previously described as being high affinity TGF-β receptors [22]. These receptor molecules were identified on all cell lines sensitive to TGF-β growth inhibition. When the TGF-β-insensitive cell lines were examined, six out of nine had no detectable TGF-β receptors, while the other three had reduced numbers of receptors. These results suggest that the loss of TGF-β receptor expression might play a role in the growth of some leukemias by allowing cells to escape negative growth regulation by TGF-β.

SUMMARY

We have found that TGF-β is a potent selective inhibitor of early murine [23,24] and human [25] hematopoietic progenitor cell growth. It inhibits CFU-GEMM,

BFU-E, and CFU-GM colony formation but has no effect on CFU-G, CFU-E, and CFU-M colony formation. Using purified IL-3-induced Thy-1+ progenitors, we have shown that multipotent but not unipotent cell growth is directly inhibited by TGF-β [26]. In addition, all three classes of the HPP progenitor cells, which are the earliest measurable cells in vitro, were inhibited by TGF-β. Since HPP-CFC-1 cells are quiescent and have to be induced to enter the cell cycle in vitro by IL-1, it seems likely that TGF-β acts by interfering with the ability of cells to enter the cell cycle. Two forms of TGF-β, TGF-β1, and TGF-β2 were found to have identical effects in all the assays employed.

In summarizing the effects of TGF-β on leukemic cell growth, it appears that sensitivity to TGF-β is determined by the state of cellular differentiation and not the growth factor responsiveness of a particular cell type. For example, while G-CSF- and GM-CSF-induced normal bone marrow proliferation and colony formation were not inhibited by TGF-β1, G-CSF- and GM-CSF-driven NFS-60 cell proliferation was inhibited. In addition, it was found that the ability to respond to TGF-β was not due to growth factor dependence or independence. The TGF-β1-resistant leukemic erythroid and macrophage cells may have been transformed at a stage in their maturation where they are insensitive to TGF-β1. On the other hand, the resistance to TGF-β inhibition of some leukemic T and B lymphocytes as well as HL-60 is due to the absence of detectable receptors for TGF-β. Thus, it appears that some leukemic cells can gain a growth advantage by escaping from the negative growth regulation of TGF-β.

The capacity to inhibit selectively early marrow progenitor cells as well as leukemic cell proliferation could have clinical application. A variety of hematopoietic tumors may be responsive in vivo to TGF-β inhibition, which could be a useful adjunct to cytoreductive chemotherapy. In addition, if growth-arrested, quiescent marrow stem cells prove to be less sensitive to the effects of cycle-active drugs, TGF-β could be used as a protective agent to ameliorate the dose-limiting toxicity of many chemotherapeutic agents.

ACKNOWLEDGMENTS

By acceptance of this article, the publisher or recipient acknowledges the right of the U.S. Government to retain a nonexclusive, royalty-free license in and to any copyright covering the article.

This project has been funded at least in part with Federal funds from the Department of Health and Human Services under contract NO1-CO-74102. The content of this publication does not necessarily reflect the view or policies of the Department of Health and Human Services, nor does mention of trade names, commercial products, or organizations imply endorsement by the U.S. Government.

REFERENCES

1. Sporn MB, Roberts AB, Wakefield LM, de Crombrugghe B: J Cell Biol 105:1039, 1987.
2. Massagué J: Cell 49:437, 1987.
3. Sporn MB, Roberts AB: Nature 313:745, 1985.
4. Sporn MB, Roberts AB, Wakefield LM, Assoian RK: Science 233:532, 1986.
5. Ellingsworth LR, Brennan JE, Fok K, Rosen DM, Bentz H, Piez KA, Seyedin SM: J Biol Chem 261:12362, 1986.

6. Metcalf D: Science 229:16, 1985.
7. Ohta M, Greenberger JS, Anklesaria P, Bassols A, Massagué J: Nature 329:539, 1987.
8. Ottman OG, Pelus LM: J Immunol 140:2661, 1988.
9. Ruscetti SK: JNCI 77:241, 1986.
10. Burgess AW, Metcalf D: Blood 56:947, 1980.
11. Keller JR, Ihle JN: J Immunol 135:1864, 1985.
12. Ihle JN, Keller J, Oroszlan S, Henderson LE, Copeland TD, Fitch F, Prystowsky MB, Goldwasser E, Schrader JW, Palaszynski E, Dy M, Lebel B: J Immunol 131:282, 1983.
13. Bradley TR, Hodgson GS: Blood 54:1446, 1979.
14. McNiece IK, Williams NT, Johnson GR, Kriegler AB, Bradley TR, Hodgson GS: Exp Hematol 15:972, 1987.
15. McNiece IK, Bradley TR, Kriegler AB, Hodgson GS: Exp Hematol 14:856, 1986.
16. Quesenberry P, Zenqxuan S, McGrath E, McNiece I, Shadduck R, Waheed A, Baber G, Kleeman E, Kaiser D: Blood 69:827, 1987.
17. Greenberger JS, Eckner RJ, Sakakeeny M, Marks P, Reid D, Nabel G, Hapel A, Ihle JN, Humphries KC: Fed Proc 42:2672, 1983.
18. Dexter TM, Garland J, Scott D, Scolnick E, Metcalf D: J Exp Med 152:1036, 1980.
19. Ihle JN, Morse HC, Keller J, Holmes K: Curr Top Microbiol Immunol 113:85, 1984.
20. Kehrl JH, Wakefield LM, Roberts AB, Jakowlew S, Alvarez-Mon M, Derynck R, Sporn MB, Fauci AS: J Exp Med 163:1037, 1986.
21. Kerhl JH, Roberts AB, Wakefield LM, Jakowlew S, Sporn MB, Fauci AS: J Immunol 137:3855, 1986.
22. Cheifetz S, Weatherbee JA, Tsang ML-S, Andersen JK, Mole JE, Lucas R, Massagúe J: Cell 48:409, 1987.
23. Ishibeshi T, Millee SL, Burnstein SA: Blood 69:1937, 1987.
24. Keller JR, Mantel C, Sing GK, Ellingsworth LR, Ruscetti SK, Ruscetti FW: J Exp Med 168:737, 1988.
25. Sing GK, Keller JR, Ellingsworth LR, Ruscetti FW: Blood In press, 1988.
26. Keller JR, Ellingsworth LR, McNiece IK, Quesenberry PJ, Sing GK, Ruscetti FW: Submitted, 1988.

Journal of Cellular Biochemistry 39:411–420 (1989)
Growth Factors and Their Receptors: Genetic
Control and Rational Application 163–172

Modulation of Mitogenic Activity and Cellular Binding of Basic Fibroblast Growth Factor by Basic Proteins

M.C. Dauchel, J. Courty, A. Mereau and D. Barritault

*Laboratoire de Biotechnologie des Cellules Eucaryotes, Université Paris XII
94010 Créteil, France*

Polycationic molecules were studied either for their ability to displace the binding of basic fibroblast growth factor (bFGF) to high- and low-affinity membrane interaction sites and/or to modulate bFGF-induced proliferation of fibroblasts. Heparin-binding polypeptides, such as polylysine, protamine, histones, and thrombin-displaced [^{125}I]bFGF bound to bovine brain membrane receptors. The most displacing polypeptides were those with the strongest affinity to heparin. Two of these polypeptides, protamine and polylysine, inhibited (at 5 μM) by more than 90% the mitogenic effect induced by bFGF on Chinese hamster lung fibroblast cells (CCL39). At the same dose, no effect was observed with basic proteins that do not bind to heparin, such as cytochrome C and lysozyme. An interesting observation was that protamine at 1 μM potentiated by 1.5-fold the mitogenic activity of bFGF, while it acted as an inhibitor at higher concentration.

Key words: **protamine, heparin-binding proteins, cell proliferation, radioreceptor assay, growth inhibition**

Protamine has been reported to inhibit mitogenic activity of several growth factors, such as acidic and basic fibroblast growth factors (a- and bFGF) [1,2] and platelet-derived growth factor (PDGF) [3], as well as to inhibit angiogenesis in vivo [4]. As protamine and these growth factors both have a strong affinity for heparin, they could interact with target cells with a heparin-like structure. Indeed, proteoglycans play a role in many different cellular activities such as cell adhesion and cellular growth, as well as regulation of receptor function [5].

A wide variety of cells are highly sensitive to the proliferation effect induced in vitro by factors belonging to the fibroblast growth factor family (a- and bFGF), eye-derived growth factor (EDGF), brain-derived growth factor (BDGF), and endothelial cell growth factor (ECGF) [6–10]. Moreover, these mitogenic polypeptides appear to play an important role in angiogenesis [11,12] and wound healing in vivo [13,14].

Biological activity of FGFs is mediated through specific high-affinity membrane receptors that have been characterized in vitro on cells grown in tissue culture, such as

Received February 26, 1988; accepted October 12, 1988.

fibroblasts [15], muscle myoblasts [16], endothelial and epithelial cells [17,18], as well as on plasma membrane preparation from bovine brain [19].

Several molecules have been reported to affect the biological activity of a-and/a bFGF by either inducing a potentiation such as heparin [10,20,21] or an inhibition such as gamma-interferon (gamma-IFN) [22], beta-transforming growth factor (beta-TGF) [23], interleukin-1 (IL-1) [24], tumor necrosis factor (TNF) [25], suramine, and protamine [2,21,26]. Protamine inhibition is of particular interest, since protamine has been reported to be an inhibitor of angiogenesis in vivo [4].

In an attempt to study the specificity of this inhibitory effect, we tested the modulation of bFGF activity induced by various heparin-binding and other basic proteins on the proliferation of fibroblasts (CCL39) and the interference of protamine on the binding of bFGF to high- and low-affinity membrane interaction sites.

MATERIALS AND METHODS

Materials

Chinese hamster lung fibroblasts (CCL39) were obtained from the American Type Culture Collection (ATCC). Protamine (P4005), polylysine (P4408), cytochrome C (C7752), lysozyme (L7001), thrombin (T6759), and DNA type III (D1626) were obtained from Sigma Corporation. Protamine was further purified by reverse-phase high-performance liquid chromatography (HPLC) using a C_4 Vydac column eluted at 17% acetonitrile-O.1% TFA. Chicken erythrocyte histones were prepared from purified nuclei by salt extraction, ultracentrifugation, and chromatography [27]. Histones were free from DNA, and an electrophoretic profile indicated that H2a, H2b, H3, and H4 were present but not H1. In order to simplify the presentation of our results, we used 20,000 daltons as the average molecular weight for the whole histone preparation. Heparinase (50,000 U/ml) was a generous gift from J. Choay, Institut Choay (France). Bovine brain bFGF was purified as previously described using heparin-Sepharose chromatography [28]. Purity of bFGF was checked on overloaded silver-stained SDS-PAGE, on reverse-phase HPLC, and with biological assay.

Cell Culture and Mitogen Assay

CCL39 cells were grown in 24-well plates (Falcon, Oxnard, CA) in Dulbecco's Modified Eagle's Medium (DMEM) containing 10% fetal calf serum (Boehringer, Mannheim, FRG) and 1 mM glutamine, penicillin (100 U/ml) and streptomycin (0.1 mg/ml) (GIBCO) at 37°C in a humidified atmosphere containing 5% CO_2.

Cells were plated at 2×10^4 cells per cm^2 in 0.5 ml of culture medium. When the cells reached confluency, they were arrested for 24 h in serum-free medium and then incubated with 0.5 ml serum-free DMEM containing increasing concentrations of growth factors and/or reagents. After incubation at 37°C for 20 h, the cells were pulsed with 1 μCi methyl [^{3}H] thymidine (Oris, France) per well for 4 h, and the cell monolayers were treated with 0.5 ml 10% trichloracetic acid for 15 min at 4°C, washed with tap water, and dissolved in 0.5 ml of 0.1 M NaOH for counting in a scintillation counter. Assays were performed in triplicate.

The stimulation unit was defined as the dose of bFGF that induced 50% of the maximal methyl [^{3}H] thymidine incorporation (ED_{50}) and obtained for CCL39 cells with 70 pM of bFGF; for other cells, such as endothelial cells, ED_{50} could be reached with 6 pM.

Radioiodination of bFGF

Briefly, 3 μg of bFGF were diluted in 0.1 M phosphate buffer (pH 7.4) containing 1% of polyethylene glycol (PEG) 1,000 and incubated with 1 mCi of Na[^{125}I] (Oris, France) and 20 μM of chloramide T at room temperature. After addition of 100 mM of N-acetyl-L-tyrosine, iodinated bFGF was purified by heparin-Sepharose chromatography. Biological activity was tested after iodination, and [^{125}I]bFGF was stored at −80°C. The specific activity of labeled bovine brain bFGF obtained was usually 80,000 cpm/ng.

Bovine Brain Membrane Preparation

Bovine brain membranes were prepared by a modification of the procedure described by Neville [29]. A batch of two freshly slaughtered bovine brains was dissected free of blood clots and meninges at 4°C and cut into small pieces. Homogenization was performed with 2 volumes of 20 mM Hepes (N-2-hydroxyethylpiperazine-N′-2-ethanesulfonic acid) (pH 7.4) containing 5 mM EDTA, 1 μg/ml leupeptine, 1 μg/ml pepstatin, 5 UK/ml aprotinin, and 0.1 mM phenylmethane sulfonyl fluoride (PMSF) (buffer A) using a blender.

All procedures were performed at 4°C. The homogenized tissue was centrifuged (14,300*g* for 60 min), and the pellet was suspended with 2 volumes of buffer A using a Teflon-glass homogenizer. After centrifugation at 1,000*g* for 30 min, 2 volumes of buffer A containing 0.3 M saccharose werc added to the pellet and centrifuged at 3,000*g* for 15 min. The resulting supernatant was centrifuged at 40,000*g* for 30 min. This pellet was resuspended with 10 volumes of buffer B (buffer A containing 1.2 mM $MgCl_2$, 1.2 mM $CaCl_2$, 0.1 M NaCl, and no EDTA) and washed by three more centrifugations (40,000*g* for 30 min). The final resulting pellets, containing crude membrane preparation, were taken up in buffer B to give a final protein concentration of about 10 mg/ml. Membrane preparations were stored frozen at −80°C until use.

Radioreceptor Assays

Radioreceptor assays were performed as described [19]. Briefly, bovine brain membrane preparations containing 40 μg of proteins were incubated in a total volume of 0.5 ml of buffer B with 23 nM of [^{125}I]bFGF supplemented or not with competitors as indicated and incubated for 1 h at 4°C. These suspensions were centrifuged at 5,000*g* for 5 min. The clear supernatant was aspirated, and the radioactivity of the pellets was counted.

Nonspecific binding reached 20–30% of the specific binding. All assays were carried out in duplicate. The specificity of the binding of bFGF to its receptor using the conditions described was demonstrated by crosslinking experiments [19].

Heparin-Sepharose Chromatography

Proteins were loaded onto a heparin-Sepharose column in PBS buffer (6.5 mM Na_2HPO_4, 1.46 mM KH_2PO_4, 3 mM KCl, 0.15 M NaCl, pH 7.4). After washing, retained proteins were eluted by a gradient of 0.15–3 M NaCl. Cytochrome C and lysozyme were eluted at 0.15 M, thrombin 0.30 M, histones 1.1 M, protamine 1.4 M, and polylysine 1.6 M NaCl.

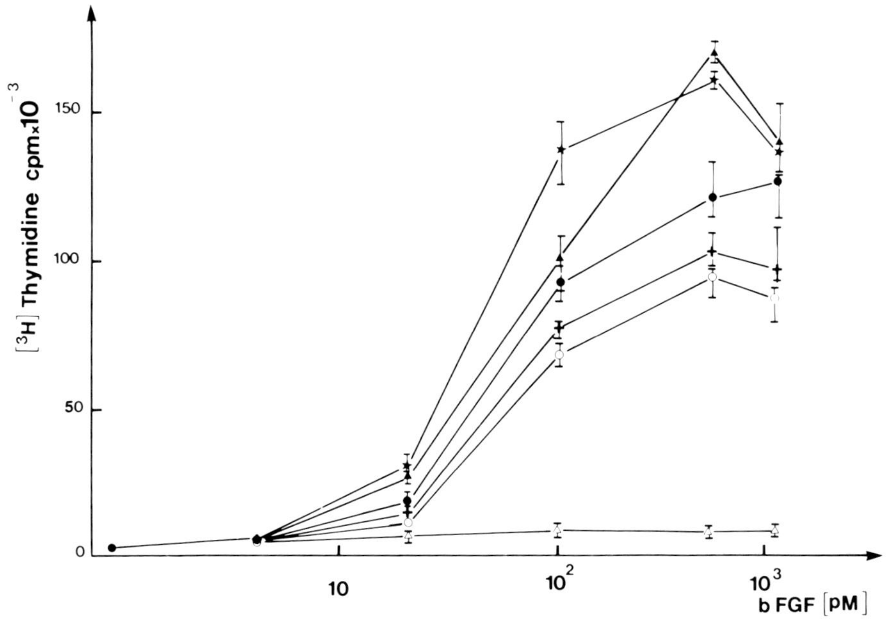

Fig. 1. Effect on biological activity of bFGF. Methyl [^{3}H]thymidine incorporation induced by bFGF at various doses in CC139 cells in the absence (+) and presence of protamine 0.5 μM (●), 1 μM (∗), 1.5 μM (▲), 2.5 μM (○), and 5 μM (△).

RESULTS

Dose Effect of Protamine on Mitogenic Activity of bFGF

As previously described for in vitro proliferation of endothelial and 3T3 cells [21], the stimulation of CCL39 cells by bFGF was modulated by the addition of protamine to the culture medium (Fig. 1).

At low doses of protamine (0.5, 1, and 1.5 μM), a potentiation effect on the incorporation of [^{3}H]thymidine was observed for different concentrations of bFGF. Hence, in the presence of 1–1.5 μM of protamine, the maximum [^{3}H]thymidine incorporation obtained with bFGF alone was increased ranging from 1.5- to 2-fold. At higher doses, an inhibitory effect was observed. Almost complete inhibition of the bFGF stimulation was already obtained with 5 μM.

In order to study these effects further, bFGF was used at a dose inducing half of the maximum [^{3}H]thymidine incorporation (ED_{50} = 70 pM) with protamine. As shown in Figure 2, maximum potentiation was also obtained with 1 μM of protamine. For this concentration, a twofold increase of [^{3}H]thymidine incorporation was observed, and the value reached corresponded to the maximum incorporation induced by bFGF alone at higher doses (=10 ED_{50}). Complete inhibition was also obtained at 5 μM.

Dose Effect of Other Basic Proteins on Mitogenic Activity of bFGF

In order to analyse the specificity of the opposite effects induced by protamine, other basic proteins were tested. As shown in Figure 2, cytochrome C, histones, and

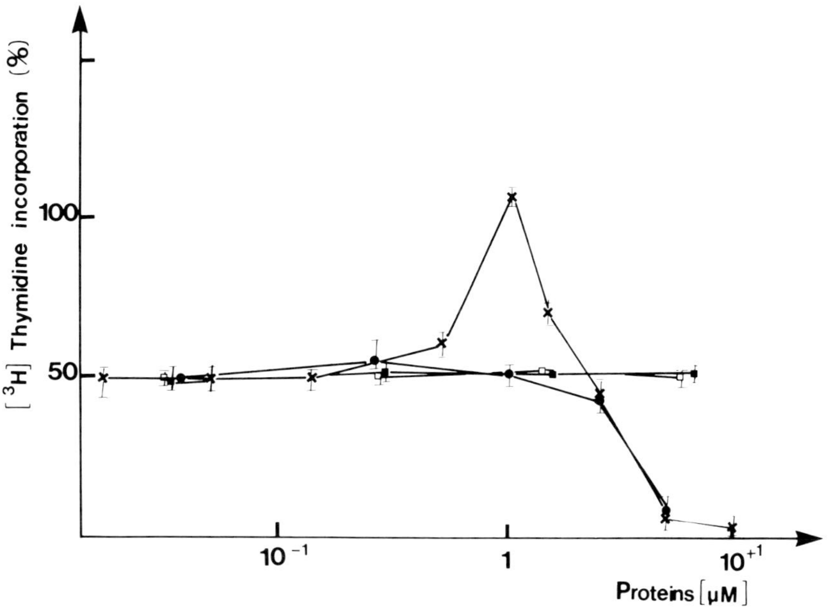

Fig. 2. Effect of various basic proteins of bFGF-stimulated cells. Methyl [^{3}H]thymidine incorporation stimulated by bFGF at a dose corresponding to ED_{50} (70 pM) in the presence of various concentrations of protamine (×), polylysine (●), cytochrome C (□), lysozyme (□), and histones (■).

lysozyme used in the same range of molarity as protamine did not affect the mitogenic activity of bFGF. In contrast, a synthetic polypeptide (polylysine) only acted as an inhibitor.

Reversibility of the Inhibitory Effect of Protamine by DNA

Protamine is a member of a family of proteins that binds strongly to DNA, giving soluble complexes [30]. When quiescent CCL39 cells were incubated with bFGF at doses corresponding to ED_{50} with 5 μM of protamine, increasing amounts of DNA added simultaneously to the culture medium reversed the inhibitory effect of protamine. This reversal was over 90% in the presence of 1 mg/ml of DNA, as shown in Figure 3. However, no effect was observed in the presence of DNA when CCL39 cells were stimulated by bFGF in the absence of protamine (results not shown).

The reversion of the inhibitory action of protamine by DNA suggested that protamine used at a dose of 5 μM was not cytotoxic for CCL39 cells, at least in these conditions.

Interference With [^{125}I]bFGF Receptor Binding by Protamine and Other Basic Proteins

The effect of protamine and other basic proteins on the binding of [^{125}I]bFGF to bovine brain membrane receptors was investigated. A constant amount of [^{125}I]bFGF 23 nM was incubated with bovine brain membranes (40 μg) in the presence of different basic proteins ranging from 10^{-4} to 10 μM. As shown in Figure 4, curve a, over 90% of displacement by unlabeled bFGF occurred at concentrations ranging from

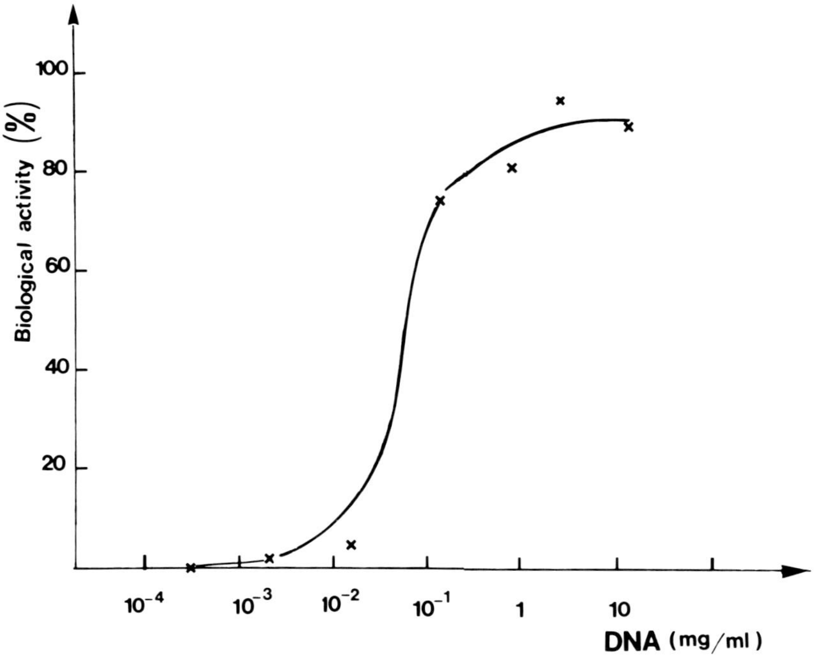

Fig. 3. Reversibility of protamine inhibition by DNA of bFGF-stimulated cells. Reversibility of the inhibitory effect of protamine at 5 μM on CC139 cells stimulated by bFGF at a dose corresponding to ED_{50} (70 pM) in the presence of various concentrations of DNA.

10^{-4} to 10^{-1} μM, suggesting that binding involved more than one type of site. Bound [^{125}I]bFGF was not displaced by cytochrome C or lysozyme used at concentrations up to 10 μM (Fig. 4).

However, protamine and polylysine displaced [^{125}I]bFGF with a half maximum displacement occurring respectively for 0.5 and 0.1 μM, while histones or thrombin induced a similar displacement at 1 μM.

To study if protamine or histones could interfere with the receptors and/or directly with bFGF, membrane preparations were preincubated with increasing concentrations of protamine or histones for 1 hr at 4°C and washed several times before adding [^{125}I]bFGF. Similar results were obtained whether preincubation was done or not, and a 90% displacement was achieved by adding 1 μM of protamine (not shown). These results indicated that protamine or histones, once they interacted with the membrane, could prevent the binding of [^{125}I]bFGF to its membrane receptor. They also suggested that protamine or histones did not interact with bFGF alone.

Heparinase Treatment of Membranes

In order to study the type of receptors interacting with [^{125}I]bFGF, a heparinase treatment was performed. [^{125}I]bFGF binding to receptors was measured after treatment with increasing concentrations of heparinase, as indicated in Figure 5 (insert). At 5 heparinase units, more than 75% of bound [^{125}I]bFGF was released from the membrane. For more reproductible results, the dose of 25 units was then chosen for

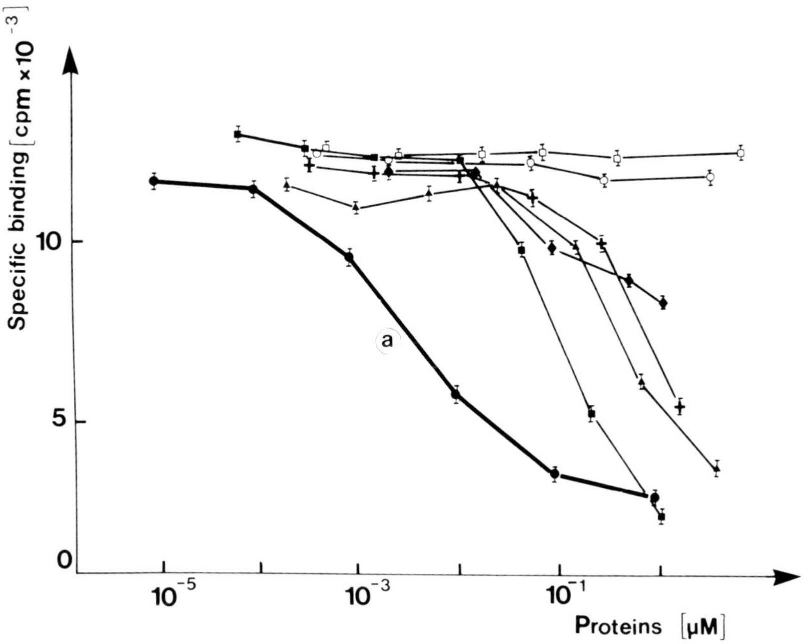

Fig. 4. Displacement of bFGF on bovine brain membranes by basic proteins. Competition binding curves of [^{125}I]bFGF (23 nM, 2,000 cpm/pmole) on bovine brain membranes, in the presence of increasing concentrations of bFGF (curve a) and basic proteins: (▲), protamine; (+), histones; (■), polylysine; (♦) thrombin; (○), cytochrome C; (□), lysozyme.

further experiments. Addition of increasing amounts of protamine displaced the remaining [^{125}I]bFGF still bound specifically to these membranes.

DISCUSSION

The data presented in this report suggest that the ability of proteins to bind to heparin could also reflect an ability to inhibit the mitogenic activity induced by bFGF. Protamine and polylysine strongly bound to heparin (elution from heparin Sepharose column at 1.4 M and 1.6 M, NaCl respectively) and were the most efficient inhibitors of [^{3}H]thymidine incorporation at 10 μM. However, histones that also bound to heparin (elution at 1.1 M NaCl) were not able to inhibit cell proliferation in the concentration range used (up to 100 μM). However, purified histone H2a at 75 μM has been described to completely inhibit bFGF-induced cell proliferation [1]. It is interesting to note that cytochrome C and lysozyme, which are basic proteins (with no affinity to heparin), did not inhibit the biological activity of bFGF, suggesting that the inhibitory effect may not be linked to ionic charges but to affinity to heparin or heparin-like structures.

Besides the presence of high-affinity specific receptors for bFGF, Scatchard analyses of binding data of radiolabeled bFGF on target cells revealed the presence of lower-affinity sites [31,32] whose interactions could be partially destroyed by hepa-

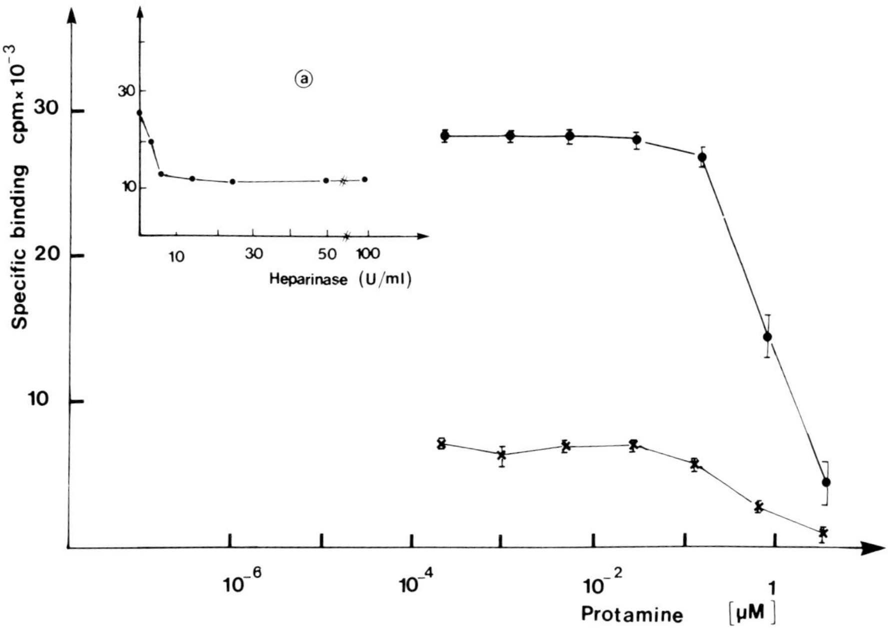

Fig. 5. Effect of heparinase treatment of bFGF binding on membranes in presence of proteins. Competition binding curves of [^{125}I]bFGF (0.023 μmole) on bovine brain membranes (●) or heparinase-treated bovine brain membranes 25 U (×) in the presence of various doses of protamine. Curve a: concentration effect of heparinase treatment on bovine brain membranes.

rinase treatment. This latter property suggests that this class of sites could be related to a heparin-like molecule whose biological significance remains unclear. Radioreceptor assays indicated that all the heparin-binding proteins tested could displace [^{125}I]bFGF bound to membrane receptors in a dose-dependent manner. The most efficient displacing polypeptides being those with the highest affinity to heparin, the results obtained suggest that heparin-binding proteins would compete for these low-affinity sites with bFGF. Indeed, heparinase treatment of bovine brain membranes reduced by 75% the total binding of bFGF.

In bovine brain membranes prepared as described here, our estimation of the ratio between the number of high- and low-affinity receptors is 1/500 [19]. One would therefore expect that heparinase treatment could reduce to a nondetectable level the total binding of bFGF if degradation by heparinase completely abolished low-affinity binding. The residual 25% of bFGF bound after heparinase treatments could simply indicate that heparinase does not degrade heparane sulfate molecules completely or that this degradation reveals other FGF binding sites. Since the residual bound bFGF could still be displaced by protamine, the first possibility seems more likely.

The most intriguing result is the potentiation of the bFGF-induced mitogenic activity in the presence of low concentrations of protamine (1 μM), as shown in Figures 1 and 2. When bFGF was used at ED_{50}, i.e., 70 pM for CCL39, 1 μM of protamine induced a 100% [^{3}H]thymidine incorporation (Fig. 2), as obtained with tenfold more bFGF alone. As already mentioned, this result cannot be explained. The potentiation of the mitogenic activity of aFGF by heparin has been previously described

[10]. It was explained by a direct interaction of aFGF with heparin [21], stabilizing aFGF and increasing its affinity to its cellular receptors.

However, the potentiation observed here with protamine and bFGF is unlikely to be due to a direct interaction of bFGF with protamine (Interference with [^{125}I]bFGF Receptor Binding by Protamine and Other Basic Proteins, in the Results section). Protamine has been shown to potentiate the biological activity of epidermal growth factor (EGF) [1]. Thus protamine might act by itself at cellular sites that are not associated with EGF or FGF receptors. When higher concentrations were used, this positive effect might be masked by its interactions with heparin-like molecules affecting directly or not FGF binding to specific cell surface receptors.

ACKNOWLEDGMENTS

The authors would like to acknowledge J. Badet (Institut National de la Santé et de la Recherche Médicale) for critical comments and help with the preparation of the manuscript, C. Gamby and A. Marez for their assistance, and E. Monti for secretarial help. This work was supported by the "Ministère de l'Education Nationale," the "Inserm" (contrat externe no. 872002), the "C.N.A.M." (contrat no. 86254E), the "Association de la Recherche sur le Cancer," the "Ligue National du Cancer," and the "Fondation de la Recherche Médicale." J. Courty is "chargé de recherche" from C.N.R.S. (Centre National de la Recherche Scientifique).

REFERENCES

1. Neufeld G, Gospodarowicz D: J Cell Physiol 132:287–294, 1987.
2. Huang JS, Huang SS, Kuo MD: J Biol Chem 261:11600–11607, 1986.
3. Huang JS, Nishimura J, Huang SS, Deuel TF: J Cell Biochem 26:205–220, 1984.
4. Taylor S, Folkman J: Nature 297:307–312, 1982.
5. Fransson LA: Trends Biochem Sci 12:406–411, 1987.
6. Gospodarowicz D, Neufeld G, Schweigerer L: Cell Differ 19:1–17, 1986.
7. Baird A, Esch F, Ling N, Gospodarowicz D: Regul Pept 12:201–213, 1985.
8. Gospodarowicz D, Neufeld G, Schweigerer L: Mol Cell Endocrinol 46:187–204, 1986.
9. Barritault D, Arruti C, Courtois Y: Differentiation 18:29–41, 1981.
10. Schreiber AB, Kenney J, Kowalski WJ, Thomas KA, Gimenez Gallego G, Rios-Candelore M, Di Salvo J, Barritault D, Courty J, Courtois Y, Moenner M, Loret C, Burgess WH, Mehlman T, Friesel R, Johnson W, Maciag T: J Cell Biol 101:1623–1626, 1985.
11. Thompson P, Arruti C, Maurice D, Plouet J, Barritault D, Courtois Y: In Clayton R (ed): "Problems of Normal and Genetically Abnormal Retinas." Orlando, FL: Academic Press, 1982, pp 63–78.
12. Folkman J, Klagsbrun M: Science 235:442–447, 1987.
13. Fourtanier AY, Courty J, Muller E, Courtois Y, Prunieras M, Barritault D: J Invest Dermatol 87:76–79, 1986.
14. Thompson P, Desbordes JM, Giraud J, Barritault D, Pouliquen Y, Courtois Y: Exp Eye Res 34:191–199, 1982.
15. Neufeld G, Gospodarowicz D: J Biol Chem 260:13860–13868, 1985.
16. Olwin BB, Hauschka SD: Biochemistry 25:3487–3492, 1986.
17. Friesel R, Burgess W, Mehlman T, Maciag T: J Biol Chem 261:7581–7584, 1986.
18. Moenner M, Chevallier B, Badet J, Barritault D: Proc Natl Acad Sci USA 83:5024–5028, 1986.
19. Courty J, Dauchel MC, Mereau A, Badet J, Barritault D: J Biol Chem 263:11217–11220, 1989.
20. Schreiber AB, Kenney J, Kowalski WJ, Friesel R, Mehlman T, Maciag T: Proc Natl Acad Sci USA 82:6138–6142, 1985.
21. Gospodarowicz D, Cheng J: J Cell Physiol 128:475–482, 1986.
22. Friesel R, Komoriya A, Maciag T: J Cell Biol 104:689–696, 1987.

23. Baird A, Durkin T: Biochem Biophys Res Commun 138:475–482, 1986.
24. Norioka K, Hara M, Kitani A, Hirose T, Hirose W, Hakigai M, Suzuki K, Kawakami M, Tabata H, Kawagoe M, Nakamura H: Biochem Biophys Res Commun 145:969–975, 1987.
25. Schweigerer L, Malerstein B, Gospodarowicz D: Biochem Biophys Res Commun 143:997–1004, 1987.
26. Coffey R, Leof E, Shipley G, Moses H: J Cell Physiol 132:143–148, 1987.
27. Van der Broek HW, Nooden LD, Sevall JS, Bonner J: Biochemistry 12:949–969, 1973.
28. Courty J, Chevallier B, Moenner M, Lagente O, Courtois Y, Barritault D: Biochem Biophys Res Commun 136:102–108, 1986.
29. Neville O: Biochim Biophys Acta 154:540–552, 1968.
30. Li HJ, Chang C, Weiskopf M: Biochemistry 12:1763–1771, 1973.
31. Moscatelli D: J Cell Physiol 131:123–130, 1987.
32. Moenner M, Badet J, Chevallier B, Tardieu M, Courty J, Barritault D (eds): "Current Communication in Molecular Biology: Angiogenesis." New York: Cold Spring Harbor Press, 1987, pp 52–57.

Journal of Cellular Biochemistry 39:421–428 (1989)
Growth Factors and Their Receptors: Genetic Control and Rational Application 173–180

Growth Factor Modulation of Melanoma Growth Stimulatory Activity mRNA Expression in Human Malignant Melanoma Cells Correlates With Cell Growth

R. Bordoni, G. Thomas, and A. Richmond

V.A. Medical Center and Department of Medicine, Emory University School of Medicine, Atlanta, Georgia 30033

This report demonstrates that the expression of melanoma growth stimulatory activity (MGSA) mRNA can be modulated in a positive fashion in the Hs294T human melanoma cell line by PDGF and MGSA. There is close correlation between MGSA expression and the pattern of cell growth in Hs294T cells.

Key words: MGSA/KC/gro, melanoma cells, expression modulation, cell proliferation

Cellular proliferation is controlled by external factors, among which the most important are hormones and growth factors (GFs) [1–3]. One of the most potent serum derived GFs identified is platelet-derived growth factor (PDGF) [4]. A variety of mesenchymal cells require PDGF for optimum growth in vitro [5, for review]. However, under growth restrictive conditions, PDGF by itself is not an efficient mitogen. A second set of GFs contained in the platelet-poor plasma fraction of the serum described as "progression factors" function synergistically with PDGF to promote the optimum mitogenic response [6,7].

PDGF treatment induces the mRNA of genes corresponding to the "PDGF competence family," including c-myc, c-fos, JE, JB, and KC [8,9]. Tumorigenic cells commonly have lower requirements for GFs than do the untransformed cells from which they are derived [10,11]. One cause of this decreased GF requirement is the production of "tumor-derived growth factors" by some transformed cells [12].

Melanoma growth stimulatory activity (MGSA) is an endogenous GF that is produced by human melanoma tumors and certain cultured nevus cells [13,14]. MGSA has been well characterized and purified from serum-free (SF) conditioned medium (CM) from the human melanoma cell line Hs294T [14–16]. Amino terminus sequence data have been obtained, MGSA cDNA clones have been isolated, and the nucleotide sequence of MGSA has been determined, demonstrating that the oligonucleotide sequence of MGSA is identical with the human gro gene [17] and is closely

Received April 1, 1988; accepted October 31, 1988.

related to, if not identical with, the human KC gene (R. Derynck, personal communication). A cDNA probe for MGSA has been developed that can be used to evaluate MGSA mRNA expression in a variety of cell types [17]. In this report we describe the effects of MGSA and PDGF on expression of MGSA mRNA and cell growth in malignant melanoma cells.

MATERIALS AND METHODS

Cell Culture

The melanoma cell line Hs294T was cultured as previously described [14].

Source of Growth Factors and Other Materials

MGSA was obtained from the CM of Hs294T cells as described before [18]. PDGF was purchased from AMGen Biologicals, Thousand Oaks, California, as human, over 95% pure PDGF.

Experimental Design

Hs294T cells were transferred from T-150 stock flasks to 100 × 20-mm plastic petri dishes and kept on F-12 medium containing 10% fetal bovine serum (FBS) until they reached 60–70% confluency. The Hs294T cells were then partially synchronized by culturing on SF media for 24 hours. MGSA (0.3 ng/ml) was added to the SF culture media for different time periods, from 15 minutes to 24 hours. For other experiments, PDGF (2 ng/ml in SF), alone or in combination with MGSA (0.3 ng/ml) was added to the cells for 1 or 8 hours prior to extraction of RNA.

RNA Extraction

A 4 M solution of guanidine isothiocyanate was added directly to the monolayer cell culture or the cell pellet when a cell number assay was done. Total RNA was isolated by centrifugation through a cushion of 5.7 M cesium chloride (CsCl). Reextraction of RNA with phenol/chloroform was then followed by 3 M sodium acetate (NaAC) and ethanol precipitation [19].

Northern Blot Analysis

Twenty micrograms of total RNA was denatured at 65°C in the presence of formamide and formaldehyde and then subjected to electrophoresis on a 1.2% agarose gel in the presence of formaldehyde [32] and thereafter blotted to nitrocellulose filter. Hybridizations were performed using 50% formamide-containing solution, at 42°C, overnight. The 0.73-K/bp EcoRI fragment of the MGSA cDNA insert isolated from a human term placenta cDNA library was used as a hybridization probe [17], radiolabeled with [^{32}P] using a multiple DNA labeling system. Posthybridization washes were done at 42°C in 0.2 × SSC and 1% sodium dodecyl sulfate (SDS) to remove the probe in excess. Changes in the expression of MGSA mRNA were calculated by scintilligraphic analysis of the nitrocellulose membranes after hybridization with the [^{32}P]-labeled MGSA cDNA probe. The integrity of total-RNA was analysed by ethidium bromide stain of the samples. The position of the MGSA mRNA on the gel was determined by comparison with the position of the 28S and 18S bands of ribosomal RNA (r-RNA) in the stained gel.

Hs294T Cell Number Assay

Hs294T cells were transfered from T150 stock flasks to 100 × 20-mm plastic petri dishes (5 × 10^5 cells/dish) and incubated with F12 culture medium containing 10% FBS, to 60–70% confluency. The cells were then partially synchronized by culturing on SF medium for 24 hours. The corresponding additions of GFs were done at the concentrations mentioned before, and the cells were incubated for an additional 72 hours. On the 3rd day after GF additions, cell number was determined using a hemocytometer. The experiments were done by duplicate, and the results are expressed as absolute number of cells in millions (1 × 10^6).

RESULTS

Modulation of MGSA Expression by PDGF and/or MGSA

One hour after treatment of the Hs294T cells with 2 ng of purified PDGF, or 0.3 ng of MGSA per ml of SF medium, the expression of MGSA was not significantly enhanced when compared with the cells arrested on SF. However, MGSA mRNA expression was twofold enhanced by 8 hours treatment with either MGSA or PDGF. The use of PDGF combined with MGSA for 8 hours showed a synergistic enhancement in MGSA mRNA level (Fig. 1A).

Effects of PDGF and MGSA on the Growth of Hs294T Melanoma Cells

The effects of exogenously added PDGF (2 ng/ml), MGSA (0.3 ng/ml), or PDGF (2 ng/ml) plus MGSA (0.3 ng/ml) on the growth of Hs294T melanoma cells were evaluated in a 72-hour cell number assay. Addition of PDGF produced a 90% (±15) (1.4 × 10^6 cells) stimulation in cell proliferation when compared with Hs294T cultures mantained on SF media (0.658 × 10^6 cells). The addition of MGSA produced a 95% (±12) (1.44 × 10^6 cells) stimulation in cell proliferation with respect to the control culture (0.658 × 10^6 cells). The presence of both GFs simultaneously in the cell media produced a 128% (±18) (1.682 × 10^6 cells) stimulation in cell proliferation. (Fig. 2).

Modulation of MGSA Expression by MGSA

When the Hs294T cells were treated with 0.3 ng/ml of exogenous MGSA for different periods of time, from 15 minutes to 24 hours, the expression of MGSA mRNA was enhanced after 15–30 minutes, reaching a peak at about 2 hours after continuous exposure. MGSA mRNA expression began to decline after 4 hours of MGSA exposure, reaching undetectable levels after 24 hours (Fig. 3).

DISCUSSION

Polypeptide GFs regulate cell growth by enabling the cells to progress through the G1-phase of the cell cycle to reach the S-phase. To bring these events about, GFs must bind to specific receptors on the cell surface and induce a series of signals, which are relayed to the nucleus, where ultimately nuclear genes are induced and DNA replication ensues [22,23]. PDGF is a powerful mitogen released by platelets when blood clots [4,20]. PDGF induces in mesenchymal cells a stable state, termed "competence," by causing the activation or synthesis of cytoplasmic factors, which can be blocked by inhibitors of RNA synthesis [8]. PDGF by itself does not promote cell

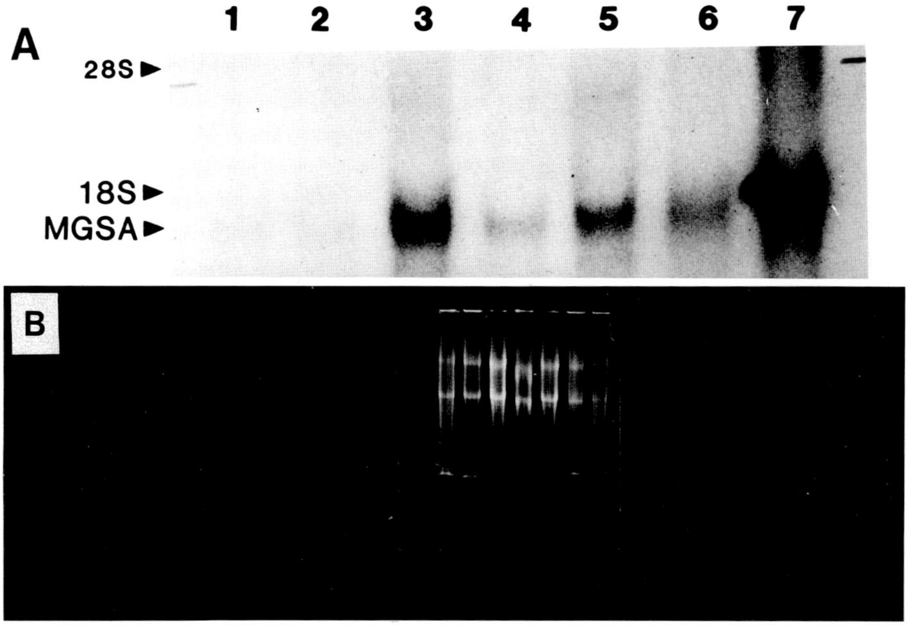

Fig. 1. MGSA mRNA detection by Northern hybridization of Hs294T melanoma cells treated with MGSA and/or PDGF. **A:** Cellular RNA was extracted by CsCl centrifugation and ethanol precipitation. Twenty micrograms of total RNA per sample was electrophoresed on a 1.2% agarose/formaldehyde gel. Levels of MGSA were determined by Northern blot analysis. The 0.73-kbp EcoRI fragment of the MGSA cDNA library was used as a hybridization probe. Hybridization was performed under high stringency conditions as described in Methods. The washed, hybridized membrane was exposed to a XAR-5 Kodak film for 48 hours to obtain the autoradiographic scan. MGSA mRNA levels in cells treated with serum-free medium (**lane 1**), MGSA (0.3 ng/ml) (**lanes 2,3**), PDGF (2 ng/ml) (**lanes 4,5**), or a combination of both (**lanes 6,7**), for 1 and 8 hours, respectively. The position of the 28S and 18S bands from the ethidium bromide stain of a duplicate gel (**B**) is indicated by the arrows. B: Ethidium bromide stain of total-RNA electrophoresed on agarose gel. A duplicate of the gel shown in A was electrophoresed and stained with ethidium bromide. The integrity of the RNA is shown by the 28S and 18S bands indicated by the arrows.

cycling, and the GFs present in platelet-poor plasma are required for cells to progress into S-phase [6,7]. Insulin-like GFs and EGF are two "progression" factors required for mouse fibroblasts to enter S-phase [24], after PDGF treatment. Several genes have been found to be responsive to PDGF stimulation in BALB/c-3T3 cells [8].

Of great interest here is the regulation of the MGSA/KC/gro gene, which is induced by PDGF and serum in fibroblasts [8], by EGF in endothelial cells and keratinocytes [25], by lectins in peripheral blood mononuclear cells [26], and by lipopolysaccharides (LPS) and colony-stimulating factor-1 (CSF-1) in macrophages [27,28]. The role of PDGF in the proliferation of melanoma cells is not clearly established [21]. Previous studies of [^{3}H]-thymidine incorporation into DNA of low-density Hs294T cultures have shown that PDGF lacked growth stimulatory activity for Hs294T cells under the conditions of the bioassay developed for MGSA [14]. However, the experiments reported here show that PDGF will promote increase in cell number in Hs294T cells in a 72-hour cell number assay. These results suggest not only

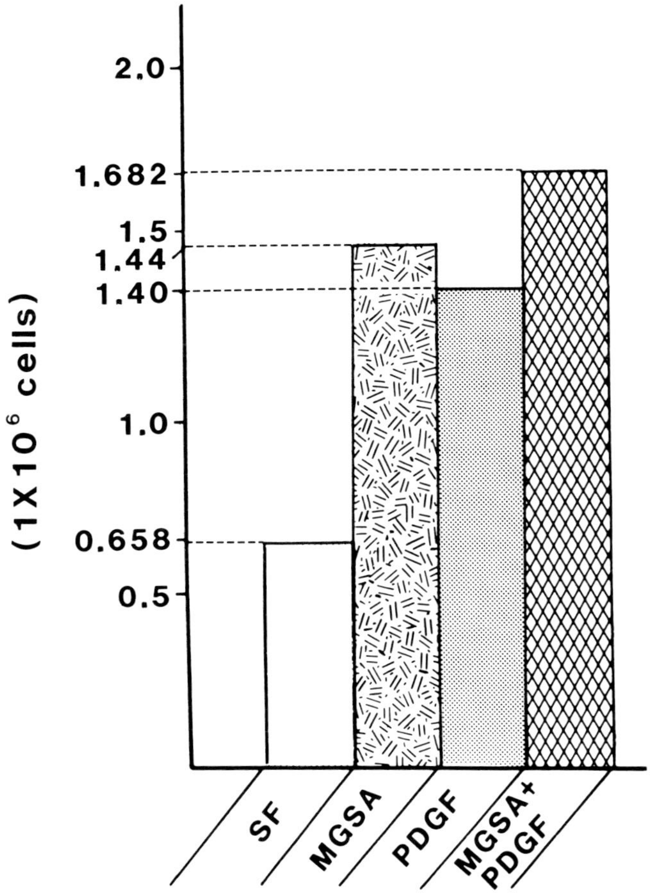

Fig. 2. Effects of MGSA, or PDGF, or both in combination on HS294T cell number: 5×10^5 Hs294T cells were plated in 100 × 20-mm plastic petri dishes and cultured to 60–70% cell density. After the cell cultures were partially synchronized by culturing on serum-free medium for 24 hours, Hs294T cells were treated with MGSA (0.3 ng/ml) alone or in combination with PDGF (2 ng/ml), in serum-free medium. After a 3-day incubation the medium was aspirated, the cells were released with trypsin, and cell counts were determined in a hemocytometer. Data are expressed as number of cells (1×10^6).

that there are functional PDGF receptors on Hs294T cells, but also that other endogenous GFs produced by the cells may be present in 72-hour Hs294T culture medium, which works in concert with PDGF to increase cell number. It appears that only suboptimal levels of these complementary endogenous growth factors are present in the 24-hour [^{3}H]-thymidine bioassay. In contrast, the increased incubation period and the greater cell density utilized in the cell number assay may provide the necessary levels of these complementary factors required for bioresponse to PDGF in Hs294T cells. IGF-I and MGSA are possible candidates for the other endogenous growth factors that are produced by Hs294T cells and work in concert with PDGF to promote Hs294T cell growth in SF culture medium during the 72-hour assay.

Other groups have reported that melanoma cells produce PDGF A-chain and to a lesser extent PDGF B-chain, but PDGF binding to PDGF receptors could not be

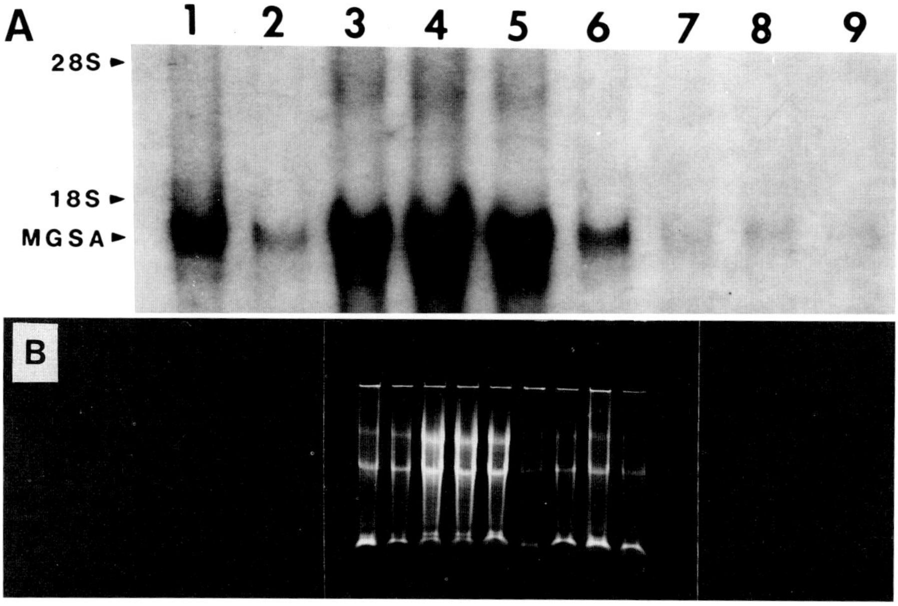

Fig. 3. MGSA mRNA detection by Northern hybridization of Hs294T melanoma cells treated with MGSA for different periods of time. **A:** Cellular RNA was extracted and processed as described in Figure 1A, and in "Materials and Methods." MGSA mRNA levels in subconfluent cell cultures kept on serum-free medium (**lane 1**), or treated with MGSA (0.3 ng/ml) for 15 or 30 minutes (**lanes 2 and 3**), or for 2, 4, 6, 8, 12, or 24 hours (**lanes 4–9,** respectively). The position of the 28S and 18S bands from the ethidium bromide stain of a duplicate gel (**B**) is indicated by the arrows. B: Ethidium bromide stain of total-RNA electrophoresed on agarose gel. A duplicate of the gel shown in A was electrophoresed and stained with ethidium bromide. The integrity of the RNA is shown by the 28S and 18S bands.

demonstrated in the melanoma cultures tested [31]. The Hs294T human melanoma cultures do contain a mRNA species, which hybridizes with the PDGF-R cDNA probe (G. Harsh and L.T. Williams, personal communication). Furthermore, antibodies that recognize the activated (tyrosine phosphorylated) form of the PDGF-R hybridize with a ~180-kD protein from serum-stimulated Hs294T cell extracts in Western blot analysis (G. Harsh, E. Balentien, and Lewis T. Williams, personal communication).

The significant stimulation of MGSA mRNA expression when the Hs294T cells are treated with PDGF suggests that one mode for PDGF induction of melanoma growth is through induction of MGSA mRNA, which is subsequently translated and secreted; MGSA would then bind to its receptor and exert a growth stimulatory effect on melanoma cells.

The pattern of MGSA message expression in the malignant melanoma cell line Hs294T is not significantly different from the profile observed in fibroblasts, where the early (1 hour) KC mRNA peak is followed by a rapid decline in the mRNA level at 4 hours [8].

Both PDGF, and TGF-alpha, have the ability to induce their own mRNA expression [29,30]. When Hs294T human melanoma cells were treated with exogenous MGSA, a significant increase of MGSA mRNA level was observed with a significant induction at 2 hours, which dropped to basal levels after 6 hours of continuous exposure. The results imply the existence of an MGSA autoinduction, with MGSA being capable of amplifying the mitogenic response in melanoma cells. In the experiments reported here, there was variation in the timing of MGSA mRNA autoinduction. This variation is probably due to the inherent difficulty in synchronization of tumor cells at a specific point in the cell cycle. The enhanced expression of MGSA mRNA when the Hs294T cells were treated with MGSA correlated with a twofold MGSA stimulation of cell proliferation. In a previous report [17] we introduced cycloheximide (CHX) in the Hs294T cell culture to assess the mechanism of MGSA mRNA autostimulation. These experiments showed a superinduction in the message in response to MGSA when the cell culture was pretreated with CHX. These findings make it unlikely that de novo synthesis of proteins is involved in MGSA message autostimulation.

In this report we have shown that MGSA mRNA expression is induced in Hs294T melanoma cells by the addition of either exogenous MGSA or PDGF. This autoinduction is time dependent and correlates with cell growth. Further efforts will be directed toward the elucidation of the significance of MGSA mRNA in association with the complex molecular network leading to cell proliferation.

ACKNOWLEDGMENTS

These studies were supported by NIH grant CA34590, a Veterans Administration Merit Award to Ann Richmond and NIH Grant DK07298-08. The excellent technical assistance of Michael Kelleher and Frances Kennedy is greatly appreciated. We are indebted to Eddy Balentien for his help in preparation of the MGSA cDNA probe.

REFERENCES

1. Temin HM: In Defendi V, Stoker M (eds): "Growth Regulating Substances for Animal Cells in Culture. Philadelphia: Wistar Institute Press, Symp Monogr, Vol 7, 1967, p 103.
2. Paul D, Lipton A, Klinger I: Proc Natl Acad Sci USA 68:634, 1971.
3. Holley RW, Kiernan JA: In Wolstenholme GEW, Knight J (eds): "Growth Control in Cell Cultures." London: Churchill and Livingstone, Ciba Foundation Symposium, 1971, pp. 3–10.
4. Antoniades HN, Scher CD, Stiles CD: Proc Natl Acad Sci USA 76:1809, 1979.
5. Ross R, Raines EW, Bowen-Pope DF: Cell 46:155, 1986.
6. Pledger WJ, Stiles CD, Antoniades HN, Scher CD: Proc Natl Acad Sci USA 74:4481, 1977; Symposium 116:87–97, 1985.
7. Stiles CD, Pledger WJ, Tucker RW, Martin RG, Scher CD: J Supra Mol Struct 13:489, 1980.
8. Cochran BH, Reffel A, Stiles CD: Cell 32:939, 1983.
9. Rollins BJ, Morrison ED, Stiles CD: Science 238:1269, 1987.
10. Dulbecco R: Nature 227:802, 1970.
11. Holley RW, Kiernan JA: In Wolstenholme GEW, Knight J (eds): "Ciba Foundation Symposium on Growth Control in Cell Cultures." London: Churchill and Livingstone, 1971, p 3.
12. Todaro GJ, De Larco JE, Marquardt H, Bryant ML, Sherwin SA, Sliski AH: Cold Spring Harbor Conf Cell Prolif 6:113, 1979.
13. Richmond A, Lawson D, Nixon D: Cold Spring Harbor Conf Cell Prolif 9:885, 1982.
14. Richmond A, Lawson DH, Nixon DW, Chawla RK: Cancer Res 43:2106, 1983.
15. Richmond A, Lawson DH, Nixon DW, Chawla RK: Cancer Res 45:6390, 1985.

16. Richmond A, Thomas HG: J Cell Physiol 129:375, 1986.
17. Richmond A, Thomas HG, Flaggs G, Barton DE, Spiess, J, Balentien E, Bordoni R, Francke U, Derynck R: EMBO J 7:2025, 1988.
18. Thomas HG, Richmond A: Mol Cell Endocrinol 57:69–76, 1988.
19. Chirgwin J, Przybyla AE, MacDonald R, Rutter WJ: Biochemistry 18:5294, 1979.
20. Stiles CD: Cell 33:653, 1983.
21. Richmond A, Thomas HG: J Cell Biochem 36:185, 1988.
22. Gustin MS, Leof EB, Shipley GD, Moses HL: Cancer Res 46:1015, 1986.
23. Weinberg R: Science 230:770, 1985.
24. Leof EB, Wharton W, Van Wyk JJ, Pledger WJ: Exp Cell Res 141:107, 1982.
25. Takehara K, Le Roy EC, Grotendorst GR: Cell 49:415, 1987.
26. Gibson CW, Rittling SR, Hirschorn RR, Kaczmarek L, Calabretta B, Stiles CD, Baserga R: Mol Cell Biochem 71:61, 1986.
27. Introna M, Bast R, Tannenbaum C, Hamilton T, Adams D: J Immunol 138:3891, 1987.
28. Orlofsky A, Stanley ER: EMBO J 6:2947, 1987.
29. Paulsson Y, Hammacher A, Heldin CH, Westermark B: Nature 328:715, 1987.
30. Coffey RJ, Derynck R, Wilcox JN, Bringman TS, Goustin AS, Moses HL, Pittelkow MR: Nature 328:817, 1987.
31. Tominaga S, Lengyel P: J Biol Chem 260:1975, 1985.
32. Maniatis T, Fritsch EF, Sambrook J: Molecular Cloning: A Laboratory Manual. Cold Spring Harbor, New York: Cold Spring Harbor Laboratory, 1982.

Journal of Cellular Biochemistry 39:429–441 (1989)
Growth Factors and Their Receptors: Genetic Control and Rational Application 181–193

Cascade of Autophosphorylation in the β-Subunit of the Insulin Receptor

Morris F. White and C. Ronald Kahn

Research Division, Joslin Diabetes Center, Department of Medicine, Brigham and Women's Hospital and Harvard Medical School, Boston, Massachusetts 02215

Insulin stimulated autophosphorylation of the β-subunit of the insulin receptor purified from Fao hepatoma cells or purified from Chinese hamster ovary (CHO/HIRC) or Swiss 3T3 (3T3/HIRC) cells transfected with the wild-type human insulin receptor cDNA. Autophosphorylation of the purified receptor occurred in at least two regions of the β-subunit: the regulatory region containing Tyr-1146, Tyr-1150, and Tyr-1151, and the C-terminus containing Tyr-1316 and Tyr-1322. In the presence of antiphosphotyrosine antibody (α-PY), autophosphorylation of the purified receptor was inhibited nearly 80% during insulin stimulation. Tryptic peptide mapping showed that α-PY inhibited autophosphorylation of both tyrosyl residues in the C-terminus and one tyrosyl residue in the regulatory region, either Tyr-1150 or Tyr-1151. Thus, a bis-phosphorylated form of the regulatory region accumulated in the presence of α-PY, which contained Tyr(P)-1146 and either Tyr(P)-1150 or 1151. In intact Fao, CHO/HIRC, and 3T3/HIRC cells, insulin stimulated tyrosyl phosphorylation of the β-subunit of the insulin receptor. Tryptic peptide mapping indicated that the regulatory region of the β-subunit was mainly (>80%) bis-phosphorylated; however, all three tyrosyl residues of the regulatory region were phosphorylated in about 20% of the receptors. As the phosphotransferase was activated by tris-phosphorylation but not bis-phosphorylation of the regulatory region of the β-subunit (White et al.:*Journal of Biological Chemistry* 263:2969–2980, 1988), the extent of autophosphorylation in the regulatory region may play an important regulatory role during signal transmission in the intact cell.

Key words: transmembrane signal, protein phosphorylation, tyrosine kinase, signal transmission, phosphorylation cascade

The initial molecular events following insulin binding involve autophosphorylation of several tyrosyl residues in the β-subunit of the insulin receptor [1–4]. The first phosphorylation sites include Tyr-1146,[1] Tyr-1150, and Tyr-1151 [2–4]. In addition, two tyrosyl residues in the C-terminus, Tyr-1316 and Tyr-1322, undergo autophosphorylation [2,6]. The phosphorylation of three tyrosyl residues in the juxtamembrane

[1]The numbering sequence used in this paper was described by Ullrich et al. [5] and is based on the position of the amino acids in the precursor of the human insulin receptor.

Received May 18, 1988; accepted October 28, 1988.

region of the β-subunit, including Tyr-953, Tyr-960, and Tyr-967, is uncertain [2,4,7]. Autophosphorylation activates a phosphotransferase in the β-subunit, which catalyzes the phosphorylation of tyrosyl residues in various peptides and proteins [3,8]. Autophosphorylation of Tyr-1146, Tyr-1150, and Tyr-1151 appears to activate the phosphotransferase of the β-subunit [3], whereas phosphorylation in the C-terminus is not required for phosphotransferase activity [6]. Tyr-960 in the juxtamembrane region may be important for signal transmission, although it is probably not phosphorylated in the intact cell [4,7].

Autophosphorylation carried out at 0°C suggests that Tyr-1146 is the initial site of autophosphorylation in the β-subunit [4]. The second step appears to be the phosphorylation of Tyr-1150 or Tyr-1151. This bis-phosphorylated region can be trapped and immunopurified with antiphosphotyrosine antibody; bis-phosphorylation does not appear to activate the tyrosine kinase [3]. In contrast, tris-phosphorylation of this region may be necessary to activate the phosphotransferase fully [3]. Thus, we provisionally call the region of the insulin receptor including tyrosyl residues 1146, 1150, and 1151 the regulatory region of the β-subunit. Since tris-phosphorylation is barely detected during insulin stimulation of Fao hepatoma cells, the state of phosphorylation of Tyr-1146, Tyr-1150, and Tyr-1151 may be an important site of kinase regulation in vivo [3]. In this report, we extend these observations to the human insulin receptor. We confirm that antiphosphotyrosine antibody inhibits the autophosphorylation cascade of the purified human insulin receptor, and we show that the bis-phosphorylated regulatory region of the β-subunit predominates during insulin stimulation of Chinese hamster ovary cells and Swiss 3T3 cells expressing the human insulin receptor.

EXPERIMENTAL PROCEDURES

Cell Culture and Expression of the Human Insulin Receptor

Experiments were performed with Fao rat hepatoma cells [9] or with Chinese hamster ovary cells (CHO/HIRC) [10] or NIH/3T3 cells (3T3/HIRC) transfected with wild-type human insulin receptor cDNA [11]. The Fao cells are a well-differentiated and insulin-sensitive rat hepatoma cell line that possesses a high concentration of insulin receptors and many insulin-stimulated responses [9]. The 3T3/HIRC cells were a gift from J. Whittaker, Howard Hughes Medical Institute at The University of Chicago. These cells, grown in Dulbecco's minimum essential medium (MEM) containing 10% FBS were obtained by cotransfection of NIH/3T3 cells with pSVEneo and a bovine papilloma virus/insulin receptor cDNA construct [11]. CHO/HIRC cells, grown in F12 medium containing 10% FBS and 400 μg/ml of G418, were obtained by cotransfection of CHO cells with pSVEneo and an SV40/insulin receptor construct [10]. All cell lines were maintained at 37°C in a humidified atmosphere composed of 95% air and 5% CO_2.

Autophosphorylation of the Partially Purified Insulin Receptor

The insulin receptor was partially purified from confluent cells by wheat germ agglutinin affinity chromatography [12]. Ten 15-cm dishes of cells were each solubilized at 22°C with 3 ml of 50 mM HEPES (pH 7.4) containing 1% Triton X-100, 0.1 mg/ml aprotinin, and 2 mM phenylmethyl sulfonylfluoride (PMSF). Following centrifugation to remove the insoluble material, the cell extract was passed over a 2 ml

WGA[2]-agarose column, and the insulin receptor was eluted with 1–2 ml of 0.3 M N-acetylglucosamine in 50 mM HEPES (pH 7.4), 0.1% Triton X-100 as previously described [3].

WGA-purified insulin receptor (0.24 μg/μl) was incubated at 22°C for 10 min in a solution (100 μl) containing 50 mM HEPES and 5 mM $MnCl_2$ in the absence or presence of 100 nM insulin [13]. Phosphorylation was initiated by adding 25 μM [γ-^{32}P]ATP (2.5 mCi/ml) and was continued for the time intervals indicated in the figure legends. In certain experiments, 3 μg of α-PY was added to the reaction mixture before the addition of the [γ-^{32}P]ATP. The phosphorylation reaction was terminated by adding a 0.5-ml portion of 50 mM HEPES containing 0.1% Triton X-100, 100 mM sodium fluoride, 10 mM sodium pyrophosphate, 5 mM EDTA, and 2 mM sodium vanadate (Aldrich) at 4°C. The phosphorylated insulin receptor was immunoprecipitated from each reaction with α-PY, separated by SDS-PAGE using 7.5% resolving gels and identified by autoradiography [10,12].

[^{32}P]Phosphorylation of the Insulin Receptor in the Intact Cells

Confluent cells in 15-cm dishes were incubated at 37°C for 2 h with 5 ml of phosphate-free and serum-free RPMI 1640 medium (GIBCO) containing carrier-free [^{32}P]orthophosphate (0.5 mCi per ml) [10]. Insulin was added, and the incubation was continued at 37°C for 5 min and then stopped quickly by removing the incubation medium and freezing the cell monolayers with liquid nitrogen. The monolayers were thawed and homogenized immediately at 4°C with 2 ml of a solution containing 50 mM HEPES (pH 7.4), 1.0% Triton X-100, 10 mM sodium pyrophosphate, 100 mM sodium fluoride, 4 mM EDTA, 2 mM sodium vanadate, 1 mg/ml aprotinin, and 2 mM phenylmethylsulfonyl fluoride [14]. The phosphotyrosine-containing insulin receptor was immunoprecipitated from the eluate with α-PY and analyzed as described above.

HPLC Separation of Tryptic Phosphopeptides and Identification of the Phosphorylated Tyrosine Residues in the β-Subunit

The β-subunit, separated by SDS-PAGE, was digested with trypsin as previously described [3,13]. Tryptic phosphopeptides were separated with a Waters HPLC system equipped with a wide-pore C_{18} reverse-phase column (Bio-Rad, RP-318). The phosphopeptides were applied to the column, which was washed for 5 min at a flow rate of 1.1 ml/min with 0.05% trifluoroacetic acid, 95% water, and 5% acetonitrile. The peptides were eluted by a gradient of acetonitrile increasing linearly from 5% to 25% during 85 min. Greater than 90% of the radioactivity in the gel fragment was routinely recovered from the reverse-phase HPLC column.

The identity of the tryptic phosphopeptides was determined by immunoprecipitation with antiphosphotyrosine antibodies and antibodies that reacted with specific regions of the β-subunit: αPep-1 (residues 1314–1324), αPep-3 (residues 1143–1152), and αPep-4 (residues 952–962) [6]. The fractions containing the peptides were dried

[2]Abbreviations used: DTT, dithiothreitol; FBS, fetal bovine serum; HEPES, 4-(2-hydroxyethyl)-1-piperazine-ethanesulfonic acid; HPLC, high-performance liquid chromatography; PAGE, polyacrylamide gel electorphoresis; α-PY, antiphosphotyrosine antibody; SDS, sodium dodecyl sulfate; WGA, wheat germ agglutinin.

in vacuo and the residue was dissolved in 0.5 ml of 50 mM HEPES containing 0.1% Triton X-100. Antibody (1 μg) was incubated with each sample for 12 h at 4°C, and then the antibody was precipitated by addition of 50 μl of a 10% suspension of Pansorbin [3]. The precipitate was washed three times with a solution containing HEPES (50 mM, pH 7.4), Triton X-100 (1%), SDS (0.1%), NaCl (150 mM), NaF (100 mM), and Na_3VO_4 (2 mM). The radioactivity in the precipitate was measured by Cerenkov counting.

RESULTS

Identification of the Autophosphorylation Sites in the β-Subunit of the Insulin Receptor

Insulin stimulates autophosphorylation of the β-subunit of the WGA-purified rat insulin receptor (data not shown). This reaction is rapid and occurs on at least five

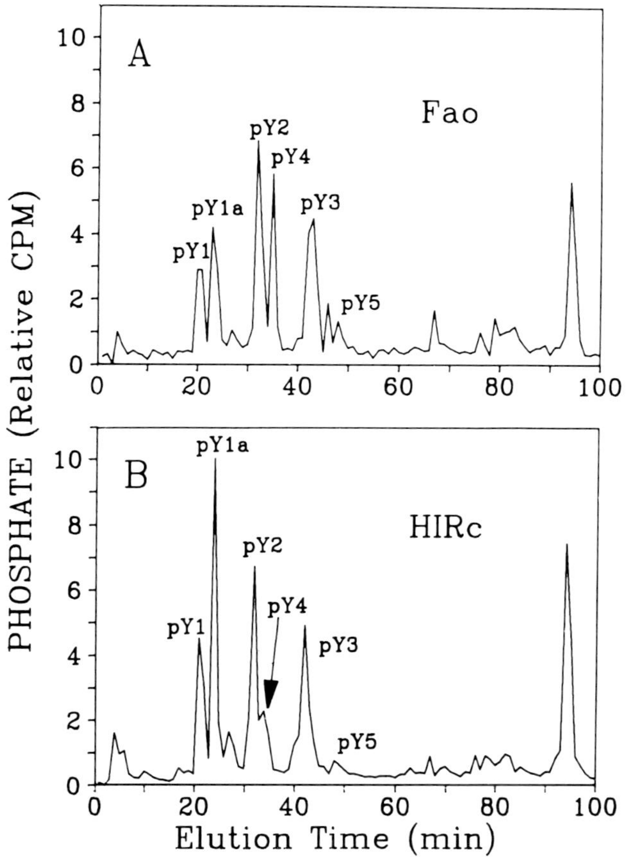

Fig. 1. Tryptic phosphopeptides of the β-subunit of the insulin receptor. Insulin receptor was solubilized from Fao cells (**A**) or CHO/HIRC cells (**B**) and partially purified on immobilized WGA. Insulin-stimulated autophosphorylation was carried out for 30 min, and the proteins were immunoprecipitated, reduced with DTT, separated by SDS-PAGE, digested exhaustively with trypsin, and the fragments were separated by HPLC.

tyrosyl residues [3]. A similar reaction occurs with the human insulin receptor [4,15]. Tryptic phosphopeptides were obtained from the β-subunit of the WGA-purified rat and human insulin receptor phosphorylated to steady state with [γ-^{32}P]ATP during insulin stimulation. Several phosphopeptides, labeled pY1, pY1a, pY2, pY3, pY4, and pY5 were obtained from the β-subunits (Fig. 1A,B). The HPLC elution profiles for each receptor were superimposable, suggesting that the phosphorylation sites in the β-subunit of the purified human and rat insulin receptor were identical.

The minimal composition of the tryptic peptides from the rat insulin receptor were deduced previously by a combination of techniques, and their identities are summarized in Table I. To determine whether the tryptic phosphopeptides obtained from the human insulin receptor purified from CHO/HIRC cells were correctly assigned, we immunoprecipitated the tryptic phosphopeptides with α-PY or with peptide-specific antibodies targeting presumed regions of autophosphorylation [3]. All of the tryptic phosphopeptides resolved from the human insulin receptor were immunoprecipitated with α-PY, indicating that each peptide contained phosphotyrosine. The pY1, pY1a, and pY4 were specifically immunoprecipitated with α-Pep3, the antibody that recognizes residues 1143–1152 of the β-subunit. This experiment verified that these peptides were derived from the regulatory region of the β-subunit, which contains tyrosyl residues 1146, 1150, and 1151. Previous studies indicated that both pY1 and pY1a contained three Tyr(P) residues but migrated differently on HPLC because of distinct tryptic cleavages at each C-terminus [3,4]. The pY4 was distinct because it contains only two Tyr(P) residues (Table I) [3]. In contrast, pY2 and pY3 were specifically precipitated with α-Pep1, the antibody directed against the C-terminus of the β-subunit, which includes Tyr(P)-1316 and Tyr(P)-1322 [6]. The α-Pep4, which recognizes amino acid residues 953–965 of the β-subunit, did not react with any of the major tryptic fragments, suggesting that none of them was derived from the Tyr-960 region of the receptor (Fig. 2); however, the final peak on our HPLC profile eluting at 95 min is composed of many minor components that have not been characterized and may contain peptides derived from the juxtamembrane region of the receptor. Thus, migration during reverse-phase HPLC and immunogenicity of the tryptic phosphopeptides obtained from the human insulin receptor indicated that the in vitro phosphorylated domains of the human insulin receptor were identical with those reported previously for the rat insulin receptor (Table I).

TABLE I. The Structure of the Tryptic Phosphopeptides Obtained From the β-Subunit of the Insulin Receptor*

Peptide	Amino acid residues	Structure
pY1	1143–1153	AspIleTyr(P)GluThrAspTyr(P)Tyr(P)ArgLys
pY1a	1143–1152	AspIleTyr(P)GluThrAspTyr(P)Tyr(P)Arg
pY2/pY3	1312–1330	ArgSerTyr(P)GluGluHisIleProTyr(P)ThrHisMetAsnGlyGlyLysLys
pY4	1143–1152	AspIleTyr(P)GluThrAsp[TyrTyr](P)Arg
pY5	1143–1153	AspIleTyr(P)GluThrAsp[TyrTyr](P)ArgLys

*The identity of these tryptic fragments was determined previously [3]. Amino acids are numbered according to Ullrich et al. [5]. The exact length of the pY2 and pY3 peptide has not been determined. The amino acids in italics indicate variable trypsin cleavage sites that give rise to the various forms of these peptides.

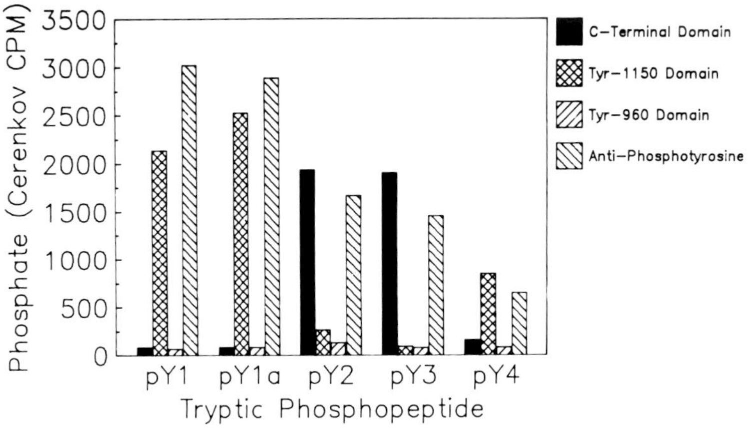

Fig. 2. Identification of tryptic phosphopeptides of the human insulin receptor. The insulin receptor was phosphorylated for 30 min, and then immunoprecipitated with α-PY, reduced with DTT, separated by SDS-PAGE, and digested exhaustively with trypsin. Tryptic phosphopeptides were separated by reverse-phase HPLC and identified by the elution position. Each phosphopeptide was immunoprecipitated with the indicated domain-specific antibody as described in "Material and Methods." pY5 was not tested.

The phosphorylation of the various sites in the rat and human insulin receptor follow distinct time courses [1,3,4,15]. After 1 min of insulin-stimulated autophosphorylation of the WGA-purified human insulin receptor, bis-phosphorylation of the regulatory region (pY4 + pY5) was near its maximal level (Fig. 3A). However, phosphorylation of the C-terminus (pY2 + pY3) and accumulation of the tris-phosphorylated form of the regulatory region (pY1 + pY1a) reached less than 25% of their steady-state level. After 30 min of autophosphorylation the aggregate intensity of pY4 and pY5 increased only 1.4-fold suggesting that the bis-phosphorylation of the regulatory region did not increase substantially (Fig. 3B). In contrast, tris-phosphorylation of the regulatory region and phosphorylation of the C-terminal sites increased five- and eight-fold, respectively (Fig. 3B). Thus, bis-phosphorylation of the regulatory region in the human insulin receptor was rapid, as previously shown for the rat receptor [3], and was the earliest detectable autophosphorylated form of the β-subunit, whereas tris-phosphorylation of this region and phosphorylation of the C-terminus lag behind.

Inhibition of Autophosphorylation of the WGA-Purified Human Insulin Receptor With Antiphosphotyrosine Antibody

Anti-phosphotyrosine antibody partially inhibits autophosphorylation of the β-subunit of the purified rat insulin receptor by binding to the initial sites of autophosphorylation; this association blocks the cascade of subsequent phosphorylations [3]. The α-PY has a similar effect on the human insulin receptor purified on immobilized WGA from CHO/HIRC cells (data not shown). In the presence of 1 μg/ml α-PY, autophosphorylation decreased 50–60% owing to the inhibition of phosphorylation of pY1, pY1a, pY2, and pY3 (Fig. 4). The pY4 (and sometimes pY5)[3] was the major

[3]The pY5 was not always detected; however, our previous report indicates that it is pY4 + Lys (see Table I).

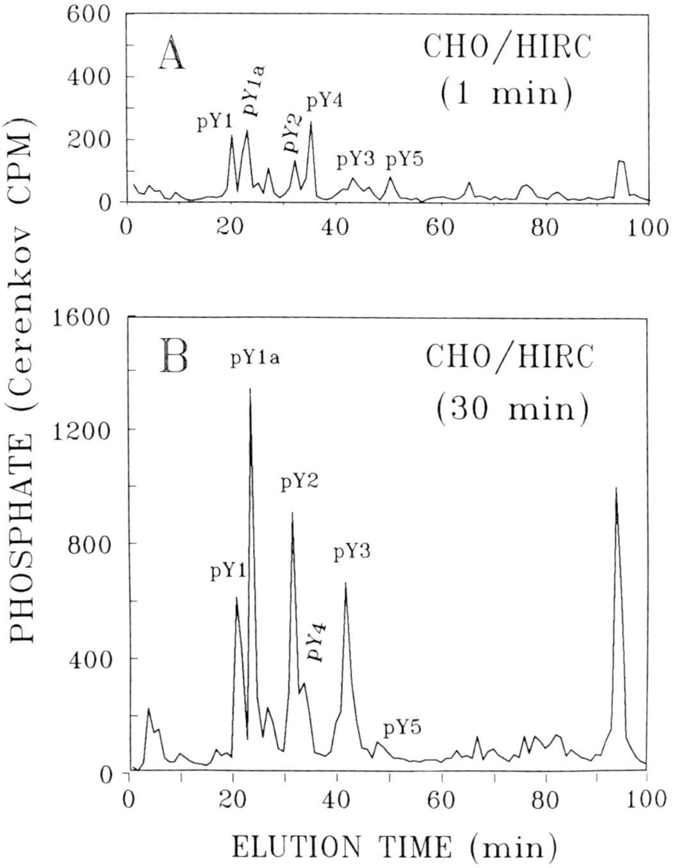

Fig. 3. The effect of time on the tryptic phosphopeptides of the β-subunit of the human insulin receptor. The WGA-purified human insulin receptor from CHO/HIRC cells was phosphorylated for 1 min (**A**) or 30 min (**B**) in the presence of 100 nM insulin. Insulin-stimulated autophosphorylation was carried out for 30 min, and the proteins were immunoprecipitated, separated by SDS-PAGE, digested exhaustively with trypsin, and separated by HPLC.

phosphopeptide detected by HPLC during insulin-stimulated autophosphorylation in the presence of α-PY. The pY4 corresponded exactly to pY4, which was obtained during phosphorylation in the absence of α-PY [3]. These data suggest that α-PY binds to the initial tyrosine phosphorylation sites in the β-subunit, which are found in pY4 (Tyr-1146 and either Tyr-1150 or Tyr-1151), and inhibits autophosphorylation at the other tyrosine residues. Thus, consistent with the time course of autophosphorylation, bis-phosphorylation of the regulatory region preceded the other phosphorylation events.

Based on our identification of the tryptic phosphopeptides from the regulatory region (i.e., pY1 and pY1a are tris-phosphorylated peptides, and pY4 and pY5 are bis-phosphorylated peptides) [3], the relative stoichiometry of autophosphorylation in this region was calculated from HPLC profiles (Table II). In the absence of α-PY, about 77% of the phosphorylated regulatory regions were tris-phosphorylated, whereas the remaining 23% were bis-phosphorylated. In contrast, during inhibition

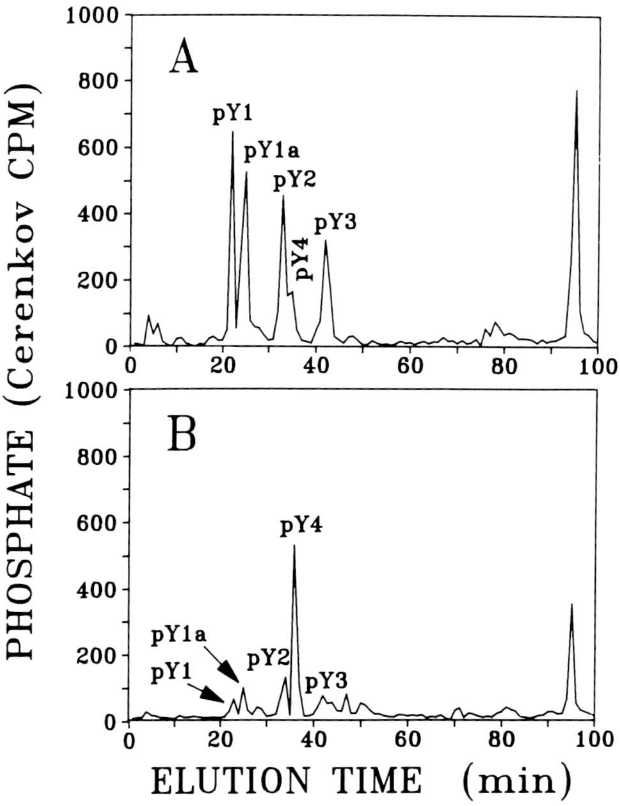

Fig. 4. The effect of antiphosphotyrosine antibody on the tryptic phosphopeptides of the β-subunit of the human insulin receptor. The WGA-purified insulin receptor from CHO/HIRC cells was stimulated with 100 nM insulin and phosphorylated for 30 min in the absence (**A**) or presence (**B**) of 0.03 μg/μl α-PY. The proteins were immunoprecipitated, reduced with DTT, and separated by SDS-PAGE, digested exhaustively with trypsin, and the phosphopeptides were separated by HPLC.

with α-PY, 83% of the regulatory regions were bis-phosphorylated, and only 17% reach the trisphosphorylated state. Moreover, during α-PY inhibition about one-half of the Tyr-1150 regions were bis-phosphorylated compared to the total phosphorylated in the absence of α-PY (Table II, compare the normalized regulatory region (Total) in the absence of α-PY to the bis-phosphorylated regulatory region during α-PY inhibition). These results suggest that only one β-subunit in the insulin-stimulated heterotetramer undergoes autophosphorylation in the presence of α-PY.

Bis-Phosphorylation Around Tyr-1150 Predominated During Insulin Stimulation of Intact Cells

Insulin-stimulated tyrosine autophosphorylation of the human insulin receptor in intact cells is quantitatively different from the in vitro reaction (Fig. 5). Tris-phosphorylation of the regulatory region was barely detected, as the recovery pY1 and pY1a was uniformly low (Fig. 5B). In contrast to in vitro autophosphorylation, only

TABLE II. The Stoichiometry of β-Subunit Phosphorylation in the Absence and Presence of Antiphosphotyrosine Antibody *

	No α-PY		α-PY Inhibited	
	CPM	Normalized CPM	CPM	Normalized CPM
Regulatory region				
Tris-phosphorylated				
pY1	743	248	96	32
pY1a	889	296	132	44
Total (pY1 + pY1a)	1,632	544 (77%)	228	76 (17%)
Bis-phosphorylated				
pY4	269	134	668	334
pY5	56	28	84	42
Total (pY4 + pY5)	325	162 (23%)	752	376 (83%)
Regulatory region total	1,957	707	980	452
C-terminal region				
pY2	478	239	217	108
pY3	574	287	167	83
C-terminal region total	1,052	526	384	191

*The WGA-purified insulin receptor from CHO/HIRC cells was stimulated with 100 nM insulin and incubated for 30 min in the absence (No α-PY) or presence (α-PY Inhibition) of 0.03 μg/μl α-PY. The β-subunit was immunopurified, separated by SDS-PAGE, and digested exhaustively with trypsin. The radioactivity (cpm) in each tryptic phosphopeptide was determined. Based on the number of Tyr(P) residues in each peptide, the normalized cpms were also calculated assuming 3 Tyr(P) residues in pY1 and pY1a (tris-phosphorylated), 2 Tyr(P) residues in pY4 and pY5 (bis-phosphorylated), and 2 Tyr(P) residues in pY2 and pY3 (C-terminus). The parenthetical values are the percentages of the normalized cpm's in the bis- or tris-phosphorylated Tyr-1150 region relative to the total.

19% of the regulatory regions of the β-subunits in intact Fao, CHO/HIRC, and 3T3/HIRC cells were tris-phosphorylated during insulin stimulation (Table III). This result was consistently observed in Fao, CHO/HIRC, and 3T3/HIRC cells (Table III). However, insulin stimulated the phosphorylation of tyrosine residues in a major phosphotyrosine-containing tryptic peptide, which migrated at the same position as pY4, suggesting that bis-phosphorylation of the regulatory region predominated in vivo (Fig. 5B). Moreover, phosphorylation of the C-terminus was also relatively low in vivo (Fig. 5). Thus, the in vivo milieu inhibited the complete autophosphorylation of the insulin-stimulated β-subunit in a way that was nearly identical with antiphosphotyrosine antibody inhibition of in vitro autophosphorylation.

DISCUSSION

Insulin-stimulated autophosphorylation is identical for the WGA-purified insulin receptor of rat or human origin. This relation was shown by an exact correspondence between the HPLC profiles of tryptic phosphopeptides obtained from β-subunit of each species phosphorylated to steady state. Moreover, the identity of each tryptic phosphopeptide determined by specific immunoprecipitation was the same; pY1, pY1a, and pY4 were derived from the regulatory region between Asp-1144 to Arg-

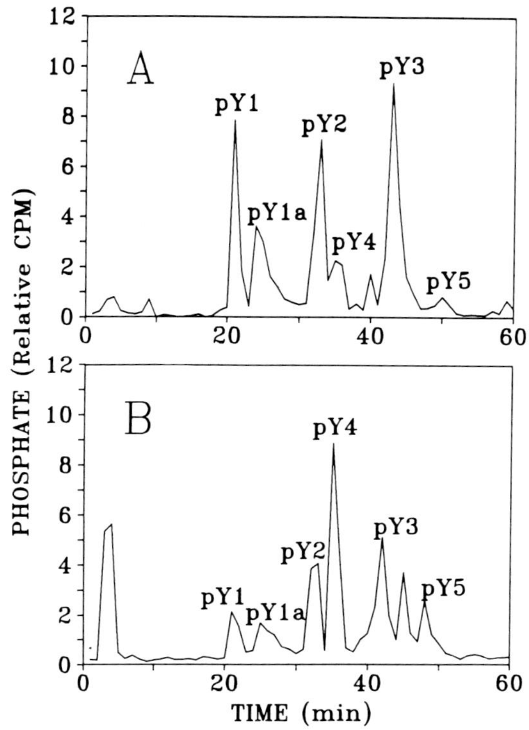

Fig. 5. A comparison of the tryptic phosphopeptides obtained from the human insulin receptor of 3T3/HIRC cells labeled in the intact cell with [^{32}P]orthophosphate (**B**) or labeled in vitro with [γ-^{32}P]ATP after WGA purification (**A**). The proteins were immunoprecipitated, reduced with DTT, and separated by SDS-PAGE, digested exhaustively with trypsin, and the phosphopeptides were separated by HPLC.

1152/Lys-1153, and pY2 and pY3 were derived from the C-terminal region between Lys-1313/Arg-1314 to Lys-1329/Lys-1330. The exact tryptic cleavage sites are quite variable, as previously discussed [3,4]. By analogy to our previous work, five tyrosyl residues in these two regions of the β-subunit undergo autophosphorylation: 1) Tyr-1146, Tyr-1150, and Tyr-1151, and 2) Tyr-1316 and Tyr-1322. These results agree with the findings of others [2,4], and are consistent with the report that the primary amino acid sequence around these tyrosyl residues is the same for the human and rat insulin receptor (R. Lewis, M. Tepper, and M.P. Czech, personal communication). However, we do not find evidence for phosphorylation of Tyr-960 or the adjacent residues Tyr-953 or Tyr-967, in contrast to previous reports [4,16,17].

As shown previously for the rat insulin receptor [3], autophosphorylation of the human insulin receptor is an ordered reaction that begins in the regulatory region at Tyr-1146 and either Tyr-1150 or Tyr-1151. Phosphorylation of two (bis-phosphorylation) of these three (tris-phosphorylation) tyrosyl residues is detected immediately after incubation of the WGA-purified insulin receptor with [γ-^{32}P]ATP and insulin. A distinct initial step in the cascade of autophosphorylation is confirmed by the fact that antiphosphotyrosine antibody (α-PY) traps the bis-phosphorylated form of the insulin receptor (found in pY4 and pY5) and actually causes its accumulation, whereas α-PY inhibits autophosphorylation of additional tyrosyl residues in the β-subunit. Thus, α-

TABLE III. Recovery of Bis- and Tris-Phosphorylated Tyr-1150 Region Phosphorylated In Vivo and In Vitro*

	Tris-phosphorylated regulatory region		Bis-phosphorylated regulatory region	
Cell line	In vitro (%)	In vivo (%)	In vitro (%)	In vivo (%)
Fao	72	12	28	88
CHO/HIRC	78	23	22	77
3T3/HIRC	74	23	26	77
Average	75	19	25	81

*The insulin receptor was phosphorylated in the intact Fao, CHO/HIRC, or 3T3/HIRC cells (in vivo), or after solubilization and WGA purification of the receptor (in vitro). The tryptic phosphopeptides were separated by reverse-phase HPLC, and pY1, pY1a, and pY4 were identified. Phosphate-labeled cells were stimulated with insulin (100 nM) for 5 min, whereas the purified receptor was incubated with insulin (100 nM) and 25 μM [γ-^{32}P]ATP for 30 min; steady-state labeling was achieved in each case. The phosphorylated receptor was immunoprecipitated with α-PY, separated by SDS-PAGE, and digested exhaustively with trypsin. The phosphopeptides were separated by HPLC, and the radioactivity in pY1 + pY1a and pY4 + pY5 was measured. The normalized amount of the tris-phosphorylated regulatory region was calculated as (pY1 + pY1a)/3, and the amount of the bis-phosphorylated regulatory region was calculated as (pY4 + pY5)/2; the percent of the Tyr-1150 region in the bis- and tris-phosphorylated forms was calculated from these values.

PY prevents tris-phosphorylation of the regulatory region (pY1 and pY1a), and inhibits autophosphorylation of Tyr-1316 and Tyr-1322 (pY2 and pY3) in the C-terminus of the molecule. A monophosphorylated form of the Tyr-1150 region cannot be detected and must exist only briefly during the initiation of the cascade or be inaccessible to α-PY. At 0°C, Tyr-1146 appears to be the first site of autophosphorylation [4]. Based on the relative stoichiometry of autophosphorylation, one-half of the available Tyr-1150 regions become bis-phosphorylated in the presence of α-PY. These results suggest that autophosphorylation of both β-subunits in the $\alpha_2\beta_2$ oligomer is not a concerted reaction and must begin in one of the two β-subunits and propagate to the other.

The cascade of autophosphorylation in the β-subunit is significant as tris-phosphorylation of the regulatory region appears to be required to activate the tyrosyl-specific phosphotransferase [3]. Our previous report indicated that bis-phosphorylation of the regulatory region obtained in the presence of α-PY did not activate the receptor kinase eluted from the antibody, and assayed under conditions that inhibit additional autophosphorylation, it was inhibited [3]. In contrast, Tornqvist et al. suggested that autophosphorylation occurs randomly and that less than complete autophosphorylation of the regulatory region activates the kinase [16]. However, substitution of Tyr-1150 alone with phenylalanine inhibited almost completely the activation of the phosphotransferase [18]. Thus, we conclude provisionally that phosphorylation of all three tyrosyl residues is necessary to activate the kinase [3].

The physiological significance of the autophosphorylation cascade is emphasized by the fact that the bis-phosphorylated Tyr-1150 region predominates in the intact cell during insulin stimulation [3,19]. Only about 20% of the phosphorylated regulatory regions are tris-phosphorylated, whereas 80% are bis-phosphorylated. This stoichiometry is found in Fao cells, as well as the CHO/HIRC and 3T3/HIRC cells, which express the human insulin receptor, suggesting that inhibition of tris-phosphorylation is a general characteristic in vivo. Although insulin binding initiates the autophosphorylation cascade in vivo [14], some other factors in the intact cell prevent the accumulation of the tris-phosphorylated β-subunit. The inhibition is removed during WGA-purification as tris-phosphorylation around Tyr-1150 predominates by 4 to 1 in the β-subunit during in vitro phosphorylation. Inhibition of in vivo autophosphorylation may be due to phosphatases present in the intact cell, which are removed during WGA chromatography. Whereas this explanation is likely for a hepatocyte-derived cell line (Fao), it is less likely for CHO or 3T3 cells, which should contain less Tyr(P) phosphatase activity [20]. Seryl phosphorylation of the β-subunit, presumably mediated through a protein kinase C pathway, may also regulate the cascade of autophosphorylation [21]; however, seryl phosphorylation appears to inhibit the cascade completely, rather than inhibit the phosphorylation of specific residues selectively [22].

It is interesting that in vivo tyrosine autophosphorylation of the β-subunit is very similar to in vitro autophosphorylation during inhibition with the α-PY. Partial autophosphorylation in vivo may result from the presence of binding proteins, which specifically interact with the bis-phosphorylated regulatory region. These binding proteins could inhibit additional autophosphorylation while simultaneously transducing the insulin signal. However, transmission of the insulin signal may occur in cells by tyrosyl phosphorylation of cellular substrates. In this case, inhibition of tris-phosphorylation could play a regulatory role in signal transmission in vivo.

In conclusion, the autophosphorylation cascade in the β-subunit demonstrated for the rat insulin receptor [3] also applies to the human insulin receptor. The first site of autophosphorylation is probably Tyr-1146, followed immediately by either Tyr-1150 or Tyr-1151. At this stage, bis-phosphorylation can be trapped by antiphosphotyrosine antibodies [3]. In the absence of α-PY, autophosphorylation of the Tyr-1150 region rapidly progresses to tris-phosphorylation and phosphorylation of Tyr-1316 and Tyr-1322 in the C-terminus of the β-subunit. Progression to the tris-phosphorylated form appears necessary for full activation of the phosphotransferase during in vitro assays and may play an important regulatory role in vivo [3].

ACKNOWLEDGMENTS

This work has been supported in part by NIH Grants DK38712 (to M.F.W.), AM31036, and AM33201 (to C.R.K.). M.F.W. is a scholar of the PEW Foundation, Philadelphia.

REFERENCES

1. White MF, Haring HU, Kasuga M, Kahn CR:J Biol Chem 259:255, 1984.
2. Tornqvist HE, Pierce MW, Frackelton AR, Nemenoff RA, Avruch J: J Biol Chem 262:10212, 1987.
3. White MF, Shoelson SE, Keutmann H, Kahn CR: J Biol Chem 263:2969, 1988.
4. Tavare JM, Denton RM: Biochem J 252:607, 1988.
5. Ullrich A, Bell JR, Chen EY, Herrera R, Petruzzelli LM, Dull TJ, Gray A, Coussens L, Liao Y-C, Tsubokawa M, Mason A, Seeburg PH, Grunfeld C, Rosen OM, Ramachandran J: Nature 313:756, 1985.
6. Goren HJ, White MF, Kahn CR: Biochemistry 26:2374, 1987.
7. White MF, Livingston JN, Backer JM, Lauris V, Dull TJ, Ullrich A, Kahn CR: Cell 54:641, 1988.

8. Rosen OM, Herrera R, Olowe Y, Petruzzelli LM, Cobb MH: Proc Natl Acad Sci USA 80:3237, 1983.
9. Crettaz M, Kahn CR: Endocrinology 113:1201, 1983.
10. White MF, Stegmann EW, Dull TJ, Ullrich A, Kahn CR: J Biol Chem 262:9769, 1987.
11. Whittaker J, Okamoto AK, Thys R, Bell GI, Steiner DF, Hofmann CA: Proc Natl Acad Sci USA 84:5237, 1987.
12. Kasuga M, White MF, Kahn CR: Methods Enzymol 109:609, 1985.
13. White MF, Haring HU, Kasuga M, Kahn CR: J Biol Chem 259:255, 1984.
14. White MF, Takayama S, Kahn CR: J Biol Chem 260:9470, 1985.
15. Herrera R, Rosen OM: J Biol Chem 261:11980, 1986.
16. Tornqvist HE, Avruch J: J Biol Chem 263:4593, 1988.
17. Herrera R, Lebwohl D, Garcia de Herreros A, Kallen RG, Rosen OM: J Biol Chem 263:5560, 1988.
18. Ellis L, Clauser E, Morgan DO, Edery M, Roth RA, Rutter WJ: Cell 45:721, 1986.
19. Tornqvist HE, Gunsalus JR, Nemenoff RA, Frackelton AR, Pierce MW, Avruch J: J Biol Chem 263:350, 1988.
20. Shriner CL, Brautigan DL: J Biol Chem 259:11383, 1984.
21. Takayama S, White MF, Lauris V, Kahn CR: Proc Natl Acad Sci USA 81:7797, 1984.
22. Takayama S, White MF, Kahn CR: J Biol Chem 263:3440, 1988.

Journal of Cellular Biochemistry 39:443–454 (1989)
Growth Factors and Their Receptors: Genetic
Control and Rational Application 195–206

Cell Type and Tissue Distribution of the Fibroblast Growth Factor Receptor

Bradley B. Olwin and Stephen D. Hauschka

Department of Biochemistry, University of Wisconsin, Madison, Wisconsin 53706 (B.B.O.); Department of Biochemistry, SJ-70, University of Washington, Seattle, Washington 98195 (S.D.H.)

A receptor for fibroblast growth factor (aFGF, bFGF) was partially characterized in intact cell cultures, cell plasma membranes, and tissue plasma membrane preparations. Analysis of 24 different cell types from four species identified a 165-kDa FGF receptor present on the cell surface of most mesodermal and neuroectodermal cells. Chemical crosslinking of ^{125}I-aFGF to its cell surface receptor was specifically blocked by a 100-fold molar excess of either aFGF or bFGF. In contrast to the similar molecular weight of FGF receptors, different cell types exhibited significant variations in binding of ^{125}I-aFGF to intact cultures with low values of 8 pM and 700, to high values of 60 pM and 30,000, for the K_d and receptor number per cell, respectively. A binding assay was developed for quantitation of ^{125}I-aFGF binding to cell- and tissue-derived membrane preparations. Membranes prepared from baby hamster kidney cells exhibited a K_d of 55 pM, while a similar K_d of 67 pM was determined for intact baby hamster kidney cells. Although ten different adult bovine tissue membrane preparations and human term placental membranes exhibited no specific binding of ^{125}I-aFGF, FGF receptor was detected in embryonic murine tissues (17 days gestation). These results support the existence, in a variety of cells, of either a common FGF receptor that binds both aFGF and bFGF or closely related FGF receptors that cannot be distinguished by molecular weight. The differential binding of FGF to its receptor in embryonic vs. adult tissues suggests a potentially broad role for FGF in embryonic development and a more restrictive role in the adult.

Key words: proliferation, differentiation, growth factor receptors, embryogenesis

A number of biological activities have been demonstrated for acidic and basic fibroblast growth factors [aFGF, bFGF; reviewed in Gospodarowicz et al., 1986; Baird et al., 1986; Folkman and Klagsbrun, 1987]. The recent identification of oncogenes with extensive homology to both bFGF and aFGF [Dickson and Peters, 1987; Delli-Bovi et al., 1987; Yoshida et al., 1987; Smith et al., 1988] and the discovery that FGF may play a key role in amphibian development [Slack et al., 1987; Kimmelman and Kirschner, 1987], has implicated FGF in a diverse number of biological roles. In contrast, little is known concerning receptors for FGF. A single FGF receptor has been

Received May 25, 1988; accepted October 5, 1988.

identified and partially characterized in a few cell types [Neufeld and Gospodarowicz, 1985; Friesel et al., 1986; Huang et al, 1986; Moenner et al., 1986; Olwin and Hauschka, 1986; Moscatelli, 1987]. Based on results from analysis of FGF receptors in a single cell line, BHK (baby hamster kidney), the existence of two receptors has been proposed [Neufeld and Gospodarowicz, 1986]. In addition, both aFGF and bFGF competed for ^{125}I-FGF binding, suggesting that aFGF and bFGF compete for a single cellular receptor [Olwin and Hauschka, 1986] or two cellular receptors [Neufeld and Gospodarowicz, 1986]. Although these studies suggest aFGF and bFGF mediate their actions through a common receptor, they were performed on a limited number of cell types, and FGF binding to FGF receptor in vivo was not examined.

In this study, we investigated the binding characteristics and sodium dodecyl sulfate-polyacrylamide gel electrophoresis (SDS-PAGE) migration patterns of FGF receptor from 24 different cell types that exhibit biologically diverse responses to FGF. A comparison of ^{125}I-aFGF binding to intact BHK cells and BHK cell membrane preparations was also undertaken to identify optimum conditions for examination of FGF receptors in tissue membrane preparations. Embryonic, but not adult membranes expressed significant numbers of FGF receptor. The results suggest that aFGF and bFGF bind to a common FGF receptor or a group of closely related receptors, which mediate the biological actions of FGF.

MATERIALS AND METHODS

Materials

Na^{125}I was purchased from New England Nuclear, Enzymobeads from BioRad Laboratories, and disuccinimidyl suberate from Pierce Chemical Co. Hydrophilic Durapore filters (HVLP, 0.45 μm) were purchased from Millipore, Inc.

Methods

Purification of aFGF and bFGF was as previously described [Olwin and Hauschka, 1986, 1988] with modifications described as follows. After separation of aFGF and bFGF by heparin affinity chromatography, each polypeptide was isolated by FPLC on a mono-S column (0.5 $\times$ 5 cm; Pharmacia, Inc.) developed with a 0–1 M sodium chloride gradient in 0.05 M sodium phosphate, pH 7.0. High-performance liquid chromatography (HPLC) analysis [Olwin and Hauschka, 1986] confirmed that detectable crosscontamination of FGF in either preparation was less than 1% of the total FGF present. As an added control, recombinant human bFGF, kindly supplied by Zymogenetics, Inc., was used to confirm results obtained with purified bovine bFGF. ^{125}I-aFGF was prepared as previously described [Olwin and Hauschka, 1986] except that FPLC-purified bovine aFGF was used for ^{125}I-labeling. Intact cell binding and crosslinking of ^{125}I-aFGF using 0.15 mM disuccinimidyl suberate were as described previously [Olwin and Hauschka, 1986].

Mouse myoblast MM14, MM14 DD-1, C2C12, and rat myoblast L6E9 cells were grown as described by Olwin and Hauschka [1988]. Rat pheochromocytoma (PC-12) cells were grown in Dulbecco's modified Eagle medium (DMEM) containing 10% horse serum, 5% fetal bovine serum, and 1% antibiotics (10,000 IU/ml penicillin, 0.5 mg/ml streptomycin sulfate). Mouse hybridoma cells were grown in RPMI 1640 buffered with 10 mM Hepes, pH 7.4, and containing 10% horse serum, 5% fetal bovine serum, and 1% antibiotics. All other cells were grown in DMEM supplemented with 10% defined calf serum (Hyclone laboratories) and 1% antibiotics.

BHK cell membranes were prepared by rinsing three confluent 15-cm plates three times with sodium phosphate, pH 7.4, containing 0.1 M NaCl (PBS), scraping the cells in 3 ml of HB (homogenization buffer: 20 mM N-2-hydroxyethylpiperazine-N'-2-ethanesulfonic acid, pH 7.4, 0.25 M sucrose, 1 mM ethylenediaminetetraacetic acid, 1 mM ethylene glycol bis(-amino-ether-)-N,N,N',N'-tetraacetic acid, 1 μg/ml leupeptin, 20 Kallikrein inhibitory Aprotinin units/ml) including 1 mM phenylmethylsulfonylflouride, and centrifuging for 5 min at 3,000*g*. The cell pellet was resuspended in 3 ml HB and homogenized 20 strokes by hand in a dounce homogenizer with a tight-fitting pestle, centrifuged at 100,000*g* for 30 min, homogenized, and centrifuged again. The membrane pellet was then resuspended in 0.2 ml HB, frozen in liquid N_2 and stored at -70°C. ^{125}I-aFGF binding on freshly prepared and frozen membrane preparations demonstrated no loss of binding activity upon freezing (unpublished data). Similar procedures were used to prepare tissue membrane samples except that Polytron homogenization was utilized to disrupt samples instead of dounce homogenization. Both cell and tissue membranes were adjusted to a protein concentration of 10 mg/ml before they were frozen in liquid N_2 and stored at -70°C.

Detection of particulate membrane FGF receptor was performed as follows. First, membranes were resuspended in PBSMB (PBS containing 2 mM $MgCl_2$ and 0.2% bovine serum albumin) by trituration through a 25 gauge needle. Unless otherwise specified, binding assays included 200 pM ^{125}I-aFGF and 20 μg membrane protein in a final volume of 100 μl of PBSMB for 30 min at 22°C. To terminate the assay, samples were diluted with 2.5 ml PBSMB and applied to a Durapore filter (Millipore, Inc.) under vacuum; the tubes were rinsed twice with 2.5 ml of PBSMB, and the filters were rinsed three times with the same volume of PBSMB. The filters were then removed, and ^{125}I-aFGF bound to the particulate membrane fraction was determined by γ-counting. Typically, 0.5% of the total applied ^{125}I-cpm bound irreversibly to the filter and was subtracted from the total cpm to determine the percent of nonspecific binding.

Protein concentrations were determined by the method of Bradford [1976] using Pentex Grade V bovine serum albumin (Miles Scientific) as a standard.

RESULTS

Analysis of FGF Receptor in Cultured Cells

The distribution and partial characterization of a receptor for FGF in a variety of cultured cell lines, primary cell cultures, and tissues was examined by identification of high-affinity FGF binding sites. FGF receptors were initially identified by covalently crosslinking specifically bound ^{125}I-aFGF to intact cells and analyzed by SDS-PAGE and autoradiography. Further FGF receptor characterization included the determination of the K_d and the relative FGF receptor number by Scatchard analysis of equilibrium binding of ^{125}I-aFGF to intact cells or tissue membrane preparations.

Examination of FGF receptor in 24 different cell lines from human, mouse, hamster, and rat revealed a strikingly similar pattern of migration on SDS polyacryl-

TABLE I. Characteristics of FGF-Receptor in Cell Lines*

Cell type	FGFR per cell	K^d(pM)	FGFR MW (kDa)
Mouse myoblast MM14	700	8	165
Mouse myoblast MM14 DD-1	12,000	11	165
Mouse myoblast C2C12	~5,000	nd	165
Mouse myoblast BC3H1	7,000	15	165
Mouse swiss 3T3	20,000	20	165
Mouse C3H10T1/2	30,000	20	165
Mouse neuroblastoma N-18	nd	nd	165
Mouse endocarcinoma F-9	nd	nd	165
Mouse fibroblast L-cell	nd	nd	165
Mouse hybridoma MF-20	—	—	—
Mouse hybridoma FOX/NY	—	—	—
Human A431	10,000	30	165
Human smooth muscle	nd	nd	165
Human endothelial	nd	nd	140
Human fibroblast (SK-5)	nd	nd	165
Human osteosarcoma MG-63	nd	nd	165
Human fibroblast W1-38	nd	nd	165
Human astrocytoma 1321N1	nd	nd	165
Human melanoma A875	nd	nd	165
Human HeLa	nd	nd	165
Rat myoblast L6	—	—	—
Rat neuronal PC-12	3,000	20	165
Hamster BHK	13,000	60	165
Hamster CHO	nd	nd	165

*Data for K_d and FGF receptor number per cell were obtained by extrapolation of Scatchard plots, and FGF receptor MW was determined by relative mobility of ^{125}I-aFGF-FGF receptor complexes on SDS-polyacrylamide gels (15 kDa was subtracted from the labeled complex for the estimation of FGF receptor MW). nd = Not determined; —= not detectable.

amide gels, suggesting the existence of a single, or closely related FGF receptors (Table I; Fig. 1). Furthermore, in all cell types examined, both unlabeled aFGF and bFGF competed specifically for the binding of ^{125}I-aFGF to a FGF receptor. Although the MW of the FGF receptor was nearly identical for most cells, the K_d for ^{125}I-aFGF binding and receptor number per cell varied significantly (Fig. 1; Table I). Most mesoderm and neuroectoderm-derived cell lines as well as primary cultures expressed significant specific binding for ^{125}I-aFGF. The rat L6 skeletal muscle cell line (Table I), was the only mesodermal cell type tested that did not bind ^{125}I-aFGF. In addition, cell lines of hematopoeitic origin, including two myeloma lines (Sp2/0 and FOX/NY) and two different hybridoma cell lines, did not bind FGF (Table I).

In all cell lines examined, the presence of a high MW band is seen on the gel, which runs at the interface of the 4% stacking gel and the 7.5% separating gel. Some of these bands are visible in the autoradiograms illustrated in Figure 1. Presumably, they represent higher-order crosslinked polypeptides incapable of entering the separating gel. In addition, a few cell lines, notably the F-9, C3H10T1/2, and 132 1N1, exhibit a band migrating at less than 180 kDa that may represent a proteolytic degradation product (see "Discussion").

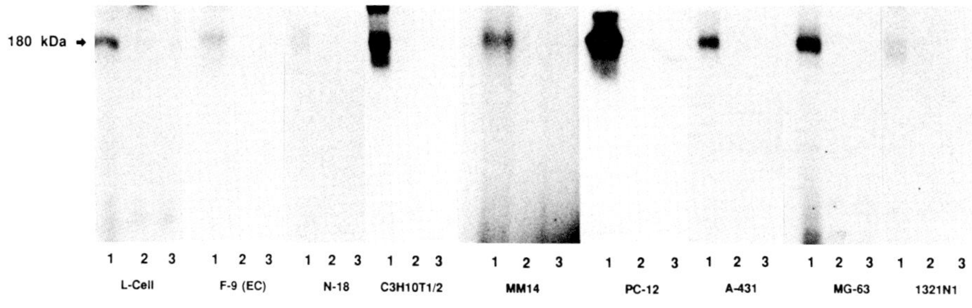

Fig. 1. Crosslinking of ^{125}I-aFGF to intact cells. Intact cell cultures were incubated for 1 h at 22°C with 100 pM ^{125}I-aFGF in the absence (**1**) and presence of a 100-fold molar excess of either unlabeled aFGF (**2**) or bFGF (**3**) and then processed as described in "Materials and Methods." Names of cell types correspond to those given in Table I. These are composites of several 7.5% SDS-polyacrylamide gels. The arrow marks a relative molecular weight of 180,000 as determined by the R_f of the crosslinked product in each gel. The bands appearing at the top of some of the lanes represent material that did not enter the running gel and is presumably composed of aggregated, crosslinked complexes. The SDS-polyacrylamide gels were analyzed by autoradiography at −70°C between 4 and 14 days.

Diverse Biological Actions of FGF Appear to Be Mediated by a Common Cell Surface Receptor

Addition of FGF to murine C3H10T1/2 cells [Reznikoff et al., 1973], rat pheochromocytoma (PC-12) cells [Schubert et al., 1974], and murine skeletal muscle (MM14) myoblasts [Hauschka et al., 1979] causes cell division in C3H10T1/2 cells (unpublished observations), stimulation of neurite outgrowth in PC-12 cells [Togari et al., 1985; Wagner and D'Amore, 1986; Walikie et al., 1986], and repression of myogenic differentiation in mouse MM14 myoblasts [Linkhart et al., 1981], respectively. Examination of FGF receptor by crosslinking of ^{125}I-aFGF to these three intact cells identified a similar 165-kDa receptor (Fig 1). However, significant variations of both the affinity for ^{125}I-aFGF and the receptor number per cell are observed from equilibrium binding studies (Fig 2; Table I). Consistent with our previous observations, both aFGF and bFGF competed for specific binding of ^{125}I-aFGF to intact C3H10T1/2, PC-12, and MM14 cells, respectively (Fig. 1).

Comparison of ^{125}I-aFGF Binding to Intact BHK Cells and BHK Membranes

Because the binding and crosslinking of ^{125}I-bFGF has been well documented in BHK cells [Neufeld and Gospodarowicz, 1985; Moscatelli, 1987], these cells were chosen for a comparison of ^{125}I-aFGF binding to intact cells and cell membrane preparations. Scatchard analysis of ^{125}I-bFGF binding to intact BHK cells has identified the presence of two distinct binding sites, a low-affinity site sensitive to ionic strength and heparinase and a high-affinity site insensitive to these treatments [Moscatelli, 1987]. Our analysis of ^{125}I-aFGF binding to intact BHK cells identified only a single class of noninteracting sites (Fig. 3a), possibly because of the reduced affinity of aFGF for heparin when compared to bFGF. To determine if high-affinity ^{125}I-aFGF binding

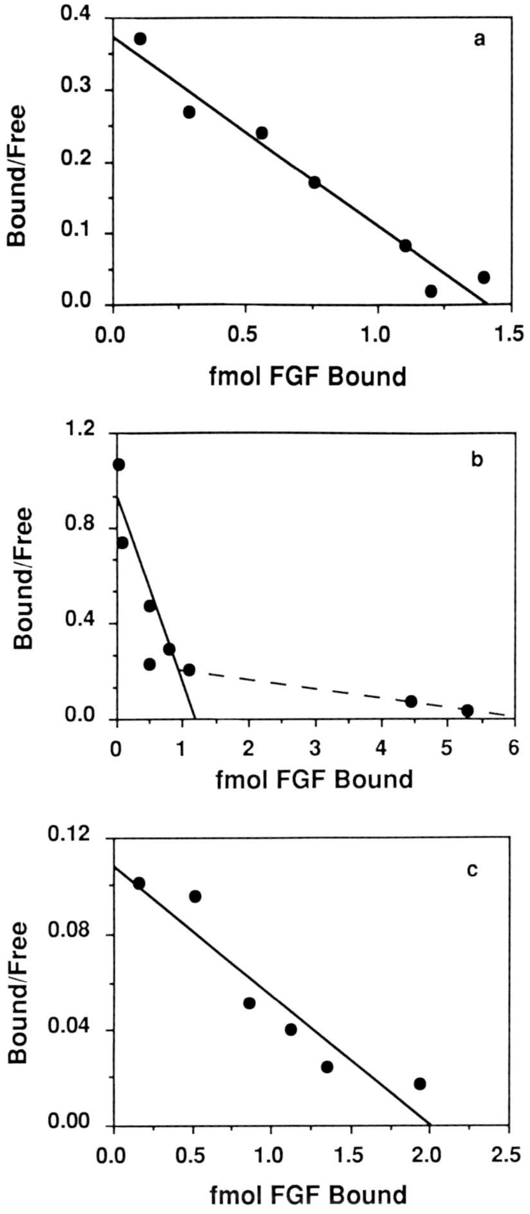

Fig. 2. Equilibrium binding and crosslinking of ^{125}I-aFGF to intact C3H10T1/2, PC-12, and MM14 cells. Increasing concentrations of ^{125}I-aFGF were incubated with intact PC-12 (**a**), C3H10T1/2 (**b**), or MM14 (**c**) cell cultures and processed as described previously [Olwin and Hauschka, 1986]. The data are plotted according to Scatchard [1949]. Cell numbers for each point were 2.5×10^5, 3×10^4, and 1.4×10^6 for PC-12 (a), C3H10T1/2 (b), and MM14 (c), respectively. Each point represents the mean from triplicate determinations. Nonspecific binding at saturation did not exceed 20% of the total bound ^{125}I-cpm for all three cell types. These experiments were repeated two times with similar results.

was retained in plasma membrane preparations, an assay was developed for membrane-associated FGF receptor. Scatchard analysis of equilibrium binding to intact BHK cells and BHK plasma membranes yielded similar K_d values of 67 pM for intact cells and 55 pM for membranes with 13,000 rcceptors per cell and 45 fmol of ^{125}I-aFGF bound per mg membrane protein, respectively (Fig. 3).

Analysis of FGF Receptor From Tissue Membrane Preparations

Detection of BHK membrane-associated high-affinity ^{125}I-aFGF binding suggested that a similar assay may prove successful for analysis of FGF receptor in tissue-derived membranes. Specific binding of ^{125}I-aFGF was not detected in ten different adult bovine tissues or human term placental membranes (Table II). Estimations of the sensitivity for the membrane binding assay demonstrate that the limit of FGF receptor detection in these assays is at least 0.05–0.1 fmol per 20 μg membrane protein. The FGF receptor content of these membrane preparations is thus less than 2.5–5 fmol per mg membrane protein (see "Discussion").

In contrast to adult tissues, embryonic murine tissues bound significant amounts of ^{125}I-aFGF (Table II). Binding of ^{125}I-aFGF to murine embryonic FGF receptor was specifically blocked by a 100-fold molar excess of unlabeled aFGF and bFGF. The FGF receptor in embryonic membranes binds FGF with high affinity, exhibits a 180-kDa crosslinked ^{125}I-aFGF-FGF receptor complex (manuscript in preparation), and thus is analogous to a previously identified cellular receptor [Olwin and Hauschka, 1986], which is distinct from low-affinity heparin binding sites [Moscatelli, 1987].

TABLE II. FGF-Receptor Content in Mammalian Membrane Preparations

Tissue	FGFR content (fmol ^{125}I-aFGF bound per mg protein)
Bovine brain	<2.5[a]
Bovine pituitary	<2.5
Bovine atrium	<2.5
Bovine ventricle	<2.5
Bovine uterine smooth muscle	<2.5
Bovine adrenal	<2.5
Bovine kidney	<2.5
Bovine liver	<2.5
Bovine lung	<2.5
Bovine testis	<2.5
Mouse 17 day embryo body	6.5 ± 1.8
Mouse 17 day embryo head	10 ± 2
Mouse 17 day embryo placenta	11 ± 4
Human term placenta	<2.5

[a]Where measurable, values are reported as the mean ± standard deviation from three independent experiments. When not detectable, the values are reported as less than 2.5 fmol/mg membrane protein, a conservative detection estimate. Membranes (20 μg) prepared as described in "Materials and Methods" were incubated 1 h at 22°C with 200 pM ^{125}I-aFGF. Nonspecific binding did not exceed 15% of the total specifically bound ^{125}I-cpm for those values reported as measurable.

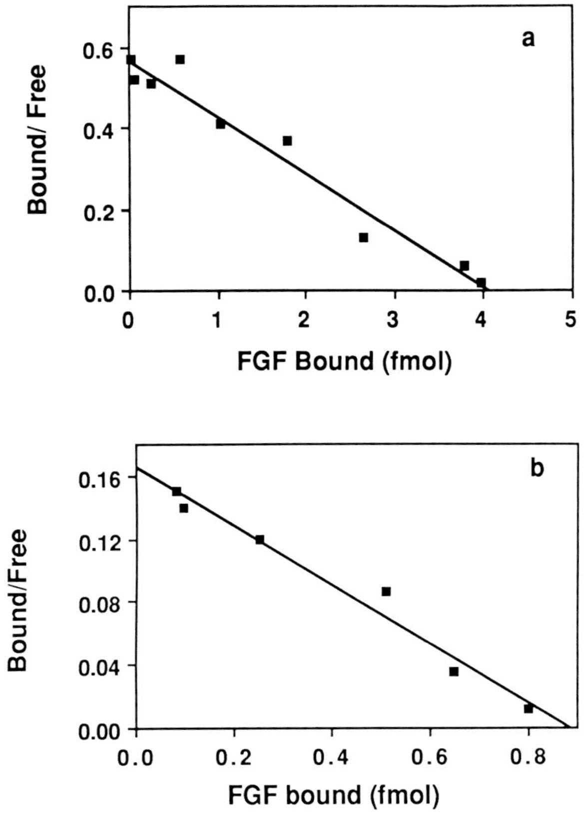

Fig. 3. Comparison of ^{125}I-aFGF binding to intact cells and cell plasma membrane preparations. **a:** Scatchard analysis of ^{125}I-aFGF binding to intact BHK cultures was as described previously [Olwin and Hauschka, 1986]. Each data point is the mean of triplicate determinations. For each point, 2.5×10^5 cells were present. The experiment was performed three times with similar results. **b:** Analysis of FGF receptor in BHK membrane preparations was performed as described in "Materials and Methods." The equilibrium binding data from a single representative experiment was plotted according to Scatchard [1949]. Each data point is the mean of triplicate determinations. At saturation, the nonspecific binding to the membrane preparation did not exceed 10%. These experiments were repeated two times with similar results.

DISCUSSION

High-affinity FGF binding sites are present on a wide variety of mesoderm, neuroectoderm, and tumor-derived cells. In many cell types, FGF functions primarily as a mitogen. All cells examined that express high-affinity ^{125}I-aFGF binding sites also express a 165-kDa FGF receptor. Furthermore, in all cell types examined, bovine aFGF, bFGF, and recombinant human bFGF compete for ^{125}I-aFGF binding to a cell surface receptor (Fig. 1) (unpublished data). Thus, FGF receptors identified by cross-linking of ^{125}I-aFGF to intact cells appear to interact with both acidic and basic forms of FGF. It is unlikely that the competition observed for either aFGF or bFGF is due to contamination of these preparations with the corresponding FGF. The contamination of bFGF in the aFGF preparation is 1% or less (see "Materials and Methods"), and thus inclusion of a 100-fold molar excess of bFGF would contain at most aFGF concentrations equivalent to the ^{125}I-aFGF concentration in the binding assay. Since a 100-fold molar excess of FGF is required for complete reduction of ^{125}I-aFGF binding sites, the contaminating FGF concentration is not sufficient for complete competition. Furthermore, as demonstrated in previous studies, equivalent concentrations of aFGF and bFGF compete for ^{125}I-aFGF binding to Swiss 3T3 cells [Olwin and Hauschka, 1986] and to a similar extent in BHK cells [Neufeld and Gospodarowicz, 1986], suggesting that both factors bind to a common receptor. Other growth factor receptors also bind multiple ligands. For example, epidermal growth factor receptor binds both epidermal growth factor and transforming growth factor α [reviewed in Carpenter, 1987].

The lower MW band present in F-9, C3H10T1/2, and 1321N1 cells most likely represents a proteolytic fragment of the FGF receptor. Similar bands are occasionally seen in other cell types, including BHK cells. These results are consistent with those originally observed for the epidermal growth factor receptor [reviewed in Carpenter, 1987], which is rapidly degraded by a Ca^{2+}-activated protease. Although the existence of more than one polypeptide cannot be absolutely discounted, the presence of lower MW bands most likely represents proteolytic degradation of the FGF receptor from cells possessing greater endogenous protease activity. Experiments designed to address this question were performed by coculturing BHK cells exhibiting two distinct crosslinked complexes with A431 cells exhibiting one product. The cocultures and control cultures of each individual cell type were crosslinked to ^{125}I-aFGF and analyzed by SDS-PAGE and autoradiography. These experiments invariably revealed only one crosslinked product in the cocultures, one in the A431 cultures, and two in the BHK cultures (unpublished data). Until further molecular detail is available for the FGF receptor, the existence of multiple receptor types will remain unresolved.

The biological activities of FGF in vitro are extensive [for reviews, see Gospodarowicz et al., 1986; Baird et al., 1986; Folkman and Klagsbrun, 1987]. Actions of FGF distinct from mitogenesis include stimulation of angiogenesis [for a review, see Folkman and Klagsbrun, 1987], stimulation of neuronal differentiation [Togari et al., 1985; Walickie et al., 1986; Wagner and D'Amore, 1986; Morrison et al., 1986], and repression of skeletal muscle differentiation [Linkhart et al., 1981; Lathrop et al., 1985; Clegg et al., 1987]. Previous studies indicated that the 165-kDa FGF receptor, identified by homobifunctional crosslinking of ^{125}I-aFGF to intact Swiss 3T3 cultures and murine MM14 myoblasts, binds FGF with affinities consistent with its function as a FGF receptor and mediator of FGF's biological activities [Olwin and Hauschka, 1986, 1988]. To determine whether the biological actions of FGF in this study were mediated by distinct cell surface receptors, we examined the binding characteristics and FGF receptor MW in these three cell lines. Although the K_d and FGF receptor number per cell varies between these cell types, they were within the range exhibited by other cells (Table I). Also consistent with other cell types, an apparent receptor MW of 165,000 is observed for all three cell lines (Fig. 1). Both aFGF and bFGF are biologically active for all three cell types, and both forms of FGF effectively compete for ^{125}I-aFGF binding (Fig. 1). Similar results have been observed using ^{125}I-bFGF (manuscript in preparation). These data suggest that the diverse biological actions of FGF do not originate via interaction of either aFGF or bFGF with different cell surface receptors specific for either growth factor. Although the low-affinity FGF binding sites seen in C3H10T1/2 cultures were not examined further, similar low-affinity sites have been characterized in detail for ^{125}I-bFGF binding to BHK cells [Moscatelli, 1987]. Since the low affinity BHK binding sites do not correlate with the biological activity of FGF, and since the sites are sensitive to both high ionic strength and treatment with heparinase [Moscatelli, 1987], the C3H10T1/2 low-affinity sites most likely represent cell-associated heparin binding sites for FGF.

A comparison of FGF receptors from intact BHK cells and BHK plasma membrane preparations indicates the presence of high-affinity ^{125}I-aFGF binding sites in the membrane fraction (Fig. 2b). The affinity of BHK membranes for ^{125}I-aFGF is virtually identical with the affinity of receptors for ^{125}I-aFGF in intact cell cultures (Fig. 2). In addition, the concentration of added unlabeled aFGF required for displacement of 50% of bound ^{125}I-aFGF from either intact cells or membrane preparations was equal to the concentration of ^{125}I-aFGF added to the incubation (unpublished data), suggesting the presence of a single class of noninteracting binding sites.

Retention of high-affinity ^{125}I-aFGF binding in BHK membranes allowed the development of efficient assays for the detection of FGF receptor in tissue-derived membrane preparations. However, an initial examination of ^{125}I-aFGF binding to adult bovine tissues and human term placental membranes failed to reveal specific ^{125}I-aFGF binding (Table II). In contrast to adult membranes, embryonic murine membranes express high-affinity ^{125}I-aFGF binding sites. Crosslinking of embryonic murine membranes identifies a FGF receptor of 165 kDa, which binds both aFGF and bFGF (manuscript in preparation).

One potential artifact that may reduce the apparent number of high-affinity FGF binding sites is the presence of high concentrations of endogenous FGF in the membrane preparations. To test for such artifacts, the FGF content in the membrane preparations was estimated using a repression of differentiation assay for MM14 myoblasts [Olwin and Hauschka, 1986]. Since MM14 cells are absolutely dependent on FGF for cell division [Linkhart et al., 1981; Clegg et al., 1987], the assay is specific for unbound FGF in crude extracts and has a detection limit of 0.3 fmol bFGF [Seed et al., 1988]. Although the initial supernatants from the membrane preparations exhibited significant FGF biological activity, analysis of several washed, membrane preparations did not detect significant levels of biologically active FGF. Additional support for the absence of detectable FGF receptor is the observation that unlabeled aFGF or bFGF compete for bound ^{125}I-aFGF. Therefore, significant exchange should occur between the added excess of ^{125}I-aFGF and the endogenous bound FGF present. Finally, we have attempted to dissociate bound FGF from membrane preparations by treatment with acidic buffers. Such treatment, which does not reduce binding affinity, yields increases in specific ^{125}I-aFGF binding of less than 20% (unpublished data). Thus, the inability to detect FGF receptor in adult tissues does not appear to be due to the presence of endogenous FGF in the membrane preparations, and therefore we conclude that the FGF receptor content of these adult bovine membranes is less than 2.5 fmol per mg protein, the detection limit of the assay.

Since aFGF and bFGF have been implicated as mesoderm-inducing factors in the early development of the amphibian embryo [Slack et al., 1987; Kimmelman and Kirschner, 1987], a distribution of FGF receptor might be present in the developing embryo. Analysis of high-affinity sites for ^{125}I-aFGF binding to embryonic membranes reveals low levels of specific binding (Table II). It is thus tempting to speculate that the presence of relatively high levels of FGF receptor in the embryo signifies the importance of FGF for embryonic development. The relative lack of widespread FGF receptor in adult tissues examined in this study may indicate more restrictive roles for FGF action in the adult.

Cultured cell lines and embryonic tissues, but not several adult bovine tissues, express high-affinity FGF binding sites. Partial characterization of FGF receptors from several species suggests the existence of a common 165-kDa receptor that is capable of interacting with both aFGF and bFGF. Related FGF receptors are also present on cells that exhibit divergent biological responses upon addition of exogenous FGF. The recent identification of oncogenes possessing approximately 40–50% homology with bFGF [Dickson and Peters, 1987; Delli-Bovi et al., 1987; Yoshida et al., 1987] suggests the existence of a FGF gene family. Although bFGF and aFGF appear to bind to a common cell surface FGF receptor, it is unlikely that an entire FGF gene family exists for which there is only one cell surface receptor. If more than one receptor subtype exists, then the current methods used to identify FGF receptor may not distinguish between related FGF receptor subtypes. It is likely that the receptor that binds both aFGF and bFGF may not bind other members of the FGF family. Identification of such receptor subtypes, if they exist, will require further biochemical characterization of FGF receptor(s).

ACKNOWLEDGMENTS

We thank Elaine Raines and Russell Ross for the gifts of A 431, human SK-5, and WI-38 cells; Mark Bothwell for L-cells, PC-12 cells, MG-63 cells, and A875 cells. We also thank Jennifer Martin and Neil Nathanson for the gift of 1321N1 cells and Jim Florini for the rat L6 cells. Zymogenetics, Inc. of Seattle is acknowledged for providing recombinant human basic FGF. Jennifer Martin, Laura Burrus, Bruce Lueddecke, and John Doctor are thanked for their comments on the manuscript.

REFERENCES

1. Baird A, Esch F, Mormede P, Ueno N, Ling N, Bohlen P, Ying S, Wehrenberg WB, Guilleman R: Recent Prog Horm Res 42:143–205, 1986.
2. Bradford MM: Anal Biochem 72:248–254, 1976.
3. Carpenter G: Annu Rev Biochem 56:881–914, 1987.
4. Clegg CH, Linkhart TA, Olwin RB, Hauschka SD: J Cell Biol 105:949–956, 1987.
5. Delli Bovi P, Curatola AM, Kern FG, Greco A, Ittmann M, Basilico C: Cell 50:729–737, 1987.
6. Dickson C, Peters G: Nature 326:833, 1987.
7. Folkman J, Klagsbrun M: Science 235:442–447, 1987.
8. Friesel R, Burgess WH, Mehlman T, Maciag T: J Biol Chem 261:7581–7584, 1986.
9. Gospodarowicz D, Neufeld G, Schweigerer L: Cell Differ 19:1–17, 1986.
10. Gospodarowicz DJ, Wesman J, Moran JS, Lindstrom J: J Cell Biol 70:395–405, 1976.
11. Hauschka SD, Linkhart TA, Clegg C, Merril G: In Mauro A (ed): "Muscle Regeneration." New York: Raven Press, 1979, pp 311–322.
12. Huang JS, Huang SS, Kuo MD: J Biol Chem 261:11600–11607, 1986.
13. Kimmelman D, Kirschner M: Cell 51:869–877, 1987.
14. Lathrop B, Olson E, Glaser L: J Cell Biol 100:1540–1547, 1985.
15. Linkhart TA, Clegg CH, Hauschka SD: Dev Biol 86:19–30, 1981.
16. Moenner M, Chevallier B, Badet J, Barritault D: Proc Natl Acad Sci USA 83:5024–5028, 1986.
17. Morrison RS, Sharma A, DeVellis J, Bradshaw RA: Proc Natl Acad Sci USA 83:7537–7541, 1986.
18. Moscatelli D: J Cell Physiol 131:123–130, 1987.
19. Neufeld G, Gospodarowicz D: J Biol Chem 260:13860–13868, 1985.
20. Neufeld G, Gospodarowicz D: J Biol Chem 261:5631–5637, 1986.

21. Olwin BB, Hauschka SD: Biochemistry 25:3487–3492, 1986.
22. Olwin BB, Hauschka SD: J Cell Biol 107:761–769, 1988.
23. Reznikoff CA, Brankow DW, Heidelberger C: Cancer Res 33:3231–3238, 1973.
24. Scatchard G: Ann NY Acad Sci 51:660–672, 1949.
25. Schubert D, Harris J, Devine CF, Heinemann S: J Cell Biol 61:398–413, 1974.
26. Seed JS, Olwin BB, Hauschka SD: Dev Biol 128:50–57, 1988.
27. Slack JMW, Darlington BG, Heath JK, Godsave SF: Nature 326:197–200, 1987.
28. Smith R, Peters G, Dickson C: EMBO J 7:1013–1022, 1988.
29. Togari A, Dickens G, Huzuya H, Guroff G: J Neurosci 5:307–316, 1985.
30. Wagner JA, D'Amore PA: J Cell Biol 101:1363–1367, 1986.
31. Walikie P, Cowan WM, Ueno N, Baird A, Guilleman R: Proc Natl Acad Sci USA 83:3012–3016, 1986.
32. Yoshida T, Miyagawa K, Odagiri H, Sakamoto H, Little PFR, Terada M, Sugimura T: Proc Natl Acad Sci USA 84:7305–7309, 1987.

Journal of Cellular Biochemistry 39:455–465 (1989)
Growth Factors and Their Receptors: Genetic Control and Rational Application 207–217

Ligand-Induced Association of Epidermal Growth Factor Receptor to the Cytoskeleton of A431 Cells

P.M.P. van Bergen en Henegouwen, L.H.K. Defize, J. de Kroon, H. van Damme, A.J. Verkleij, and J. Boonstra

Department of Molecular Cell Biology, University of Utrecht, 3584 CH Utrecht (P.M.P.v.B.e.H., J.d.K., H.v.D., A.J.V., J.B.) and Hubrecht Laboratory, Netherlands Institute for Developmental Biology, 3584 CT Utrecht, The Netherlands (L.H.K.D.)

Recently, we have obtained evidence in favor of a structural interaction between the epidermal growth factor (EGF) receptor and the Triton X-100-insoluble cytoskeleton of epidermoid carcinoma A431 cells. Here we present a further analysis of the properties of EGF receptors attached to the cytoskeleton. Steady-state EGF binding studies, analyzed according to the Scatchard method, showed that A431 cells contain two classes of EGF-binding sites: a high-affinity site with an apparent dissociation constant (K_D) of 0.7 nM (7.5×10^4 sites per cell) and a low-affinity site with a K_D of 8.5 nM (1.9×10^6 sites per cell). Non-equilibrium binding studies revealed the existence of two kinetically distinguishable sites: a fast-dissociating site, with a dissociation rate constant (k_{-1}) of $1.1. \times 10^{-3}s^{-1}$ ($1.0–1.3 \times 10^6$ sites per cell) and a slow-dissociating site, with a k_{-1} of $3.5 \times 10^{-5}s^{-1}$ ($0.6–0.7 \times 10^6$ sites per cell).

The cytoskeleton of A431 cells was isolated by Triton X-100 extraction. Scatchard analysis revealed that ~5% of the original number of receptors were associated with the cytoskeleton predominantly via high-affinity sites (K_D = 1.5 nM). This class of receptors is further characterized by the presence of a fast-dissociating component ($k_{-1} = 2.0 \times 10^{-3}s^{-1}$) and a slow-dissociating component ($k_{-1} = 9.1 \times 10^{-5}s^{-1}$). The distribution between fast and slow sites of the cytoskeleton was similar to that of intact cells (65% fast and 35% slow sites). Incubation of A431 cells for 2 h at 4°C in the presence of EGF resulted in a dramatic increase in the number of EGF receptors associated to the cytoskeleton. These newly cytoskeleton-associated receptors appeared to represent low-affinity binding sites (K_D = 7 nM). Dissociation kinetics also revealed an increase of fast-dissociating sites. These results indicate that at 4°C EGF induces the binding of low-affinity, fast-dissociating sites to the cytoskeleton of A431 cells.

Key words: Scatchard analysis, dissociation kinetics, epidermal growth factor, binding analysis, Triton X-100 extract

Epidermal growth factor (EGF), a polypeptide growth factor, acts on its target cells first by binding to receptors located on the cell surface followed by the initiation

Received February 1, 1988; accepted November 18, 1988.

of a number of characteristic responses and, depending on the cell system, this binding leads ultimately to initiation of DNA synthesis and cell division [for reviews see 1–4].

Among the effects of EGF on its target cells are morphological changes, such as rounding up of cells [5], induction of membrane ruffling, extension of filopodia [6,7], and redistribution of actin and α-actinin [8]. Since cell morphology is largely maintained by an integrated action of cytoskeletal systems [9], these results suggest that EGF causes changes in the organization of the cytoskeleton via a direct or indirect coupling. In this respect it is of interest that the EGF-receptor kinase has been demonstrated to be associated with the Triton X-100-insoluble cytoskeletons of A431 cells [10]. Furthermore, in these cells the EGF receptor appeared to be structurally associated with the cytoskeleton, as demonstrated by a variety of electronmicroscopic methods [11].

The cytoskeleton-associated fraction of the EGF receptors represented specifically the high-affinity class of receptors, as deduced from Scatchard analysis of EGF binding [11]. These results were supported by the observations that the high-affinity class of EGF receptors of A431 cells exhibited no lateral mobility [12]. The association of growth factor receptors to the cytoskeleton has been demonstrated also in pheochromocytoma (PC12) cells for both the EGF receptor and the nerve growth factor (NGF) receptor [13–15]. In addition, other receptors have been demonstrated to be associated with the cytoskeleton; these include the N-formylated peptide receptor [16,17], the cyclic AMP receptor [18], and the fibronectin receptor [19].

It has been suggested that the cytoskeleton plays a direct intermediate role in signal transduction exerted by growth factors, ultimately leading to DNA synthesis and cell division [20,21]. Actin filaments have been suggested to be involved in endocytosis [22], while the microtubule system has been demonstrated to regulate intracellular transport of the EGF receptor [23] and the transferrin receptor [24]. The cytoskeleton may thus be involved in intracellular targeting of either the ligand or the receptor or both and may in this way influence signal transduction.

In this paper we present a further characterization of the properties of cytoskeleton-associated EGF receptors of A431 cells. It is demonstrated that high-affinity receptors are associated with the cytoskeleton. Addition of EGF to the cells resulted in an additional association of low-affinity, fast-dissociating receptors to the cytoskeleton.

MATERIALS AND METHODS

Cells and Culture Conditions

A431 human epidermoid carcinoma cells were grown in Dulbecco's modified Eagle's medium (DME) supplemented with 10% fetal calf serum (Flow Labs., Irvine, Scotland) in a 7% CO_2 humidified atmosphere at 37°C. Cells were grown routinely to a final density of 50,000–100,000 cells/cm^2.

Isolation of Cytoskeletons of A431 Cells

Cytoskeletons of A431 cells were prepared by extraction for 10 min with 0.5% Triton X-100 in 25 mM N-2-hydroxyethylpiperazine-N′-2-ethanesulfonic acid (Hepes) (pH 7.4) and 1 mM phenylmethylsulfonyl fluoride (PMSF) at 4°C. After removal of the Triton X-100-containing buffer, the cytoskeletons were washed in the same buffer but without Triton X-100.

EGF Binding Studies

Scatchard analysis. Determination of specific binding and Scatchard analysis according to the LIGAND program was performed as described in detail previously [25]. Briefly, A431 cells were grown on tissue culture clusters (16 mm well diameter, Costar Cambridge MA) and washed twice with phosphate-buffered saline (PBS) (pH 7.4). Subsequently, cells or isolated cytoskeletons were incubated in 0.5 ml DME supplemented with 0.1% bovine serum albumin and buffered with 25 mM Hepes (pH 7.4). EGF (receptor grade) was obtained from Collaborative Research Corp. (Waltham, MA) and ^{125}I-EGF from New England Nuclear (Boston, MA). The ^{125}I-EGF was >95% precipitable in trichloroacetic acid. ^{125}I-EGF (100,000 cpm/ng) was added to the cells at varying final concentrations and incubated for 2 h at 4°C, sufficient to obtain steady-state conditions for all EGF concentrations used (data not shown). The cells or cytoskeletons were then washed thrice in PBS and solved in 0.5 ml 1 M NaOH. Bound radioactivity was determined by counting in a gamma counter (Crystal 5412, United Technologies Packard, Downers Grove, IL).

The values for ^{125}I-EGF binding were corrected for non-specific binding by subtracting the amount of radioactivity bound in the presence of 500 ng/ml unlabeled EGF. At equilibrium these values ranged between 1% and 10% of the total binding of intact cells.

EGF dissociation. A431 cells or isolated cytoskeletons were labeled with ^{125}I-EGF (20 ng/ml final concentration) for 2 h at 4°C, as described above. After one wash with PBS, the cells were incubated for various periods of time in 0.25 ml DME containing 200 ng/ml EGF. After two washes with PBS, the radioactivity was determined as described above. The results presented are the mean values of three different experiments; each part of all experiments was performed in triplicate.

Dissociation of a ligand from its receptor can be described according the model of Clark [26] by $B_t = B_0 e^{-kt}$. If more receptor classes are involved, the receptor dissociation can be described by $B_t = B_{0,1} \cdot e^{-k-1,1.t} + B_{0,2} \cdot e^{-k-2,2.t} \ldots + B_0 e^{-k-1,n.t}$ [26]. $B_{0,1}, B_{0,2}, \ldots B_{0,n}$ represents equilibrium binding to each of the different components at the onset of dissociation; $k_{-1,1}, k_{-1,2} \ldots k_{-1,n}$ are the dissociation rate constants of the components 1 to n expressed in s^{-1}. Analysis of the dissociation kinetic data was performed using the EnzFitter program (Elsevier-Biosoft, Cambridge, England) or plotting ln (Bt/Bo) against time; both methods yield similar results.

Protein and Phospholipid Determination

A431 cells were grown in 150 cm^2 culture flasks and suspended in 1 ml of PBS by trypsinization. Various quantities of Triton X-100 were added to 100 μl samples, and the cells were extracted for 10 min at 4°C. After extraction the cells were centrifuged for 2 min at 10,000*g*. Protein and phospholipid determinations were performed both on pellet and supernatant. Protein determination was performed according the BCA–protein assay reagent (Pierce Chemical Company, Rockford, IL). For phospholipid isolation the pellet and supernatant were extracted with chloroform/methanol according to Bligh and Dyer [27], and the phosphate content was determined according to Rouser et al. [28].

RESULTS

Characteristics of Triton X-100 Extraction

The association of EGF receptors with the cytoskeleton has been demonstrated on glutaraldehyde-fixed A431 cells, using 1% Triton X-100 for 30 min [11]. To avoid the artificial effects of fixatives and to allow a study of the dynamics of EGF receptor–cytoskeleton association, we determined EGF binding on cytoskeletons of non-fixed A431 cells.

The effect of various concentrations of Triton X-100 on cellular protein and phospholipid contents in non-fixed A431 cells was studied. As shown in Figure 1, the phospholipid content decreases dramatically after the addition of Triton X-100 and reaches a minimum value after extraction for 10 min at 4°C with 0.5% Triton X-100. Increasing the Triton X-100 concentration only results in a small further decrease of phospholipid content. Protein extraction occurred at much lower detergent concentrations. Addition of 0.05% Triton X-100 for 10 min at 4°C was sufficient to remove 63% of the total protein content (Fig. 1). No further loss of protein was observed at increasing concentrations of Triton X-100. These data demonstrate that the presence of 0.5% Triton X-100 for 10 min at 4°C is sufficient to remove the soluble proteins and vir-

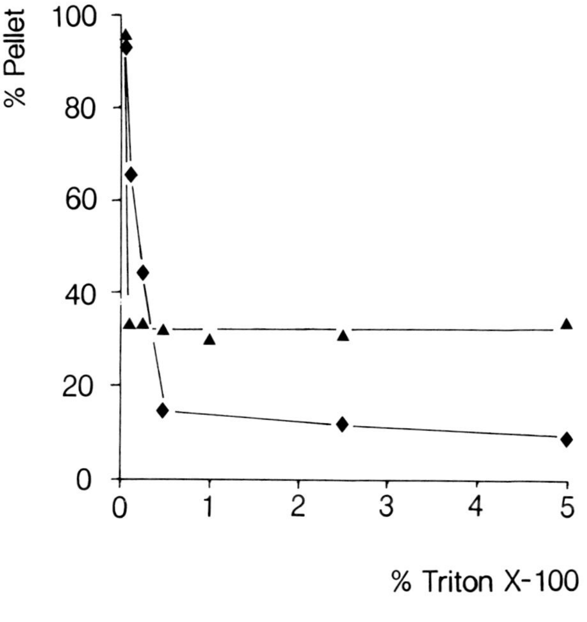

Fig. 1. Effect of Triton X-100 concentration on extraction of proteins and lipids. A431 cells were extracted with different concentrations Triton X-100 for 10 min at 4°C. After extraction, cells were pelleted by centrifugation at 10,000*g* for 2 min and protein and lipid content was determined, both in the supernatant and pellet, as described in Materials and Methods. Percentage protein present in pellet (▲—▲); percentage phospholipid present in pellet (♦—♦).

tually all phospholipids. The residue obtained is defined as the cytoskeletons of A431 cells.

EGF Binding to A431 Cytoskeletons

Binding of EGF to the cytoskeletons of A431 cells is shown in Figure 2. Within a minute of Triton X-100 treatment, a maximal decrease of EGF binding is obtained. No further decrease of EGF binding is observed during prolonged Triton X-100 treatment (Fig. 2).

A similar effect of Triton X-100 was obtained when the cells were incubated with EGF prior to the Triton X-100 extraction. However, in this case the binding of EGF was significantly higher than that measured on isolated cytoskeletons (Fig. 2). These results demonstrate that at least a part of the EGF receptor population is associated with the cytoskeleton of non-fixed A431 cells. Furthermore, the results show that incubation of intact cells with EGF at 4°C results in an increase of Triton X-100-resistant EGF-binding sites, suggesting an EGF-induced association of receptors to the A431 cytoskeleton.

Scatchard Analysis

Receptor characteristics, such as affinity and receptor numbers per cell, are usually obtained by determination of the relationship between specific binding and

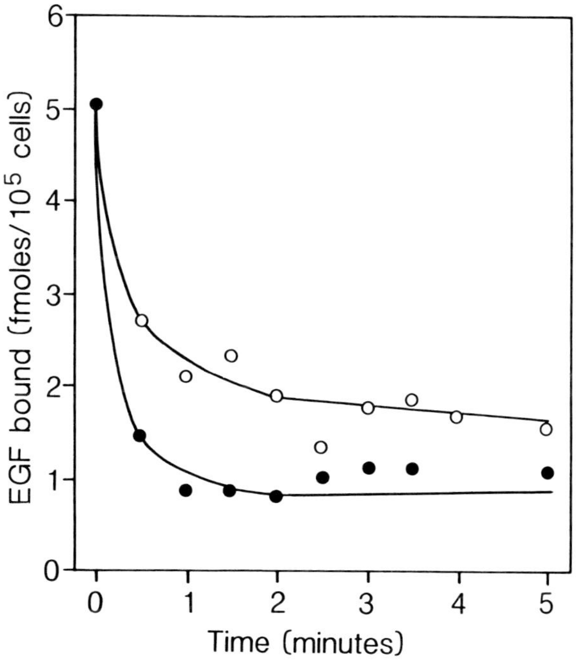

Fig. 2. Effect of Triton X-100 extraction time on EGF binding. The Triton X-100 extraction was performed, as described in Materials and Methods. Specific EGF binding was determined by incubation of intact or Triton X-100-treated cells in the presence of ^{125}I-EGF (2 ng/ml, 100,000 cpm/ng) for 2 h at 4°C, as described previously [25]. Cells treated with Triton X-100 followed by EGF binding (●—●) cells incubated with ^{125}I-EGF, followed by Triton X-100 extraction (○—○).

ligand concentration at steady-state conditions, followed by analysis via the Scatchard method. According to this procedure, A431 cells have been demonstrated to contain two classes of EGF-binding sites; that is, a high-affinity and a low-affinity binding site [11,29], although the precise determination of affinity and numbers of high-affinity sites is considerably hampered by the low number of sites compared with the total number and the relatively small difference between affinities of high- and low-affinity sites [29]. Nevertheless, the best fit of the EGF-binding data plotted following the Scatchard method is according to the two-site model, as shown in Figure 3A.

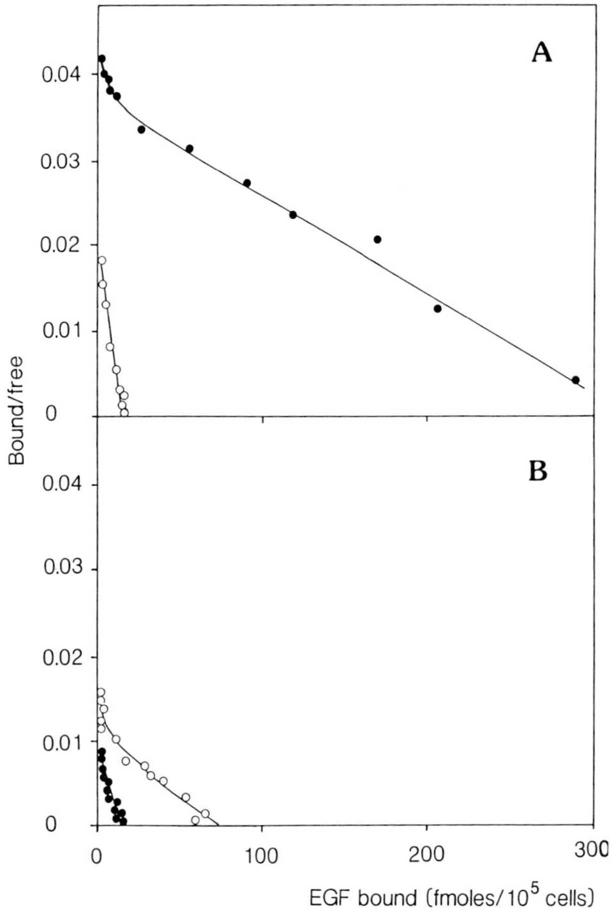

Fig. 3. Scatchard analysis of EGF binding. EGF binding was performed at 4°C for 2 h. The various concentrations of EGF were obtained by mixing ^{125}I-EGF (0.5 ng/ml; 100,000 cpm/ng) with unlabeled EGF, as described in detail previously [25]. Each point represents the mean value of two experiments each performed in triplicate. The data were analyzed using the LIGAND program [25]. **A:** Untreated cells (●—●) and cells preincubated with 40 ng/ml 2E9 (○—○), as described in detail elsewhere [30]. **B:** Cells were extracted with 0.5% Triton X-100 followed by EGF binding, as described in Materials and Methods (●—●), and cells were exposed to EGF as described above, followed by Triton X-100 extraction (○—○).

The high-affinity site is characterized by an apparent dissociation constant (K_D) of ~0.7 nM and 75,000 binding sites per cell, while the low-affinity sites have a K_D of 8.5 nM and 1.9×10^6 binding sites per cell. The reliability of the numbers obtained from Scatchard analysis is demonstrated by the usc of the monoclonal anti-EGF-receptor antibody 2E9. This antibody has been demonstrated to specifically block EGF binding to low-affinity binding sites, leaving EGF binding to the high-affinity binding site undisturbed [29]. As shown in Figure 3A, in the presence of saturating amounts of 2E9, EGF binding is characterized by a linear relationship in a Scatchard graph. The K_D is calculated as 0.7 nM, and maximal binding represents 90,000 receptors/cell. These values closely agree with the characteristics of the high-affinity binding site of untreated cells. Isolation of cytoskeletons of non-fixed A431 cells by 0.5% Triton X-100 for 10 min at 4°C, followed by EGF binding also reveals a linear relationship in a Scatchard plot (Fig. 3B), indicating that cytoskeletons exhibit only one class of EGF-binding sites. This class of binding sites is characterized by a K_D of 1.5 nM and 70,000 sites per cell, thus resembling strongly the high-affinity sites of intact cells. As such, these data agree with the presence of high-affinity EGF-binding sites in cytoskeletons of glutaraldehyde-fixed A431 cells [11].

As shown in Figure 2, incubation of intact cells in the presence of EGF, followed by Triton X-100 extraction, results in an increase of Triton X-100-resistant EGF binding, suggesting an EGF-induced association of receptors with the cytoskeleton. Therefore, the EGF-binding characteristics were determined on cytoskeletons of non-fixed cells isolated after the binding period. Analysis of the Scatchard plot clearly reveals the presence of two classes of binding site, that is, a high-affinity site with a K_D of 0.6 nM and 20,000 receptors per cell and a low-affinity site with a K_D of 7 nM and 380,000 receptors per cell (Fig. 3B). These results demonstrate that EGF causes an increase in the association of EGF receptors with the cytoskeleton, primarily due to low-affinity receptors.

Effect of Triton X-100 Extraction on EGF-Dissociation Characteristics

Steady-state binding experiments, as described above, provide important information on receptor characteristics, but information on receptor dynamics, that is, the rate at which the ligand is bound to or dissociated from the receptor is lacking. Thus we have performed a further characterization of cytoskeleton-associated EGF receptors by studying EGF dissociation.

A431 cells were incubated in the presence of 20 ng/ml ^{125}I-EGF for 2 h at 4°C, after which the cells were washed with binding buffer. EGF dissociation was initiated by adding 200 ng/ml unlabeled EGF in binding buffer. The dissociation rate was considerably increased by the addition of excess unlabeled EGF (data not shown). Since adding higher concentrations of EGF (0.5 or 1 μg/ml) did not alter the dissociation rate further, we used an EGF concentration of 200 ng/ml for all dissociation experiments. Plotting $\ln(B_t/B_o)$ against time results in a curvilinear relationship (Fig. 4), indicating that dissociation of EGF is not simply from one single class of receptors but, most likely, from two different classes. Dissociation of EGF from the fast component is characterized by a dissociation rate constant k_{-1} of $1.1 \times 10^{-3}s^{-1}$ and from the slow component by a k_{-1} of $3.5 \times 10^{-5}s^{-1}$. Furthermore, the distribution of EGF is determined as 64% and 36% in the fast and slow sites, respectively. The same distribution

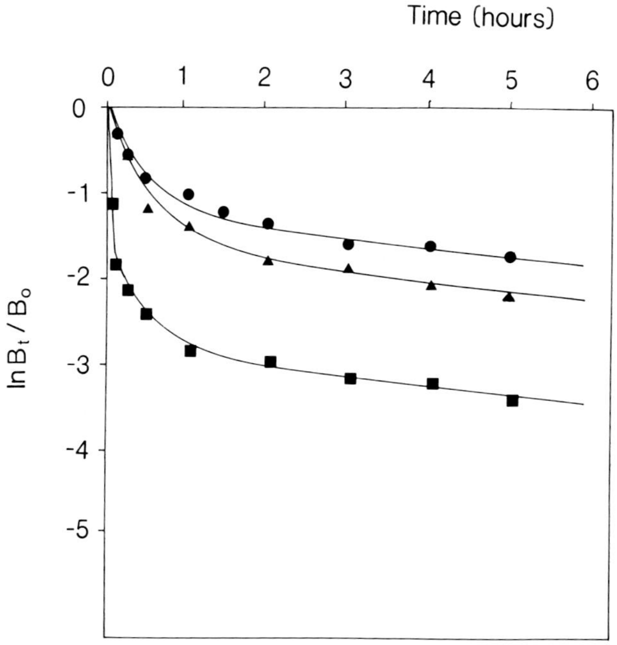

Fig. 4. Dissociation of EGF from its receptor. A431 cells were exposed to 20 ng/ml ^{125}I-EGF (10,000 cpm/ng) for 2 h at 4°C. Subsequently, the dissociation of EGF from its receptor was determined in the presence of 200 ng/ml unlabeled EGF, as described in Materials and Methods. Each point represents the mean value of at least three experiments performed in triplicate. Control cells (●—●); isolated cytoskeletons (▲—▲); dissociation from cytoskeletons, isolated after EGF binding to intact cells (■—■).

between fast and slow receptors is found when labeling is performed under saturating conditions (100 ng/ml EGF) (data not shown).

Subsequently, EGF dissociation has been determined from A431 cytoskeletons. After the Triton extraction, 0.5% Triton X-100 for 10 min at 4°C, the isolated cytoskeletons were incubated in the presence of 20 ng/ml ^{125}I-EGF for 2 h at 4°C, and EGF dissociation was determined, as described in Materials and Methods. As shown in Figure 3B, cytoskeletons contain predominantly high-affinity EGF-binding sites, but, surprisingly, the dissociation curve demonstrates clearly a curvilinear relationship comparable to control cells (Fig. 4). The rate constants of the fast- and slow-dissociation sites are determined as $2.0 \times 10^{-3}s^{-1}$ and $9.1 \times 10^{-5}s^{-1}$, respectively, while the distribution of EGF in the slow and fast compartments is 35% and 65%, respectively. These results demonstrate clearly that dissociation of EGF from cytoskeleton-associated receptors is identical to that of control cells and, furthermore, that the slow- and fast-dissociation sites are not simply identical to the high- and low-affinity receptors, respectively.

Finally EGF dissociation was determined from cytoskeletons of A431 cells isolated after the EGF binding. The EGF receptors associated to these cytoskeletons are characterized by both low- and high-affinity binding sites (Fig. 3B). Analysis of the dissociation data reveal again fast- and slow-dissociation sites; the dissociates rates are

$6.8 \times 10^{-3} s^{-1}$ and $6.1 \times 10^{-5} s^{-1}$, respectively, (Fig. 4). Compared with intact cells and isolated cytoskeletons, a clear difference is observed in the distribution of the two dissociation sites: 90% of the total receptor population associated with the fast site and 10% with the slow site, respectively. Thus isolation of cytoskeletons after EGF binding results in the appearance of low-affinity, fast-dissociating binding sites associated with the cytoskeleton (Figs. 3B and 4).

DISCUSSION

In this paper we have further characterized the properties of EGF receptors associated with the cytoskeletons of A431 cells and present evidence in favor of an EGF-induced association of receptors with the cytoskeleton.

Recently, we demonstrated that high-affinity EGF receptors were associated with the cytoskeleton of glutaraldehyde-fixed A431 cells [11]. Here we also demonstrate that non-fixed A431 cells exhibit high-affinity receptors that were resistant to Triton X-100 treatment. Binding of EGF to intact cells at 4°C followed by Triton X-100 extraction reveals that, under these conditions, additional low-affinity receptors are associated with the cytoskeleton. These observations demonstrate that EGF induces an association of EGF receptors with the cytoskeleton. These observations were obviously not obtained when the cells were pre-fixed with glutaraldehyde [11].

Dissociation of EGF from its receptors reveals the presence of two classes of sites, a fast- and a slow-dissociating site. The fast sites represent 64% and the slow sites 36% of the total receptor population independent of the EGF concentration used for pre-labeling.

It is tempting to suggest that slow-dissociating sites are identical to the high-affinity binding sites and fast-dissociating sites to the low-affinity binding sites. However, the ratio between occupied high- and low-affinity receptors at steady-state conditions depends on the EGF concentration [29]. In contrast, the ratio between occupied slow and fast receptors in intact cells appears to be independent of the EGF concentration (unpublished observations), demonstrating that the high-affinity sites obtained in a Scatchard graph are not simply identical to the slow-dissociating sites obtained in a dissociation assay.

Based on the numbers obtained in this study (~5% high-affinity receptors and 35% slow-dissociating sites) it can be concluded that the quantity of slow sites is strongly over-estimated, most likely due to a re-association of released ligand with the receptor during the dissociation assay.

In this respect, the dissociation of EGF from isolated cytoskeletons also should be considered with caution, since cytoskeletons exhibit apparently only one class of high-affinity binding sites (Fig. 3B) but clearly two classes of dissociating receptors. These conflicting results may be due to either the presence of a small fraction of low-affinity binding sites associated with the cytoskeleton (not detectable in a Scatchard graph), a re-association of released EGF to the receptors during the dissociation assay, or both. The presence of a small fraction of low-affinity sites on isolated cytoskeletons is supported by the observation that the K_D is calculated as 1.5 nM, a twofold increase compared with the K_D of the high-affinity binding site. These findings clearly demonstrate that a quantitative comparison of binding characteristics obtained from equilibrium and non-equilibrium studies requires a further detailed study of the mechanisms of association and dissociation under experimental conditions.

Finally, binding of EGF at 4°C to intact cells followed by Triton X-100 extraction and EGF dissociation demonstrates a seven- to eightfold increase in the number of fast-dissociating sites. Although a quantitative comparison between the data obtained in Figures 3 and 4 appears not to be justified, a qualitative comparison demonstrates that EGF causes association of a relatively high number of low-affinity binding sites with the cytoskeleton and a significant increase in the relative number of fast-dissociating sites. Therefore it is concluded that the novel cytoskeleton-associated EGF receptors are both low-affinity and fast-dissociating ones.

An important aspect of the interaction between the EGF receptor and the cytoskeleton involves the nature of the cytoskeletal component. Previously, it has been suggested that actin is involved in this association, as judged by the diameter of the elements involved in electron micrographs [11]. Recently, this has been supported by co-localization of EGF receptors and actin filaments by immunofluorescence microscopy (manuscript in preparation). It is tempting to suggest that this interaction between EGF receptors and the actin filaments is involved in receptor clustering, as demonstrated quantitatively in A431 cells and, consequently, specifically involved in receptor mediated endocytosis. For intracellular transport, the microtubule system—not actin—seems to be more important [23,24]. It is clear that further analysis of the characteristics of the EGF receptor, including cytoskeletal association, level of phosphorylation, and clustering, is required for a full understanding of the molecular mechanism underlying the activation of signal transduction by EGF.

REFERENCES

1. Carpenter G: Annu Rev Biochem 56:881–914, 1987.
2. Schlessinger J: J Cell Biol 103:2067–2072, 1986.
3. de Laat SW, Moolenaar WH, Defize LHK, Boonstra J, van der Saag PT: Biochem Soc Symp 50:205–220, 1985.
4. Defize LHK, Mummery CL, Moolenaar WH, de Laat SW: Cell Differ 20:87–102, 1987.
5. Chinkers M, McKanna JA, Cohen S: J Cell Biol 88:422–429, 1981.
6. Chinkers M, McKanna JA, Cohen S: J Cell Biol 83:260–265, 1979.
7. Boonstra J, Mummery CL, Feijen A, de Hoog WJ, van der Saag PT, de Laat SW: J Cell Physiol 131:409–417, 1987.
8. Schlessinger J, Geiger B: Exp Cell Res 134:273–279, 1981.
9. Ben-Ze'ev A: Biochim Biophys Acta 780:197–212, 1985.
10. Landreth GE, Williams LK, Rieser GD: J Cell Biol 101:1341–1350, 1985.
11. Wiegant FAC, Blok FJ, Defize LHK, Linnemans WAM, Verkleij AJ, Boonstra J: J Cell Biol 103:87–94, 1986.
12. Rees AR, Gregoriou M, Johnson P, Garland PB: EMBO J 3:1843–1847, 1984.
13. Schechter AL, Bothwell MA: Cell 24:867–874, 1981.
14. Vale RD, Shooter EM: J Cell Biochem 22:99–109, 1983.
15. Vale RD, Ignatius MJ, Shooter EM: J Neurosci 5:2762–2770, 1985.
16. Jesaitis AJ, Naemura JR, Sklar LA, Cochrane CG, Painter RG: J Cell Biol 98:1378–1387, 1984.
17. Painter RG, Zahler-Bentz K, Dukes RE: J Cell Biol 105:2959–2971, 1987.
18. Ludérus MEE, Van Driel R: J Biol Chem 263:8326–8331, 1988.
19. Tamkun JW, DeSimone DW, Fonda D, Patel RS, Buck C, Horwitz AF, Hynes RO: Cell 46:271–282, 1986.
20. Thyberg J: Exp Cell Res 155:1–8, 1984.

21. Rebillard M, Leibovitch S, Jullien M, Talha S, Harel L: Exp Cell Res 172:432–438, 1987.
22. Geiger B: Biochim Biophys Acta 737:305–341, 1983.
23. van 't Hof RJ, Defize LHK, Nuijdens R, de Brabander M, Verkleij AJ, Boonstra J: Eur J Cell Biol: in press.
24. de Brabander M, Nuijdens R, Geerts H, Hopkins C: Cell Motil Cytoskeleton 9:30–47, 1988.
25. Boonstra J, Mummery CL, van der Saag PT, de Laat SW: J Cell Physiol 123:347–352, 1985.
26. Boeynaems JM, Dumont JE: "Outlines of Receptor Theory." Amsterdam: Elsevier/North Holland Biomedical Press.
27. Bligh EG, Dyer WJ: Can J Biochem Physiol 37:911–917, 1959.
28. Rouser G, Fleischer S, Yamamoto A: Lipids 5:494–496, 1970.
29. Defize LHK, Arndt-Jovin DJ, Jovin TM, Boonstra J, Meisenhelder J, Hunter T, de Heij HT, de Laat SW: J Cell Biol 107:939–949, 1988.

Growth Factors and Their Receptors: Genetic
Control and Rational Application, pages 219–227

Aberrant Expression of a Murine Hemopoietic Growth Factor (Granulocyte Macrophage Colony Stimulating Factor) in Transgenic Mice

Ashley R. Dunn, Richard A. Lang, R. Andrew Cuthbertson, and Donald Metcalf

Melbourne Tumor Biology Branch, Ludwig Institute for Cancer Research (A.R.D., R.A.L.) and Walter and Eliza Hall Institute for Medical Research (D.M.), PO Royal Melbourne Hospital and Howard Florey Institute for Experimental Physiology and Medicine, University of Melbourne (A.C.), Victoria 3050, Australia

Transgenic mice aberrantly expressing the murine granulocyte macrophage colony stimulating factor gene have been generated. Mice show infiltrations of macrophages in the eyes, peritoneal cavity, and ultimately in striated muscle, leading to destruction of normal tissue and premature death. Analysis by Northern blots and in situ hybridization histochemistry indicates that the macrophages express the transgene resulting in autocrine stimulation. As a result of autostimulation by GM-CSF, the macrophages are to some extent functionally activated and responsible for local tissue destruction. Tissue damage occurs either as a result of the direct action of macrophages on normal cells or alternatively by the production of one or more growth factors or immunoregulatory molecules.

Key words: GM-CSF transgenic mice, autocrine stimulation, activated macrophages

The growth and multiplication of mammalian cells depends on the availibility of a complex array of nutrients. In addition to those molecules required for various metabolic processes, the proliferation of many specialized cell types requires activation of cell surface receptors by corresponding specific polypeptide growth factors. Upon receptor occupation a biochemical signal is transmitted to downstream targets that in turn activate substrates forming part of signal transduction circuits resulting in alterations in gene expression, initiation of DNA synthesis and cell division.

Results from in vitro studies have provided compelling evidence that the proliferation and differentiation of hemopoietic cells is dependent on a group of glycoprotein regulators, the colony stimulating factors (CSFs) [1]. In the past 4 years, the individual CSFs have been molecularly cloned and the range of biological activities displayed by the recombinant encoded regulators characterized. Intriguingly, in addition to their

Received February 1, 1988.

mitogenic properties, CSFs have also been shown to play a role in the functional activation of terminally differentiated hemopoietic cells in vitro. Thus, CSFs are likely to play a pivotal role in many aspects of normal hemopoiesis.

For the most part active hemopoiesis occurs in the bone marrow where both stem cell self-renewal as well as proliferation and development of mature progeny is achieved by interaction with stromal cells forming part of the microenvironment. While little is known of the regulatory molecules produced by the stroma, it is clear that target cells are stimulated by direct contact with the stromal layer. By contrast, hemopoietic cells in peripheral circulation are likely to be stimulated by soluble factors (possibly CSFs).

In response to increased demand such as in the course of an infection, specific CSFs as well as other immunoregulatory molecules are produced by T cells as a result of their activation by bacterial products. The T cell products interact with target hemopoietic cells, providing an important level of regulation that allows for selective amplification of certain cell compartments in response to increased demand.

In our own laboratories, we have considered the possible outcome of the aberrant expression of a hemopoietic growth factor by a hemopoietic cell also capable of responding to this regulator. This situation might be expected to have ramifications both to the regulation of normal hemopoiesis as well as to the development of the leukemic state. While the answers to some aspects of these questions have been addressed using in vitro systems we have sought to study the problem in vivo. Accordingly, we have established transgenic mice aberrantly expressing the murine granulocyte macrophage colony stimulating factor gene.

RESULTS

The transgene consisted of the murine granulocyte macrophage (GM)-CSF gene [2] containing a segment of ϕX174 sequences in its 3′ untranslated region and driven by the Moloney murine leukemia viral long terminal repeat (LTR) including promoter/enhancer elements. Transgenic SJL × C57BL/6 mice were generated using standard procedures [3]. Two founder mice (one male and one female) were certified, by Southern blotting, to contain the transgene, and a breeding programme was established using each founder animal independently as a breeding partner.

In the course of surveying potential transgenic mice, we noted a bilateral opacity of the eyes in several animals. Southern blotting of tail DNAs revealed a perfect correlation between presence of the transgene and the abnormal eye phenotype. Histological sections of affected eyes showed the presence of large numbers of macrophages whose presence was often found to be associated with varying degrees of cataract, corneal, and retinal damage [4]. In contrast to the eyes of control littermate mice, the vitreous of eyes from transgenic animals contained biologically active GM-CSF; moreover, analysis of RNA from pooled ocular tissue by Northern blotting using a ϕX174 probe revealed the presence of abundant levels of mRNA corresponding to the GM-CSF transgene. To identify the sites of expression of the transgene at the level of individual cells we have undertaken an analysis of sections of individual eyes by in situ hybridization histochemistry.

As revealed in Figure 1 when a single-stranded radiolabelled probe complementary to the transcripts of the GM-CSF/ϕX174 transgene is used, significant labelling of the macrophages in the vitreous is observed. By contrast, only background silver

A

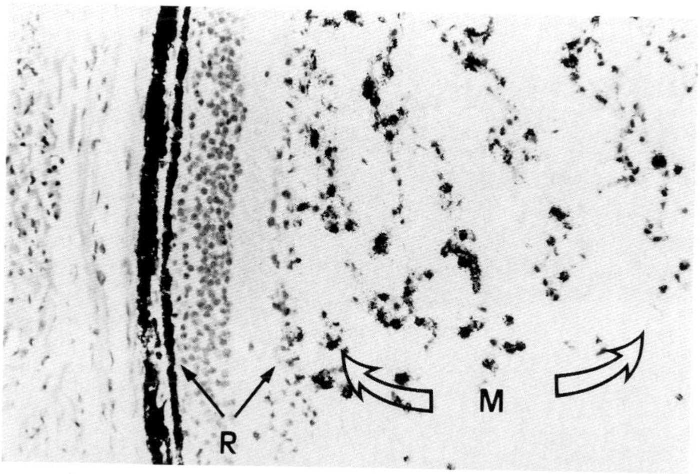

B

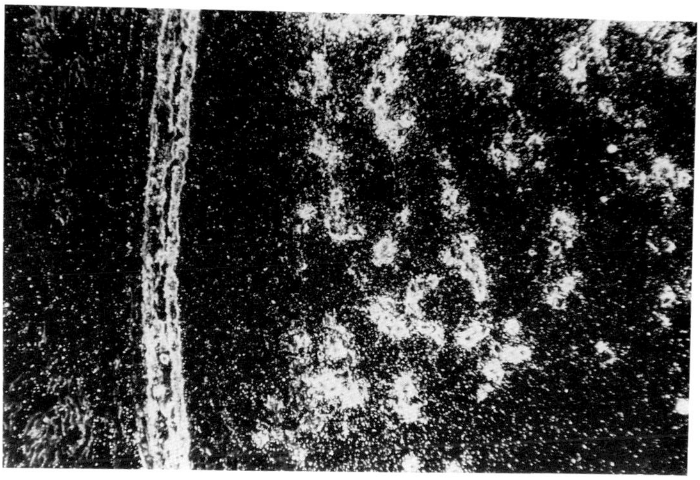

Fig. 1. In situ hybridization histochemistry. Sections of eyes of transgenic mice were annealed with a single-stranded radiolabelled [^{32}P]SP6 probe containing ϕX174 sequences complementary to the transcripts of the GM-CSF/ϕX174 transgene. Following autoradiographic exposure sections were photographed in light (**A**) and dark field (**B**). Autoradiographic grains are associated specifically with the abnormal vitreous macrophages (M) with only background labelling in the remaining normal retina (R). The dark band on the left of the section is the normal pigment epithelial layer. Hybridization histochemistry was carried out according to the procedures outlined by Penschow [5].

grains are seen in corresponding tissue sections using a mRNA synonymous probe; morever, there is no significant association of grains with the macrophages (Fig. 2). Thus, it appears that the predominant sites of synthesis of the GM-CSF transgene product are the macrophages themselves.

In addition to ocular abnormalities, a detailed analysis has revealed high levels of GM-CSF in the serum and urine of transgenic animals not evident in littermate control mice. In addition, transgenic mice contained abnormal accumulations of macrophages most pronounced in the peritoneal cavity. The cells were often multinucleate with basophilic cytoplasm; with respect to their morphology they were reminiscent of

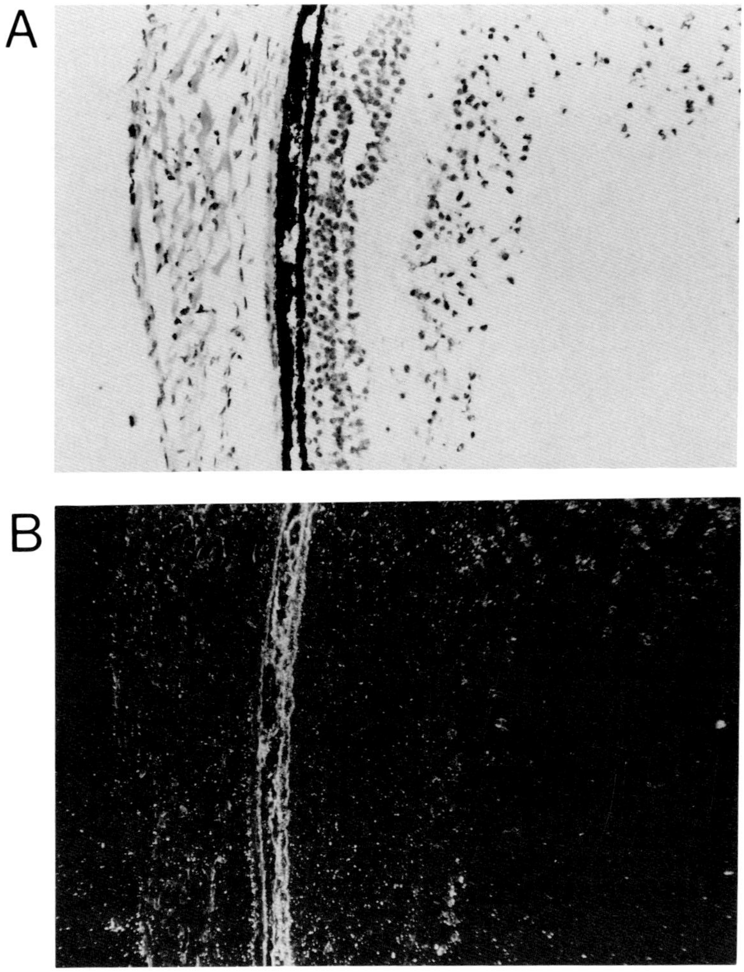

Fig. 2. In situ hybridization histochemistry. Sections of eyes of GM-CSF transgenic mice were annealed with a single-stranded radiolabelled [^{32}P]SP6 probe containing ϕX174 sequences in a synonomous configuration to transcripts corresponding to the GM-CSF/ϕX174 transgene. Only background autoradiographic grains are evident in light (**A**) or dark field (**B**) compared to the specific labelling seen using the mRNA complementary probe in Figure 1.

activated macrophages, a conclusion reinforced by their demonstrated capacity for heightened phagocytosis. Northern blot analysis of RNA isolated from cells recovered from the peritoneal cavities of transgenic mice indicated expression of the transgene [4].

With increasing age transgenic mice became less mobile, suffered weight loss, assumed a hunched posture (often involving partial paralysis of the hind limbs), and during the later stages of life exhibited an obvious tremor when handled. At this advanced stage of disease, mice usually succumbed in 2–3 days. At autopsy the skeletal muscle, and particularly the thigh muscle, showed grossly abnormal infiltrations of macrophages that formed focal lesions between the muscle fibre bundles. Northern

TABLE I. Functional Activation of Macrophages/Monocytes by GM-CSF

Acquired functional activity	Authors	Reference No.
1. Induction of antibody-dependent cellular toxicity	Lopez et al., 1983	[6]
2. Potentiates antibody responses by activation of an accessory function of murine macrophages	Grabstein et al., 1986a	[7]
3. Killing leishmania by phagocytosis	Handman and Burgess, 1979	[8]
4. Stimulation of tumoricidal action against malignant cells	Grabstein et al., 1986b	[9]
5. Induction of TNF by monocytes	Cannistra et al., 1987	[10]
6. Inhibition of neutrophil migration	Gasson et al., 1984	[11]

blot analysis of striated muscle from transgenic mice showed an age-related correlation between levels of expression of the tansgene and the degree of macrophage infiltration [4].

A major question arising from these observations is the basis for the abnormal accumulations of macrophages in the peritoneal cavity, eyes, and striated muscle of transgenic animals. Since our data support the notion that macrophages themselves express the transgene, their capacity for autostimulation might underlie both their occupation of abnormal sites as well as their apparent functionally activated state.

As noted previously, CSFs including GM-CSF have been shown to be capable of enhancing certain functional characteristics of macrophages. Some of these effects or functions are listed in Table I. Macrophages, as well as platelets and lymphocytes, are also involved in the acute-phase response to inflammation, and repair is often mediated by the production of various biologically active molecules. While many of these regulators are known to be produced in response to the interaction of macrophages and bacterial products (e.g., lipopolysaccharide [LPS]) it is less clear whether they are bona fide products of macrophages activated by normal physiological regulators like GM-CSF. Since the action of one or more macrophage products might account for some aspects of the pathology of GM-CSF transgenic mice, we have sought to identify potential regulators produced by normal bone marrow-derived macrophages in response to GM-CSF. To this end parallel cultures of normal bone marrow-derived macrophages were stimulated with LPS (for 2 and 6 h) and GM-CSF (for 2, 4, and 12 h). A series of identical Northern blots of whole cell poly-A-containing RNAs corresponding to the various time points were hybridized with murine interleukin (IL)-1α, IL-1β, and tumor necrosis factor (TNF) cDNAs, respectively. While LPS stimulated macrophages express IL-1α, IL-β, and TNF, no such transcripts were observed in GM-CSF stimulated macrophages (Fig. 3). Neither GM-CSF or LPS stimulated macrophages contained detectable GM-CSF transcripts.

DISCUSSION

A long-held belief is that the release of a cell's dependence on growth factors might represent a contributory factor in progression to malignancy [14,15]. This might be achieved by the unscheduled synthesis of a growth factor to which the same cell responds (autocrine stimulation), or alternatively, by the activation of a gene prod-

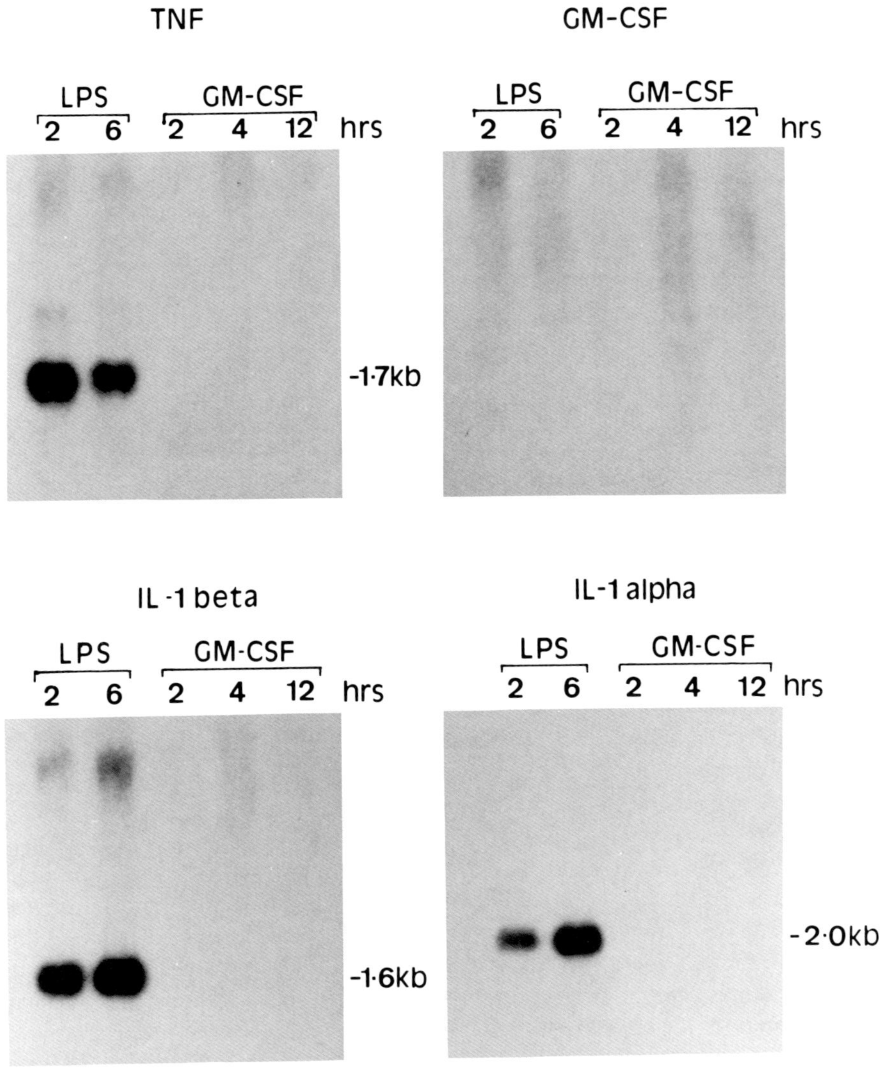

Fig. 3. Northern blots of normal bone marrow-derived macrophages. Bone marrow macrophages were prepared by the method of Tushinski et al. [12], as modified by Vairo and Hamilton [13], and stimulated with LPS (100 ng/ml) or GM-CSF (100× maximal proliferative stimulatory concentration for FDC-P1 cells) for the indicated times and whole cell poly-A-containing RNA prepared. Hybridization was performed with radiolabelled SP6 probes corresponding to murine IL-1α, IL-β, TNF, and GM-CSF.

uct normally participating in a biochemical circuit linked to a normal growth factor receptor. Platelet-derived growth factor (PDGF), tumor growth factorα (TGFα), and TGFβ have all been implicated in oncogenesis since they are produced by a variety of cancer cells which often display corresponding receptors and accordingly are subject to autocrine stimulation.

Previous work from our own laboratory [16] has shown that a nonleukemic and growth factor dependent immortalized hemopoietic cell (dependent on either GM-

CSF or IL-3) can be converted into a leukemic cell following the introduction and expression of the gene encoding GM-CSF. Thus, experimentally induced autocrine stimulation of hemopoietic cells can represent a key event in the development of the leukemic state.

This observation poses the question of whether autocrine stimulation might represent a contributing factor in the formation of a leukemic cell in vivo. Since leukemia, like other malignancies, is a multistage disease, one might have expected the GM-CSF transgenic mice to be more susceptible to leukemia requiring fewer additional genetic mutations than littermate control animals. While this possibility was one underlying reason for establishing GM-CSF transgenic mice, the question remains unfulfilled. In part this is because the mice survive for only one-quarter of their normal life span and also because the transgene is probably not expressed in a hemopoietic cell sufficiently immature to serve as the founder cell of a leukemic clone. This latter conclusion is based on the essentially normal cellularity of the bone marrow coupled with our failure to detect transcripts corresponding to the transgene in transgenic marrows.

While some of the abnormalities observed in the eyes and striated muscle of transgenic mice can be regarded as damage, the situation is reminiscent of some aspects of inflammation where inflammatory cells enter the injured area and co-ordinate the process of repair. In the normal course of an inflammatory response, macrophages and other inflammatory cells respond to hormonal and cellular signals (proteins and polysaccharides) and participate in a variety of physiological and pathological events, including phagocytosis, the release of growth factors essential for the recruitment of other cells, the formation of new matrix, angiogenesis, and the modulation of the immune response. An important issue therefore is to what extent constitutive stimulation of macrophages by GM-CSF correlates with macrophages activated in response to infection or inflammation. Clearly, macrophages derived from the peritoneal cavities of transgenic mice have a heightened capacity for phagocytosis, and this in itself might account for some of the observed destruction. Alternatively, transgenic macrophages may produce one or more of the authentic products normally associated with LPS stimulated macrophages that act in a paracrine manner to influence the behaviour of neighboring cells. Candidate molecules would include PDGF, IL-1, TNF, and TGFβ.

PDGF, while principally a product of platelets, is also synthesized by macrophages [17,18]. In addition to its well-documented mitogenic properties, PDGF is a potent chemoattractant for fibroblasts [19] and monocytes and neutrophils [20]. Moreover, PDGF is known to stimulate the production of collagenase by fibroblasts thereby contributing to the formation of matrix [21]. This activity might explain the morphology of some of the lesions observed in transgenic mice.

Activated macrophages are also known to produce IL-1 [22,23], a hormone that modulates tissue responses in inflammation but is also implicated in muscle wasting [24]. TNF (cachectin) has been shown to induce necrosis and tumor regression [25], but the mechanism by which this is achieved is unknown. Interestingly TNF, like IL-1, induces muscle wasting [26] and may be responsible for some of the damage observed in transgenic mice. In this context it is noteworthy that a macrophage cell line (U937) exposed to GM-CSF leads to the transient induction of TNF. Moreover, GM-CSF apparently enhances TNF production by monocytes stimulated with endotoxin and phorbol myristate acetate (PMA) [10].

It is also clear that TGFβ plays an important role in tissue repair, is a product of macrophages (as well as platelets and T lymphocytes), and, based on in vitro studies, controls the activities of other peptide growth factors.

Clearly, the pleiotropic activities of one or more of these regulators may account for some of the damage and muscle wasting observed in the GM-CSF transgenic mice. Our preliminary analysis, at face value, would appear to rule out IL-1α, IL-1β, and TNF as being significant products of GM-CSF stimulated normal bone marrow macrophages. This conclusion is, however, fragile since GM-CSF transgenic macrophages in vivo may well be exposed to other molecules that in concert with GM-CSF serve to enhance the production of macrophage products. Accordingly, we must undertake a survey of the products of freshly derived transgenic macrophages with probes corresponding to candidate regulators.

Whether one or more regulators plays a causal role in the abnormalities observed in transgenic mice will ultimately require the long-term infusion of specific neutralizing antibodies to potentially important regulators in an attempt to alleviate or prevent some of the pathology we have observed.

ACKNOWLEDGMENTS

We are grateful to Linda Tabone for her help in preparing the manuscript.

REFERENCES

1. Metcalf D: "The Hemopoietic Colony Stimulating Factors." Amsterdam: Elsevier, 1984.
2. Stanley E, Metcalf D, Sobieszczuk P, Gough NM, Dunn AR: EMBO J 4:2569–2573,1985.
3. Brinster RL, Chen HV, Trumbauer ME, Yagle MK, Palmiter RD: Proc Natl Acad Sci USA 82:4438–4442, 1985.
4. Lang RA, Metcalf D, Cuthbertson RA, Lyons I, Stanley E, Kelso A, Kannourakis G, Williamson DJ, Klintworth GK, Gonda TJ, Dunn AR: Cell 51:675–686, 1987.
5. Penschow JD, Haralambidis J, Aldred P, Tregear GW, Coghlan JP: Methods Enzymol 124:534, 1986.
6. Lopez AF, Nicola NA, Burgess AW, Metcalf D, Battye FL, Sewell WA, Vadas M: J Immunol 131:2983–2988, 1983.
7. Grabstein K, Mochizuki D, Kronheim S, Price V, Cosman D, Urdal D, Gillis S, Conlon P: J Mol Cell Immunol 2:199–207, 1986 (a).
8. Handman E, Burgess AW: J Immunol 122:1134–1137, 1979.
9. Grabstein KH, Urdal DL, Tushinski RJ, Mochizuki DY, Price VL, Cantrell MA, Gillis S, Conlon PJ: Science 232:506–508, 1986 (b).
10. Cannistra SA, Rambaldi A, Spriggs DR, Herrmann F, Kute D, Griffin JD: J Clin Invest 79:1720–1728, 1987.
11. Gasson JC, Weisbart RH, Kaufman SE, Clark SC, Hewick RM, Wong GG, Golde DW: Science 226:1339–1342, 1984.
12. Tushinski RJ, Oliver IT, Gilbert LJ, Tynan PW, Warner JR, Stanley ER: Cell 28:71–81, 1982.
13. Vairo G, Hamilton JA: Biochem Biophys Res Commun 132:430–437, 1985.
14. Todaro G, De Larco JE: Nature 267:526–528, 1987.
15. Sporn MB, Roberts, AB: Nature 313:745–747, 1985.
16. Lang RA, Metcalf D, Gough NM, Dunn AR, Gonda TJ: Cell 43:531–542, 1985.
17. Shimokado K, Raines EW, Madtes DK, Barrett TB, Benditt EP, Ross R: Cell 43:277–286, 1985.
18. Martinet Y, Bitterman PB, Mornex J, Grotendorst GR, Martin GR, Crystal RG: Nature 319:158–160, 1986.
19. Grotendorst GR, Seppa HE, Kleinman HK, Martin GR: Proc Natl Acad Sci USA 78:3669–3672, 1981.

20. Deuel TF, Senior RM, Huang JS, Griffin GL: J Clin Invest 69:1046–1049, 1982.
21. Bauer EA, Cooper TW, Huang JS, Altman J, Deuel TF: Proc Natl Acad Sci USA 82:4132–4136, 1985.
22. Moore RN, Oppenheim JJ, Farrar JJ, Carter CS, Waheed A, Shaduck RK: J Immunol 125:1302–1305, 1980.
23. Oppenheim JJ, Kovacks EJ, Matsushima K, Durum SK: J Immunol 7:45–56, 1986.
24. Baracos V, Rodemann HP, Dinarello CA, Goldberg AL: N Engl J Med 308:553–558, 1983.
25. Carswell EA, Old LJ, Kassel RL, Green S, Fiore N, Williamson B: Proc Natl Acad Sci USA 72:3666–3670, 1975.
26. Beutler B, Krochin N, Milsark IW, Luedke C, Cerami A: Science 232:977–980, 1986.

Growth Factors and Their Receptors: Genetic Control and Rational Application, pages 229–238

Mechanisms in Interleukin-3 Dependent Growth of Factor Dependent Myeloid Leukemia Cell Lines

Robert J. Isfort, Robert Abraham, W. Stratford May, Don A. Stevens, A. Raymond Frackelton, Jr., and James N. Ihle

Department of Molecular Mechanisms of T Cell Leukemogenesis, NCI-Frederick Cancer Research Facility, Frederick, Maryland 21701 (R.J.I., J.N.I.); Department of Immunology, Mayo Clinic Rochester, Minnesota 55905 (R.A.); The Johns Hopkins Oncology Center, Baltimore, Maryland 21231 (W.S.M., D.A.S.); Department of Medicine, Brown University, Providence, Rhode Island 02908 (A.R.F.)

Interleukin 3 (IL-3) is a hematopoietic growth factor which supports the proliferation and terminal differentiation of hematopoietic stem cells and cells committed to several lineages. IL-3 is also an obligatory growth factor for a series of hematopoietic cell lines derived from long-term bone marrow cultures or primary retrovirus induced leukemias. We have examined the effects of IL-3 stimulation on protein kinase C (PKC) activation, total protein phosphorylation, and tyrosine phosphorylation. Our results demonstrate that PKC is not involved in the IL-3 signal transduction pathway, but rather, IL-3 induces the rapid tyrosine phosphorylation of a number of cellular substrates. One of the most rapidly phosphorylated phosphoproteins is a membrane associated glycoprotein protein of 140 kd. IL-3 binding and cross-linking studies demonstrate that the 140 kd phosphoprotein is a component of the IL-3 receptor.

Key words: tyrosine phosphorylation, protein kinase C, cell growth regulation by, myeloid leukemia, receptor, myeloid leukemia cell lines, interleukin 3

Interleukin 3 (IL-3) is a T cell derived lymphokine which was initially identified and purified to homogeneity based on its ability to induce the enzyme 20α-hydroxysteriod dehydrogenase in splenic lymphocytes of athymic mice [1,2]. The genes for both human [3] and murine IL-3 [4,5] have been cloned. From the structure of the gene and biochemical studies of the purified factor, IL-3 has been shown to be a glycoprotein of approximately 28 kd and secreted IL-3 consists of a single polypeptide chain of 140 amino acids. Purified IL-3 has been demonstrated to be active in a variety of assays using hematopoietic cells due to its ability to support the proliferation and differentiation of cells which can become committed to a variety of lineages [2,6–8]. In addition to its growth factor activity on normal cells, IL-3 has been shown to be an

Robert J. Isfort and James N. Ihle are now at the Department of Biochemistry, St. Jude Children's Research Hospital, Memphis, TN 38101.

Received February 1, 1988.

essential growth factor for a series of cell lines derived from long-term bone marrow cultures [9,10] and for cell lines derived from retrovirus induced leukemias [11,12].

The mechanisms in IL-3 regulated growth are largely unknown. IL-3 dependent cell lines have been shown to express a single class of high-affinity receptors for IL-3 [13]. By cross-linking studies and in immunoprecipitation experiments, the major IL-3 binding protein has an apparent molecular weight of approximately 65 kd [14–16]. In addition to the 65 kd protein, however, a higher molecular weight cross-linked protein has been detected with an apparent molecular weight of 140 kd [16] (May and Ihle, unpublished data). The relationship between the 65 kd and 140 kd IL-3 binding proteins is not currently known.

The cellular signals which are involved in IL-3 regulated growth are unknown. Initial studies indicated that the protein kinase C (PKC) system may be involved based on the ability of IL-3 to cause the apparent translocation of PKC in factor dependent cells [17]. The reported effects of IL-3 on Ca^{++} flux have been variable [18,19] and IL-3 does not cause an increase in phosphatidylinositiol turnover [18]. A possible role for protein phosphorylation has been suggested based on the observation that IL-3 stimulation of factor dependent cells results in the threonine phosphorylation of a 68 kd protein [20]. A role for tyrosine phosphorylation has been proposed based on the observation that a series of transforming viruses, containing tyrosine protein kinases, can abrogate the requirement of factor dependent cells for IL-3 [21–25]. In the studies described here we have undertaken further to characterize the mechanisms in IL-3 regulated growth by examining the role of PKC and tyrosine phosphorylation. The results demonstrate that IL-3 uniquely induces the tyrosine phosphorylation of a series of cellular substrates. One of these substrates, a 140 kd phosphoprotein is an integral component of the IL-3 receptor complex.

MATERIALS AND METHODS

Cell Lines and Growth Factors

The origins and properties of the FDC-P1, DA-3, DA-1, and DA-2 cell lines have been previously described [9,12]. The cells were maintained as previously described [10]. Il-3 was purified to homogeneity by previously published procedures [12]. Other growth factors were obtained as indicated in the text.

Protein Kinase C Assays

Cytosolic and membrane-bound protein kinase C activity was determined by a modification of previously described methods [26]. PKC activity was partially purified by anion exchange chromatography over DE52 cellulose [27,28] at 4°C. Activity was quantitated by measuring incorporation of $^{32}PO_4$ from (γ) ^{32}P ATP into type II-S histones [27,28].

Gel Electrophoresis

One-dimensional SDS-PAGE (sodium dodecysulfate, polyacrylamide gel electrophoresis) was done according to the methods of Laemmli [29]. Two-dimensional gel electrophoresis was performed according to the methods of O'Farrel [30].

Cell Fractionation Experiments

Cell fractionations were done according to the methods of Courtneidge et al. [31]. In place of dounce homogenization for lysis of the cells, the cells were lysed by

TABLE I. Effects of IL-3 and TPA on the Localization of Ca^{++} and Phospholipid Dependent Protein Kinase Activity in FDC-P1 Cells*

		pmol/5 min/10^7 cells		
Additions	Fraction	Experiment 1	Experiment 2	Experiment 3
None	Cytosol	1,936	739	1,849
	Membrane	<10	<10	<10
IL-3	Cytosol	1,744	1,486	1,138
	Membrane	125	<10	190
TPA	Cytosol	1,344	1,016	<10
	Membrane	1,869	1,090	1,070

*Ca^{++} and phospholipid dependent protein kinase activity was assayed as described in Materials and Methods. In experiments 1 and 2 the concentration of TPA was 50 ng/ml and in experiment 3 the concentration was 100 ng/ml. In experiments 1 and 2, the cells were treated with IL-3 or TPA for 10 min, while in experiment 3 the cells were treated for 30 min. The assays were run in triplicate and the average values are given. The standard deviation in the assay was generally less than 20%. The concentration of the IL-3 used in all experiments was 240 units/ml.

sonification using a Sonifier cell disrupter with a microtip at a power setting of 2 for 10 s.

Cell Labelling, Lysate Preparation, and Immunoprecipitation

For standard immunoprecipitation analysis 1×10^7 cells were labeled in 1 ml of phosphate-free RPMI-1640 containing 1 mCi of carrier-free orthophosphate according to the procedures of Frackleton et al. [32,33]. Lysates were prepared from cells according to published procedures. Typically, 1×10^7 cells were lysed in 1 ml of lysis buffer. The lysate was then used in immunoprecipitation using the antiphosphotyrosine monoclonal antibody IG-2. Immunoprecipitations were done according to published protocols [32,33].

Phosphoamino Acid Analysis

The phosphoamino acid content was determined by the methods described by Cooper et al. [34].

RESULTS

Initially we examined the effects of IL-3 on the levels and apparent localization of PKC. For the experiments shown in Table I, IL-3 dependent cells were washed free of IL-3 and incubated in the absence of IL-3 for 1–2 h. The cells were subsequently stimulated with an excess of IL-3, and the amount of PKC activity associated with the cytosolic or membrane fractions was determined. As indicated by the results shown in Table I, there was no apparent effect of IL-3 on either the total activity of PKC or its distribution into cytosolic or membrane fractions. In contrast to the results with IL-3, treatment of the cells with phorbol myristate acetate (PMA) resulted in an apparent redistribution of PKC activity. It should be noted that although PMA had a readily demonstrable effect on PKC activity, PMA does not support the growth of IL-3 dependent cells (Ihle, unpublished data) and inhibits the IL-3 induced growth [35].

Chronic exposure of cells to PMA causes down-regulation of the enzyme and can result in the complete loss of detectable PKC activity [36,37]. We therefore exam-

TABLE II. The Effects of Long-Term TPA Treatment on PKC Activity and on the Response to IL-3 of FDC-P1 Cells*

Treatment	PKC activity (pmol/5 min/10^7) Cytosol	Membrane	IL-3 concentration (units/ml)	% maximum response
Control	1,860	213	0.1	18
			0.3	39
			1.0	68
			3.0	87
			10.0	96
			30.0	98
TPA	<10	<10	0.1	33
			0.3	68
			1.0	94
			3.0	116
			10.0	105
			30.0	108

*FDC-P1 cells were cultured in either normal media containing 30 units/ml of IL-3 or in media containing IL-3 and 50 ng/ml of TPA for 48 h prior to the assays. Cells were harvested and assayed for PKC activity as described in Materials and Methods. Aliquots of the cells were then assayed in a thymidine incorporation assay for their proliferative response to IL-3 in the absence of TPA. The percentages shown are calculated relative to the response of untreated control cells to optimal concentrations of IL-3. The results are an average of the results obtained in four separate experiments.

ined the effects of chronic exposure of cells to PMA on the levels of PKC activity and on the ability of the cells to respond to IL-3. As illustrated in Table II, cells which had been cultured in IL-3 with PMA for 48 h had no detectable PKC activity. The cells nevertheless responded to IL-3 with dose responses which were comparable or more sensitive to IL-3 compared to control, untreated cells. Taken together these data suggest that PKC does not play a major role in IL-3 signal transduction.

To examine the role of protein phosphorylations in IL-3 signal transduction we initially examined the effects of IL-3 stimulation of factor dependent cells on total phosphorylations and on tyrosine phosphorylation. For these experiments, cells were deprived of IL-3 for 6–8 h, labeled with orthophosphate, and were then stimulated with IL-3 for the indicated times. As is illustrated in Figure 1, there was no apparent effect of IL-3 stimulation on the total phosphoprotein profile. In contrast, when the phosphotyrosine containing fraction was isolated using a monoclonal antibody against phosphotyrosine, there was a readily demonstratable effect of IL-3 on the pattern of phosphoproteins. As is illustrated in Figure 1, there were at least eight IL-3 inducible phosphoproteins which were immunoprecipitated with the antiphosphotyrosine monoclonal antibody.

The properties of the phosphoproteins which are detected in response to IL-3 are summarized in Table III. In each case the phosphoamino acid content was determined from protein isolated from SDS-PAGE, the time of appearance was examined using FDC-P1 cells and short time points, the cellular localization in membrane or cytoplasmic fractions was determined by cellular fractionation, and glycosylation was assessed by the ability to bind to and to be eluted from wheat germ agglutinin WGA-sepharose. One of the most intriguing phosphoproteins has an apparent size of 140 kd, binds to wheat germ agglutinin (WGA)-Sepharose columns, is predominantly associated with

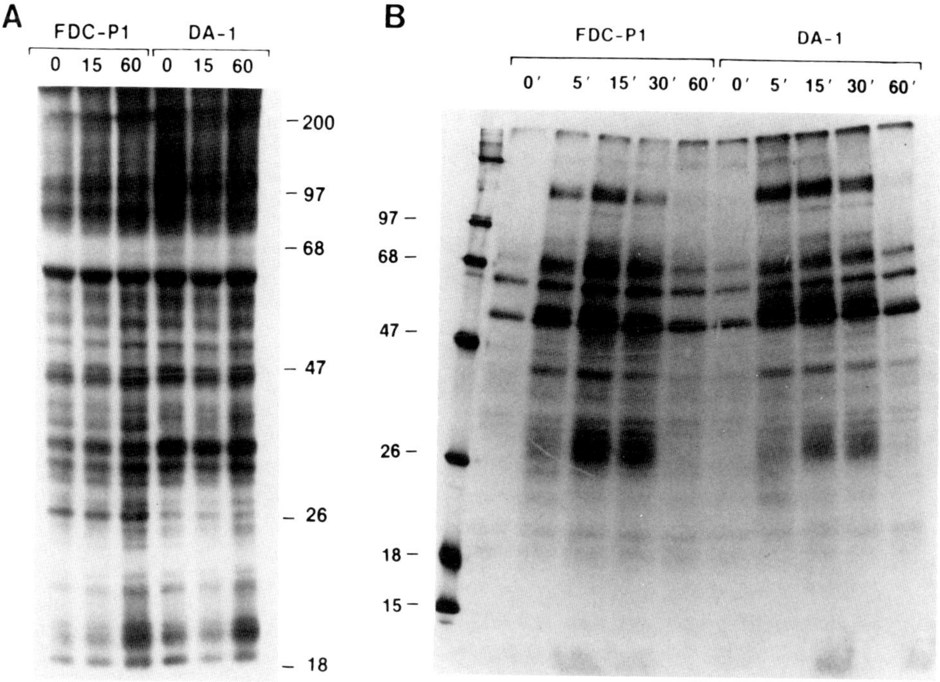

Fig. 1. Effects of IL-3 stimulation on total protein phosphorylation and on tyrosine phosphorylation. IL-3 dependent cell lines were washed free of IL-3 and maintained in the absence of IL-3 for approximately 8 h. The cells were then labeled with orthophosphate and stimulated with an excess of IL-3. At the indicated times cells were collected and lysed and the lysates either run directly on SDS-PAGE (**A**) or immunoprecipitated with a monoclonal antibody against phosphotyrosine, and the immunoprecipitates were run on SDS-PAGE (**B**). The positions of protein standards are indicated.

the membrane fraction, and is one of the earliest proteins phosphorylated in response to IL-3. The 140 kd phosphoprotein is one of the major proteins detected following stimulation with IL-3 and has been detected in all the IL-3 dependent cell lines examined. The 120 kd and 85 kd phosphoproteins are minor components which are variably detected in FDC-P1 cells but not in another IL-3 dependent cell line (DA-3).

The 56 kd phosphoprotein is a major component detectable following stimulation with IL-3 and has been observed in all the IL-3 dependent cell lines examined. It is predominately associated with the cytoplasmic fraction and does not bind to WGA-sepharose. Similarly the 38 kd phosphoprotein is a consistently detected component following IL-3 stimulation which is found in the cytosolic fraction and which does not bind to WGA-Sepharose. A minor, cytosolic phosphoprotein of 51 kd was also evident in all the IL-3 dependent cell lines examined. Lastly, FDC-P1 cells variably give a diffuse phosphoprotein band with an apparent molecular weight of 28 kd. This component is not detected in all the IL-3 dependent cell lines.

Cross-linking studies have indicated that the predominant IL-3 binding protein on FDC-P1 cells has an apparent size of 65–70 kd [14–16]. Occasionally, higher molecular weight binding proteins have been detected. In particular, as illustrated in Figure 2, when high levels of protease inhibitors are incorporated into the extraction mixture, two cross-linked components are readily detectable. After subtracting the

TABLE III. Summary of the Characteristics of Phosphotyrosine Containing Proteins Induced by IL-3*

Size (kd)	Phosphoamino acids	Time of appearance (min)	Cellular localization	Glycosylation
140	ser,tyr	0.5	Membrane	+
120	N.D.	N.D.	N.D.	N.D.
85	ser,tyr,thr	0.5	Cytoplasm	–
70	ser,tyr	2.5	Cytoplasm	–
56	ser,tyr	2.5	Cytoplasm/membrane	–
51	ser,tyr	2.5	Cytoplasm	–
38	ser,tyr	2.5	Cytoplasm	–
28	N.D.	N.D.	N.D.	N.D.

*N.D., not determined.

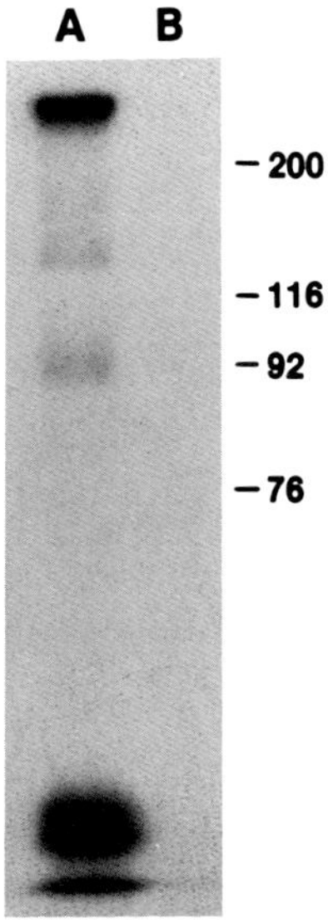

Fig. 2. Affinity labeling of the IL-3 receptor by chemical cross-linking. ^{125}I labeled IL-3 was cross-linked to FDC-P1 cells using disuccinimidyl suberate (DSS) as described in detail in a previous study [14]. Following cross-linking, the cells were collected and solubilized in 50 mM Tris-HCl and pH 7.2, 1.5% Triton-X-100 and 1.5 mM MgC12 containing 1 mg/ml bacitracin and mM phenylmethylsulfonyl fluoride. The cross-linked complex was separated by SDS-PAGE using 10% acrylamide separating gels according to the methods of Laemmli [29].

size of IL-3 the two components have apparent molecular weights of approximately 65 kd and 140 kd. These results suggest that the IL-3 receptor either contains two components and thus might have a similar structure to that of the IL-2 receptor or is a single receptor of 140 kd which can be proteolytically degraded to a smaller form which retains ligand binding activity similar to the granulocyte, macrophage specific colony stimulatory factor (GM-CSF) receptor.

The ambiguity regarding the size of the receptor and the existence of a 140 kd membrane associated protein which is rapidly phosphorylated following IL-3 stimulation suggested the possibility that the 140 kd phosphoprotein may be a component of

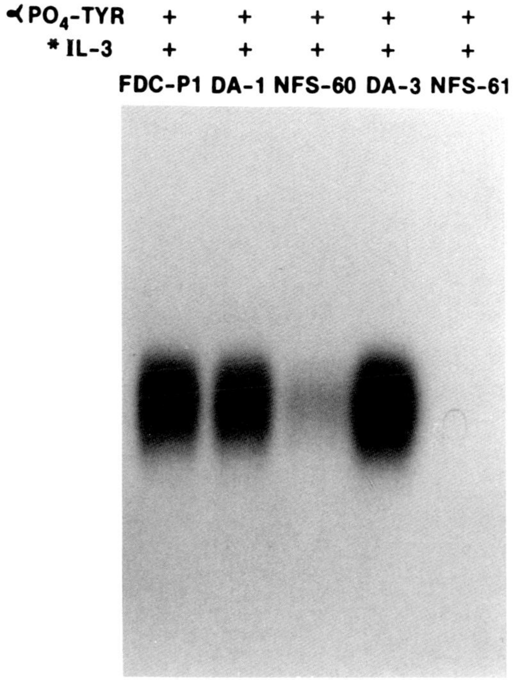

Fig. 3. Immunoprecipitation of ^{125}I labeled IL-3 in a complex using a monoclonal antibody against phosphotyrosine. The indicated cell lines were removed from IL-3 and were cultured in the absence of IL-3 for approximately 8 h. For a control, the IL-3 receptor negative, factor independent cell line was included (NFS-61) and treated comparably throughout. The cells were subsequently simulated with ^{125}I labeled IL-3 for 10 min. The cells were subsequently lysed and the phosphotyrosine containing fraction was immunoprecipitated with a monoclonal antibody against phosphotryosine. The immunoprecipitated material was eluted from the antibody-Sepharose beads by an excess of phenylphosphate, and the eluted material was analyzed by SDS-PAGE for the presence of labeled IL-3. The major radioactive band detected had the expected mobility for IL-3.

the IL-3 receptor. To examine this, we initially sought to determine whether the IL-3 binding complex contained a phosphotyrosine containing protein. For the experiments shown in Figure 3, cells were stimulated with ^{125}I labeled IL-3, the cells lysed, and the phosphotyrosine containing fraction isolated with the antiphosphotyrosine monoclonal antibody. As shown, the monoclonal antibody precipitated labeled IL-3. The precipitation was competable with cold IL-3 (data not shown) and was not detected with cells which lack receptors for IL-3. (NFS-61 in Fig. 3).

To determine the size of the IL-3 binding complex, ^{125}I labeled IL-3 was used to stimulate cells and the complex isolated with the antiphosphotyrosine monoclonal antibody. The complex was eluted with phosphotyrosine and the solubilized complex was run on glycerol gradients. As shown in Figure 4, the complex migrated in these gradients with an apparent molecular weight of 150–200 kd. In addition, the solubilized complex was cross-linked and analyzed on SDS-PAGE. Under these conditions the major IL-3 containing complex had an apparent size of 170 kd, consistent with the size of the native complex on glycerol gradients (data not shown). Taken together these results suggest that IL-3 can bind to a 140 kd protein containing phosphotyro-

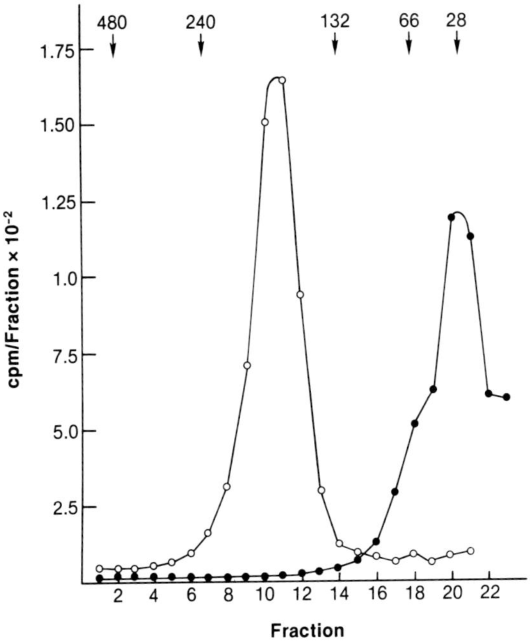

Fig. 4. Glycerol gradient analysis of IL-3 receptor complexes. FDC-P1 cells were removed from IL-3 and cultured in the absence of IL-3 for approximately 8 h. The cells were subsequently stimulated with an excess of ^{125}I labeled IL-3 for 15 min. The cells were then lysed and the phosphotyrosine containing proteins were immunoprecipitated with the IG-2 monoclonal antibody as previously described [32,33]. The immunoprecipitated materials were eluted from the monoclonal antibody Sepharose beads with phenylphosphate and the solubilized fraction immediately loaded onto a 10 ml 10–30% glycerol gradient in lysis buffer. The gradients were centrifuged at 40,000 rpm for 24 h and fractions collected by puncturing the bottom of the tubes. Markers were either run in the same gradients or in duplicate tubes and consisted of urease, bovine serum albumin, or iodinated IL-3.

sine. Whether this is the major protein which is labeled following IL-3 stimulation is not known, although preliminary experiments indicate that the 140 kd phosphoprotein comigrates with labeled IL-3 on glycerol gradients and can be cross-linked in a complex which has a comparable mobility on SDS-PAGE to the cross-linked complex detected with labeled IL-3.

DISCUSSION

Our studies have been directed at better defining the signal transduction pathway which is utilized by IL-3. The results provide no supportive evidence for the possible involvement of PKC in IL-3 associated growth. The basis for the differences between our results and previously published studies [17] are unknown. In contrast, our results demonstrate that there is a very rapid and specific pattern of tyrosine phosphorylations which occurs following stimulation of factor dependent cells with IL-3. At least eight phosphoproteins have been variably detected, of which at least three are major components which are consistently detected with all the IL-3 dependent cell lines which have been examined.

Two of the major phosphotyrosine containing proteins following IL-3 stimulation are the 56 kd and 38 kd phosphoproteins. The 38 kd component is similar in size to calpactin I or II. Consistent with this, the 38 kd component is acidic and migrates in two-dimensional (2D) gels similar to calpactin I [38]. The possibility that the 38 kd component is calpactin I is currently being assessed with antisera against calpactin I. The identity of the 56 kd component is unknown. Interestingly, a phosphotyrosine containing 56 kd component has also been shown to be commonly detected in Ph+ human CML and in Abelson transformed cells [33,39]. Experiments are currently in progress to attempt to isolate this component for further biochemical analysis.

One of the most intriguing components induced by stimulation with IL-3 is the 140 kd membrane associated phosphotyrosine containing protein. In particular, the possibility exists that this phosphoprotein may be a component of the IL-3 receptor. In cross-linking studies, some of IL-3 becomes associated with a protein which would have a predicted molecular weight of approximately 140 kd. More importantly, our studies demonstrate that IL-3 becomes associated with a phosphotyrosine containing protein and can be precipitated with a monoclonal antibody against phosphotyrosine. By cross-linking studies and by migration on glycerol gradients, IL-3 is associated with a 140 kd protein. Experiments are currently in progress to determine more definitively whether the IL-3 binding protein is the 140 kd phosphoprotein detected following stimulation with IL-3.

The possible significance of the 65 kd IL-3 binding protein is not known. Our results are not consistent with the concept that the IL-3 receptor is composed of a 65 kd and a 140 kd subunits. In particular, the size of the IL-3 binding complex on glycerol gradients is closer to 170 kd than to 235 kd as would be predicted if two subunits were involved. Similarly, cross-linking of the antiphosphotyrosine isolated complex yielded predominately a 170 kd complex consistent with the glycerol gradient results. For these reasons the data are most consistent with the concept that the 140 kd is the native, IL-3 binding protein and that this component becomes phosphorylated following stimulation with IL-3. Whether the 140 kd protein has intrinsic protein kinase activity or is phosphorylated by another kinase is not known. The 65 kd component can be speculated to be a proteolytically cleaved derivative of the 140 kd protein which remains on the cell surface and can bind IL-3. Recent studies [40] have demonstrated that the human growth hormone receptor is susceptible to proteolytic cleavage and a soluble 50 kd ligand binding is routinely detectable. Similarly it has been reported that the GM-CSF receptor is sensitive to proteases and can be degraded from a 130 kd form to a 70 kd form [41]. Experiments are currently ongoing to examine further the relationship between the 65 kd and 140 kd proteins and to characterize both proteins biochemically.

ACKNOWLEDGMENTS

We would like to thank Sonia Hutchinson for help in the preparation of this manuscript. Research sponsored by the National Cancer Institute, DHHS, under contract No. NO1-CO-74101 with Bionetics Research, Inc. The contents of this publication do not necessarily reflect the views or policies of the Department of Health and Human Services, nor does mention of trade names, commercial products, or organizations imply endorsement by the U.S. Government.

REFERENCES

1. Ihle JN, Keller J, Henderson L, Klein F, Palaszynski EW: (1982) J Immunol 129:2431–2436, 1982.
2. Ihle JN, Keller J, Oroszlan S, Henderson L, Copeland T, Fitch F, Prystowsky MB, Goldwasser E, Schrader JW, Palaszynski E, Dy M, Lebel B: (1983) J Immunol 131:282–287, 1983.
3. Yang YC, Ciarletta AB, Temple PA, Chung MP, Kovacic S, Witek-Giannotti JS, Leary AC, Kriz R, Donahue RE, Wong GG, Clark SC: Cell 47:3–10, 1986.
4. Yokota T, Lee F, Rennick D, Hall C, Arai N, Mosmann T, Nabel G, Cantor H, Arai K: Proc Natl Acad Sci USA 81:1070–1074, 1984.
5. Fung MC, Hapel AJ, Ymer S, Cohen DR, Johnson RM, Campbell HD, Young IG: Nature 307:233, 1984.
6. Spivak JL, Smith RRL, Ihle JN: J Clin Invest 76:1613–1621, 1985.
7. Suda T, Suda J, Ogawa M, Ihle JN: J Cell Physiol 124:182–190, 1985.
8. Quesenberry PJ, Ihle JN, McGrath E: Blood 65:(1):214–217, 1985.
9. Dexter TM, Garland J, Scott D, Scolnick E, Metcalf D: J Exp Med 152:1036–1047, 1980.
10. Ihle JN, Keller J, Greenberger JS, Henderson L, Yetter RA, Morse HC, III: J Immunol 129:1377–1383, 1982.
11. Holmes KL, Palaszynski E, Frederickson TN, Morse HC, III: Ihle JN: Proc Natl Acad Sci USA 82:6687–6691, 1985.
12. Ihle JN, Rein A, Mural R: Adv Viral Oncol 4:95–137, 1984.
13. Palaszynski EW, Ihle JN: J Immunol 132:1872–1878, 1984.
14. May WS, Ihle JN: Biochem Biophys Res Commun 135:870–879, 1986.
15. Nicola NA, Peterson L: J Biol Chem 261:12384–12389, 1986.
16. Park LS, Friend D, Gillis S, Urdal DL: J Biol Chem 261:205–210, 1986.
17. Farrar WL, Thomas TP, Anderson WB: Nature 315:235–237, 1985.
18. Whetton AD, Monk PN, Consalvey SD, Downes CP: EMBO J 5:3281–3286, 1986.
19. Rossio JL, Ruscetti FW, Farrar WL: Lymphokine Res 5:163, 1986.
20. Evans SW, Rennick D, Farrar WL: Blood 68:906–913, 1986.
21. Pierce JH, DiFiore PP, Aaronson SA, Potter M, Humphrey J, Scott A, Ihle JN: Cell 41:685–693, 1985.
22. Cook WD, Metcalf D, Nicola NA, Burgess AW, Walker F: Cell 41:677–683, 1985.
23. Wheeler EF, Askew D, May S, Ihle JN, Sherr CJ: Mol Cell Biol 7:1673–1680, 1987.
24. Mathey-Prevot B, Nabel G, Palacios R, Baltimore D: Mol Cell Biol 6:4133–4135, 1986.
25. Overell RW, Watson JD, Gallis B, Weisser KE, Cosman D, Widmer MB: Mol Cell Biol 7:3394–3401, 1987.
26. Abraham RT, Ho SN, Barna TJ, McKean DJ: J Biol Chem 262:2719–2728, 1987.
27. Kraft AS, Anderson WB: Nature 301:621–623, 1983.
28. Kraft AS, Anderson WB: J Biol Chem 258:9178–9183, 1983.
29. Laemmli VK: Nature 227:680–685, 1970.
30. O'Farrell PH: J Biol Chem 250:4007–4021, 1975.
31. Courtneidge SA, Levinson AD, Bishop JM: Proc Natl Acad Sci USA 77:3783–3787, 1980.
32. Frackelton AR, Ross AH, Eisen HN: Mol Cell Biol 3:1343–1352, 1983.
33. Huhn RD, Posner MR, Rayter SI, Foulkes JG, Frackelton AR: Proc Natl Acad Sci USA 84:4408–4412, 1987.
34. Cooper JA, Setrox BM, Hunter T: Methods Enzymol 99:387–405, 1983.
35. Orosz EG, Zinn NE, Olsen RG, Math LE: J Immunol 134:3396–3403, 1985.
36. May WS, Sahyoun N, Jacobs S, Wolf M, Cautrecausas P: J Biol Chem 266:9419–9426, 1985.
37. Chida K, Kato N, Kuroki T: J Biol Chem 261:13013–13018, 1986.
38. Isucke CM, Trougridge IS, Hunter T: Mol Cell Biol 6:2745–2751, 1986.
39. Frackelton AR, Jr: In Ozanne B, Stiles C (eds): "Cancer Cells 3, [Growth Factors and Transformation]." Cold Spring Harbor, NY: Cold Spring Harbor Laboratory, 1985, pp 339–345.
40. Leung DW, Spencer SA, Cachianes G, Hammonds RG, Collins C, Henzel WJ, Barnard R, Waters MJ, Wook WI: Nature 330:537–543, 1987.
41. Park LS, Friend D, Gillis S, Urdad DL: J Biol Chem 261:4177–4183, 1986.

Growth Factors and Their Receptors: Genetic
Control and Rational Application, pages 239–251

Nuclear Factor Binding Sites and Requirements for Initiation and Attenuation of Murine c-*myc* Transcription

Claude Asselin, Alain Nepveu, Jian-Qing Yang, and Kenneth B. Marcu

Biochemistry Department, SUNY at Stony Brook, Stony Brook, New York 11794-5215

To understand the role of cellular factors in the transcriptional regulation of the murine c-*myc* gene, we tested sequences in the 5′ flanking region of the gene's P_1 and P_2 transcription initiation sites for their ability to form specific DNA-protein complexes with factors present in HeLa cell nuclear extracts and for their contribution to c-*myc* transcription. DNase I footprint analysis identified five regions that bind nuclear proteins including the transcription factor Spl in addition to several others. Deletion mutagenesis of a binding domain 5′ of P_2 abolishes its activity but does not affect P_1. S1 nuclease mapping and nuclear run-on transcription analysis of deletion mutants upstream and downstream of the two *myc* promoters have identified a 116 bp sequence, residing in between them, that binds two nuclear factors and is also necessary for P_2 usage and transcriptional attenuation within the first exon. These results suggest that P_1 and P_2 are independently regulated, that the 3′ terminus of the murine c-*myc* transcriptional block maps about 150 bp 5′ of the exon 1/intron 1 splice junction, and that efficient P_2 usage is required for transcriptional blockage.

Key words: DNA binding factors, DNase I footprinting, transcriptional elongation block, nuclear run-on, S1 nuclease mapping

The c-*myc* proto-oncogene is the cellular homologue of the v-*myc* oncogene harbored by the acute transforming avian myelocytomatosis viruses. The c-*myc* gene is highly conserved in vertebrate evolution and is expressed in a wide variety of mammalian cell types. The gene is composed of three exons, the first of which is a large noncoding sequence [reviewed in 1]. The last two exons encode a nuclear DNA-binding

Alain Nepveu is now at Ludwig Institute, Montreal Branch, 687 Ave. des Pins ouest, Montreal, Canada H3A 1A1.

Jiang-Qing Yang is now at Imclone Systems Inc., 180 Varick St., New York, NY 10014.

Received February 1, 1988.

protein with an apparent molecular mass of 62–64 kilodaltons (KD). Transcription of the c-*myc* gene initiates at two major sites, P_1 and P_2. Normal c-*myc* RNAs are highly unstable in vivo and can be superinduced by growth factors in the presence of protein synthesis inhibitors, which implicates labile negative factors in their regulation [reviewed in 1].

Alterations of normal c-*myc* expression either by gene amplification, retroviral insertion, or chromosome translocation are considered important steps in the development of a variety of neoplasms [1–5]. The c-*myc* gene product may be a cellular mediator of growth factor action since its activation is believed to reflect the competency of cells to enter and progress through the cell cycle [1,6]. Indeed, the c-*myc* gene product can immortalize primary cells in culture and can cooperate with other activated oncogenes in the tumorigenic process [7]. Transgenic mice harboring a c-*myc* gene linked to tissue-specific regulatory sequences develop clonal tumors in tissues expressing the c-*myc* transgene [8–10].

The c-*myc* gene is regulated at the transcriptional and post-transcriptional levels in proliferating and differentiating cells [11–18, 30, 48, 49], and cis- and trans-acting negative and positive regulatory elements have been described within the c-*myc* locus [19–28]. One of the negative modulators of c-*myc* expression blocks the progression of RNA polymerase II through the gene's first exon [21,23,25,29]. In this report, we have identified multiple nuclear factor-binding sites within the regulatory region of the murine c-*myc* gene. Several nuclear proteins including the Sp1 transcription factor bind to c-*myc* control regions. Exon 1 sequences required for the usage of the gene's major transcription start site, P_2, are also necessary for efficient transcriptional blockage. We also show that the 3′ terminus of the transcriptional block is within exon 1, about 150 bp 5′ of the exon 1/intron 1 splice junction.

MATERIALS AND METHODS

Cell Lines and Transfection

Stably transfected HeLa cells and CV1 cells were obtained by calcium phosphate cotransfection of *myc*-CAT vectors and pSV2neo [32], followed by selection in media supplemented with G-418 (Geneticin, GIBCO).

RNA Isolation and S1 Nuclease Analysis

Total RNA was prepared by the guanidinium thiocyanate/hot phenol method [33]. S1 mapping of murine *myc*-CAT RNAs in stably transfected HeLa or CV1 cells was performed with an α^{32}P uniformly labeled *myc* exon 1 probe, mp-R*S10 (See Fig. 4) [34]. Conditions for RNA:DNA annealing, S1 nuclease digestions, and their analysis by PAGE were as previously described [34].

Nuclear Run-On Transcription

Nuclei were purified and nascent transcripts were elongated essentially as described, without pretreatment of the nuclei with RNase [17,35]. The prehybridization and hybridization procedures have been detailed elsewhere [23].

Nuclear Extract Preparation and DNase I Footprint Analysis

Nuclear extracts were prepared according to Dignam et al. [36] with 8–12 mg/ml of protein from the following tissue culture cell lines: HeLa (human cervical carcinoma), BALB/c A31 (mouse fibroblast), 18-81 (A-MuLV transformed murine pre-B lymphoma) [31], J558 (murine plasma cell tumor), and BJAB (human lymphoblastoid line). The gel retardation assay and DNase I footprint analyses were performed as described [37,38]. Affinity-purified Sp1 was a kind gift of W.S. Dynan (University of Colorado, Boulder).

Vector Constructions

The largest *myc*-CAT vector, pMBgCAT, contains a 1.71 kilobase (Kb) *myc* Bgl II fragment which encompasses 1.14 Kb of 5′-flanking sequences, the entire 563 bp first exon ending at a Bgl II site 14 bp into intron 1 [24]. pMSmBgCAT was prepared from pMBgCAT by deleting the 5′ 716 bp Bgl II-Sma I segment [24]. pMHHBg Glob contains 584 bp of *myc* sequences (commencing 7 bp 5′ of P_1 and ending at the intron one Bgl II site), attached to a rabbit β-globin cDNA in place of the CAT gene [39]. Internal deletions in the c-*myc* first exon were generated from the BamHI 5′ of P_2 site in pMBgCAT by Bal 31 exonuclease treatment. The extent of the deletions were determined by restriction enzyme mapping and are as follows: 75 bp (pMBg 2–37), 150 bp (pMBg 1–30), and 200 bp (pMBg 2–36). Additional exon one deletion mutants extending in a 3′ direction from the first exon BamHI site (+116) towards the intron one Bgl II site (+577) were also generated with Bal 31 exonuclease. These were cloned, after addition of Bgl II linkers, into the mp18 bacteriophage vector and their structures confirmed by DNA sequencing. The 5′ boundaries of deletion mutants A1-A4 were +274, +305, +374, and +449 (relative to P_1 at +1), respectively (Fig. 7).

RESULTS

Multiple Nuclear Factor Binding Sites Upstream of P_1 and P_2

A gel electrophoresis protein-DNA binding assay [37] was used to detect nuclear factors that bind to sequences upstream of the P_1 and P_2 transcription initiation sites. In this assay, DNA-protein complexes migrate slower than the free DNA probe in an nondenaturing, low salt polyacrylamide gel [37]. To reduce nonspecific binding of proteins to the labeled fragment, an excess of a double-stranded synthetic copolymer, poly(dI-dC)/poly(dI-dC), was included in the binding reaction. Nuclear extracts made from different cell lines were incubated with end-labeled probes extending either from Sca I to Hae III (−198 to −7 relative to P_1) (Fig. 1) or from Hae III to Hind III (−7 to +140 relative to P_1) (Fig. 2). Similar DNA-protein complexes were detected with nuclear extracts prepared from fibroblast, epithelial, and immature and mature B cell lines (data not shown). Competition with the homologous DNA fragments demonstrated that the complexes were specific (data not shown). We then used DNase I footprinting to identify the nuclear factor binding site(s). Three regions of protection were detected by DNase I footprinting with a c-*myc* 5′ flanking sequence from −198 to −7 (Fig. 1). The footprints are located between −188 and −156, −128 and −106, and −55 and −30 for nuclear factors 5′Mg1, 5′Mg2, and 5′Mg3, respectively (Fig. 1). Two other regions of protection were obtained within a Hae III-BamHI fragment encompassing sequences −7 to +140 (e.g., +57 to +77 and +96 to +119 for factors ME1a2 and ME1a1, respectively) which are just 5′ of P_2 at +163 (Fig. 2). The ME1a1 factor recognizes the sequence $(G)_3A(G)_4A(G)_3A$, which is sufficient for binding as an in vitro synthesized oligonucleotide (data not

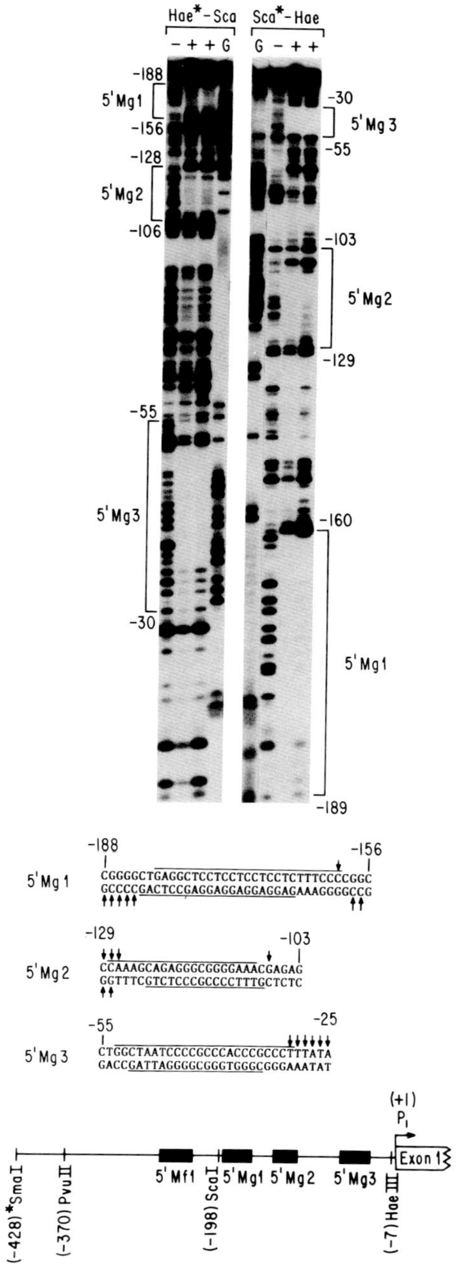

Fig. 1. DNase I footprinting of nuclear factor binding sites on the 191 bp Sca I-Hae III fragment 5′ of the c-*myc* P_1 initiation site. DNA probes were end-labeled at −198 (Sca*-Hae, coding strand) or at −7 (Hae*-Sca, noncoding strand). The end-labeled DNA probes were chemically cleaved at G residues and used as markers and were also digested with 125 or 250 ng of DNase I with (+) or without (−) prior pre-incubation with a crude HeLa cell nuclear extract. The sequences at the bottom of the figure summarize the DNase footprinting results with lines above and below the sequences delineating the regions protected from DNase I digestion by nuclear factor binding. DNase I hypersensitive sites are indicated by small arrows. The location of these binding sites relative to P_1 are shown in the map.

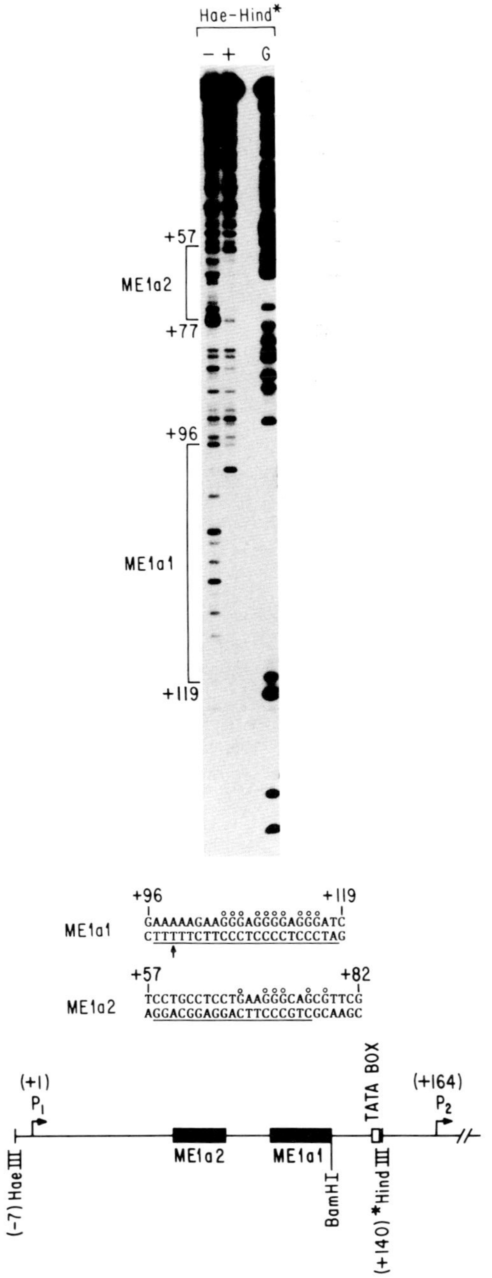

Fig. 2. DNase I footprinting of the nuclear factor binding sites on the 147 bp Hae III-Hind III fragment upstream of the P_2 initiation site. The DNA probe was end-labeled at +140 (Hae-Hind*, noncoding strand), chemically cleaved at G residues as a marker, and digested with DNase I with (+) or without (−) prior pre-incubation with HeLa cell nuclear extract. The sequences below summarize the footprinting results. Lines below the sequence represent the nuclear factor binding site protected from DNase I digestion. DNase I hypersensitive sites are designated by small arrows. Open circles represent guanines that when methylated block nuclear factor binding in a DMS interference assay [37]. The location of these binding sites relative to P_2 are indicated.

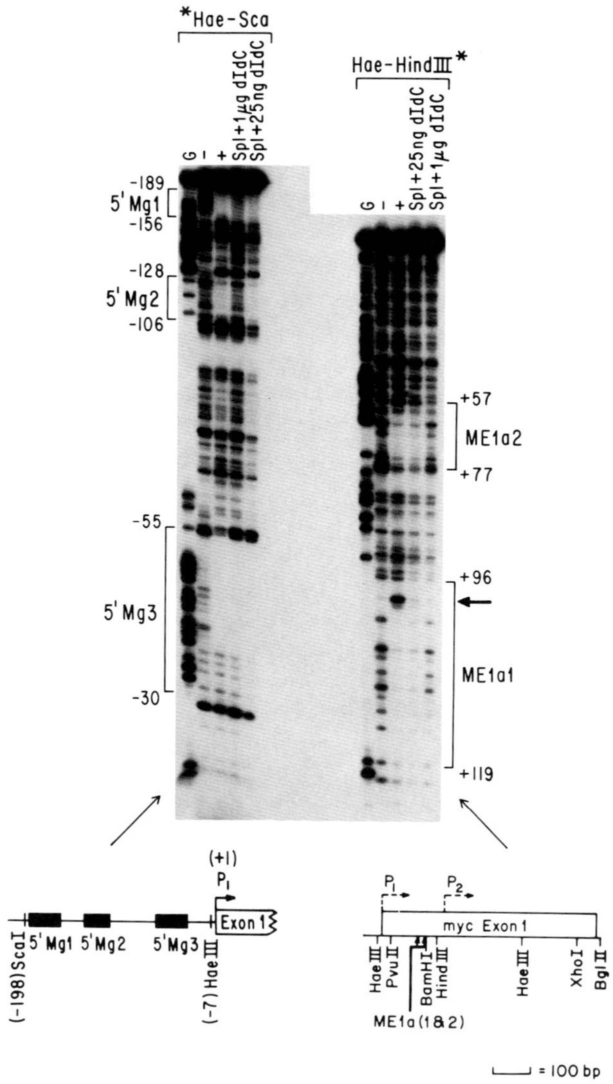

Fig. 3. DNase I protection assays using affinity-purified Spl. The DNA probes were end-labeled at −7 (Hae*-Sca, noncoding strand) for sequences upstream of P_1 and at +140 (Hae-Hind*, noncoding strand) for sequences upstream of P_2. The probes were chemically cleaved at G residues for a marker and digested with DNase I with (+) or without (−) prior incubation with HeLa nuclear extracts. Spl was incubated with the labeled probes in the presence of either 1 μg or 25 ng of the double-stranded nonspecific competitor poly (dI-dC)/poly (dI-dC). The arrow indicates a strong DNase I hypersensitive site which is only observed with the HeLa nuclear extract. The locations of the binding sites relative to P_1 and P_2 are indicated.

shown). This sequence is conserved in the same location 5′ of the human c-*myc* P_2 start site. The addition of a 100-fold molar excess of a purified unlabeled restriction fragment carrying the SV40 early promoter sequences partially abolished complex formation at these two binding sites (data not shown). It is well known that the Spl transcription factor binds to multiple GC box motifs in the SV40 early promoter [40,41].

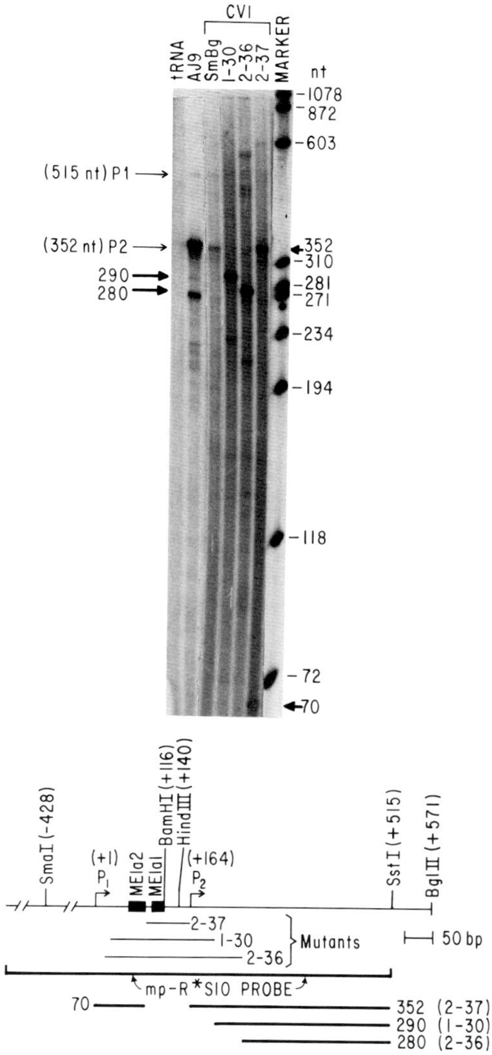

Fig. 4. S1 protection analysis of CAT RNAs expressed by CV1 cells stably transfected with *myc*-CAT vectors. S1 nuclease mapping was performed with a uniformly labeled, single-stranded *myc* exon 1 probe, mp-R*S10. AJ9 is a murine B cell line with a ~20 fold *myc* gene amplification (J. Braun, personal communication). Bands corresponding to P_1 and P_2 are indicated by arrows. S1 protected bands unique to the deletion mutants are indicated by a thick arrow and their locations and relationship to the extent of deletion are provided in the map below.

Therefore, we footprinted the c-*myc* P_1 and P_2 promoter regions with affinity-purified Sp1 along with a high (1 μg/ml) or low (25 ng/ml) concentration of the nonspecific double-stranded competitor poly(dI-dC)/poly (dI-dC). DNase I footprint analysis revealed specific Sp1 binding to the 5′ Mg1 and 5′ Mg3 sites and less specific binding to the other three sites (Fig. 3). Furthermore, a major DNase I hypersensitive site was detected at the 5′ boundary of the ME1a1 binding site with a HeLa cell nuclear extract and not with purified Sp1. In addition, comparative gel retardation assays per-

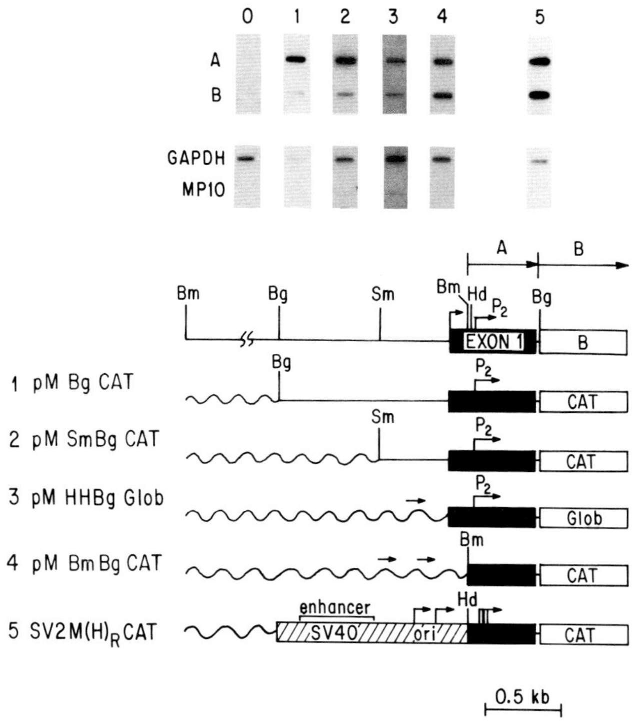

Fig. 5. Run-on transcription analysis of various CAT vectors expressed in stably transfected HeLa cells. The nascent transcripts were labeled in isolated nuclei and hybridized to an excess of c-*myc* M13 ssDNA immobilized on nitrocellulose filters. The c-*myc* M13 probe contains 423 nt of exon 1 sequences (probe A). The GAPDH probe is the Pst I insert of a rat glyceraldehyde 3-phosphate-dehydrogenase cDNA clone, pRGAPDH 13 [51]. The CAT probe is a 550 bp Bgl II-Nco I fragment from the 5′ end of the CAT gene in pCAT(ΔEP) [24]. The β-globin probe is a 437 bp Hind III-Bgl II fragment containing the first and part of the second exon of a rabbit cDNA clone. The maps of the *myc*-CAT vectors are shown below. Restriction sites are designated as follows: Bg, Bgl II; Bm, BamHI; Hd, HindIII; Sm, Sma I.

formed with crude extracts and purified Sp1 gave different band shift patterns with the P_1 and P_2 promoter segments (data not shown). Thus, these data suggest that upstream control regions of the murine c-*myc* gene between −198 and +140 are recognized by other factors in addition to Sp1.

DNA Sequences Required for Usage of the P_2 Start Site

After mapping these nuclear factor binding sites nearby P_1 and P_2, we next investigated their potential role(s) in regulating c-*myc* expression in vivo. To assess the importance of the ME1a1 region for c-*myc* expression, we introduced Bal 31 exonuclease deletions at the BamHI site, which is located at the 3′ end of the ME1a1 binding sequence, in pMBgCAT. These deletion mutants were cotransfected along with pSV2neo into CV1 cells [32]. Total cell RNAs were prepared from mixtures of 50–100 stable transformants and were analyzed for pMBgCAT expression by S1 nuclease protection with a uniformly labeled mouse exon 1 probe, mpR*S10 (Fig. 4). Mismatched hybrids formed between endogenous CV1 *myc* RNAs and this murine *myc*

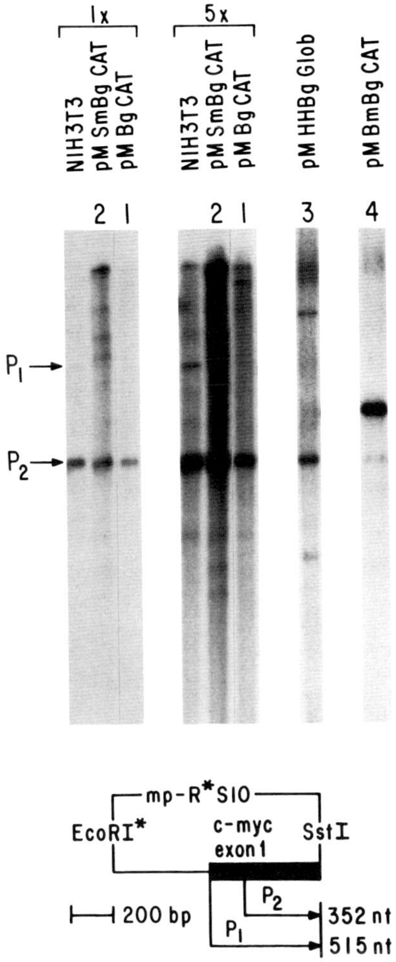

Fig. 6. S1 protection analysis of CAT RNAs expressed by stably transfected HeLa cells. S1 mapping was performed with a uniformly labeled, single-stranded *myc* exon 1 probe, mp-R*S10. 1x and 5x are 1 and 5 day exposures, respectively, with intensifier screens.

first exon probe would be degraded by S1 nuclease. P_2 is the predominant transcription start site of the pMBgCAT and pMSmBgCAT vectors (Figs. 4, 6) [24] and deletions which remove the ME1a1 binding site also abolished transcription initiation from P_2 (Fig. 4).

Strong readthrough transcription occurs in all the deletion mutants, which appears to initiate at P_1. Mutant 1-30 has a 70 bp deletion which removes the ME1a1 binding site, TATA box, and other sequences up to the P_2 start site. The presence of S1 protected bands of ~352 and 70 nt (Fig. 4) with RNAs from cells expressing mutant 1-30 would be entirely consistent with transcripts initiating from P_1. The 70 nt band would be due to P_1 transcripts initiating upstream of the deletion's 5′ boundary and the ~352 nt band would result from the continuation of these transcripts into the first exon, downstream of the deletion's 3′ boundary. Both of these bands are detected

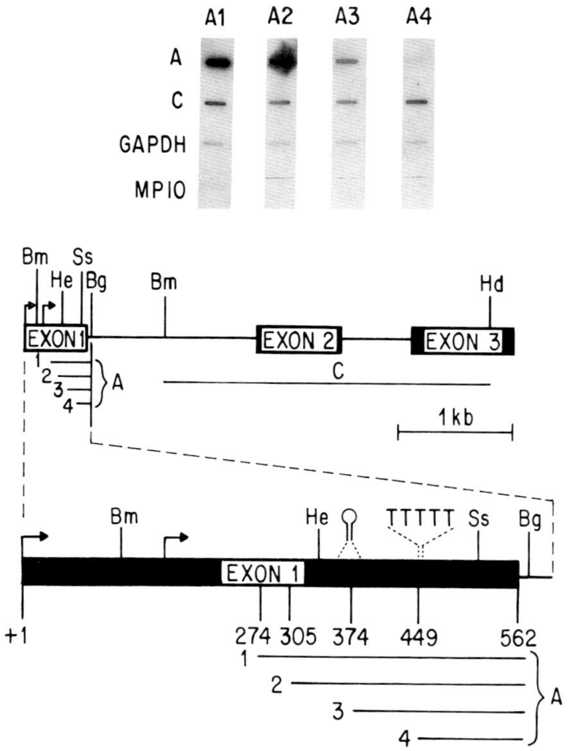

Fig. 7. Mapping of the transcriptional elongation block within the *myc* gene's first exon by nuclear run-on analysis with 54c12 nuclei and single-stranded exon 1 deletion mutants as probes. The locations of the c-*myc* DNA fragments cloned into M13 are indicated. Probe A deletion mutants A1 to A4 were derived from the gene's first exon and probe C contains 3.4 Kb of downstream sequences. The GAPDH probe is the same as described in Figure 5. If RNA polymerase is equally loaded throughout the gene, the intensity of the hybridization signal should reflect the uridine composition of the complementary RNA and probe C contains six times more adenines than probe A. 54c12 cells are an A-MuLV (P160 strain) transformant of NIH3T3 cells that contains a 19 fold amplification of the c-*myc* locus [42]. Restriction enzyme sites are as follows: Bg, Bgl II; Bm, BamHI; Hd, Hind III; He, Hae III; Ss, Sst I.

because the uniformly labeled S1 mapping probe, mpR*S10, spans the entire first exon and about 500 bp of 5′ flanking sequences. These data indicate that the P_1 and P_2 initiation sites are independently regulated by separate promoters and that the ME1a1 binding region is essential for P_2 usage.

Sequences Required for c-*myc* Transcriptional Blockage

To determine the sequences required for transcriptional attenuation within the gene's first exon, we established stable HeLa cell transformants with *myc*-CAT vectors containing different portions of the 1.71 Kb *myc* gene segment in pMBgCAT [34]. We examined the effect of these deletions on the magnitude of the transcriptional elongation block within the first exon [23]. Nuclei were isolated and incubated in vitro in the presence of [α^{32}P]UTP. Labeled nascent nuclear RNAs were hybridized to nitrocellulose filters containing an excess of single-stranded DNA probes derived from the c-*myc* first exon, or double-stranded DNA fragments of the CAT or

β-globin genes (Fig. 5). Probe A (exon 1) provides an estimation of the relative initiation rates, while probe B (either CAT or β-globin) measures the amount of RNA polymerase loading downstream of the exon one transcriptional block [23]. Transcriptional attenuation occurs with the pMHHBg Glob vector, which contains the entire first exon and 7 bp 5′ of the P_1 start site (Fig. 5, lanes 1–3). However, removal of the 5′ 123 bp Hae III-BamHI fragment (−7 to +116) of pMHHBgGlob (see pMBmBgCAT) or replacement of all c-*myc* regulatory sequences 5′ of the P_2 start site (including its TATA box) by SV40 promoter and enhancer sequences (see SV2M(H)$_R$CAT in Fig. 5) [24] completely relieves the elongation block. These data indicate that the P_2 promoter sequences located in this 123 bp Hae III-BamHI fragment are necessary for transcriptional blockage. S1 nuclease mapping in Figure 6 confirms that this 123 bp region is required for P_2 usage as expected from our other results with Bal 31 mutants that removed the ME1a1 factor binding site (Fig. 4).

We next determined the 3′ boundary of the exon one transcriptional block. A set of Bal 31 deletions were prepared in a 3′ direction from the exon 1 BamHI site (see Materials and Methods) and used as single-stranded DNA probes in a nuclear run-on experiment with nuclei from the A-MuLV transformed NIH3T3 line 54c12 (Fig. 7) [42]. Labeled RNAs were hybridized to nitrocellulose filters containing one of these first exon deletion mutants (probe A series) and the downstream portion of the reporter gene (probe C) (Fig. 7). The intensities of the hybridization signals provide a measure of the distribution of RNA polymerases on the c-*myc* gene [34]. The probe A region contains about six times fewer adenine residues than probe C. Thus, when the signals from probe A and C are of equal intensity, the rate of transcription is about sixfold higher in region A. The data in Figure 7 show a dramatic decrease in probe A signal intensity with deletion mutants A3 and A4 with the A4 probe detecting no elongation block. We conclude that the 3′ boundary of the murine c-*myc* gene's transcriptional block resides between +305 and +449, relative to P_1 at +1 (Fig. 7), about 150 bp 5′ of the exon 1/intron 1 splice junction.

DISCUSSION

To investigate the molecular requirements for c-*myc* gene expression, we have used a gel retardation assay and DNase I footprint techniques to identify nuclear protein binding sites within the c-*myc* promoter regions. Two nuclear factor binding sites were found 5′ of P_2 (ME1a1 and ME1a2) and three others 5′ of P_1 (5′Mg1, 5′Mg2 and 5′Mg3). To determine if one or more of these sites bind to any well-characterized transcription factors, we tested their ability to bind to Sp1. Sp1 selectively binds to GC-rich sequence motifs found in the 21 bp repeat elements of the SV40 early promoter [40, 43, 44], in the herpes virus thymidine kinase gene promoter [45], and in a number of cellular genes including human Ha-*ras* 1 [46]. We find that Sp1 binds with high affinity to two regions 5′ of the c-*myc* P_1 start site (defined by nuclear factors 5′Mg1 and 5′Mg3 in Figs. 1, 3). The nucleotide sequence of the 5′ Mg1 region (−55 to −25) is also well conserved nearby the human c-*myc* gene. This Sp1 binding site may be important for P_1 usage since Nishikura has shown that sequences between −60 and −37 relative to the human P_1 transcription start site are essential for its activity in microinjected *Xenopus laevis* oocytes [47]. We have also found evidence for the involvement of other trans-acting factors in addition to Sp1: i) a major DNase I hypersensitive site was detected in the ME1a1 region (+96 to +119) 5′ of the P_2 start site with a crude extract and not with purified Sp1; ii) band shift patterns with either purified Sp1 or crude nuclear extract showed different migrations on polyacrylamide gels with P_1 and P_2 promoter segments (data not shown).

We have also determined the sequence requirements for initiation at P_2. A 123 bp sequence just upstream of P_2 is necessary for its activity, implicating the ME1a1 and/or ME1a2 transcription factors in regulating P_2 usage. Deletions introduced in the ME1a1 region independently confirmed the importance of this site for P_2 activity (this paper and data not shown). However, these ME1a1 binding site deletions retain an active, upstream P_1 start site. Therefore, these data confirm that the P_1 and P_2 initiation sites are independently regulated and that the ME1a1 region plays an important role in P_2 usage.

Previous studies have revealed that regulation of c-*myc* transcription can also occur by some form of transcriptional attenuation within the gene's first exon in a variety of cell types [21,23,29]. We show here that the 3′ boundary of the elongation block is 305 to 449 bp 3′ of the P_1 start site. This region contains a GC rich hairpin structure and a stretch of Ts (see Fig. 7), elements that could be involved in premature termination of transcription [50]. We have also found that a 123 bp segment 5′ of P_2 is required for transcriptional blockage. Taken together, our data suggest that murine c-*myc* transcriptional attenuation and initiation may be coregulated or possibly dependent on each other.

ACKNOWLEDGMENTS

We thank Dr. William Dynan for generously providing us with purified Sp1 transcription factor. We also gratefully acknowledge the expert assistance of Joyce Schirmer and Chris Helmke for preparation of figures and photographs. This research was supported by NCI grant CA36246 (awarded to K.B.M.). C.A. and A.N. were supported by postdoctoral fellowships from the Canadian NCI and the Quebec F.R.S.Q., respectively.

REFERENCES

1. Marcu KB: Bioessays 6:28, 1987.
2. Leder P, Battey J, Lenoir G, Moulding G, Murphy W, Potter H, Stewart T, Taub R: Science 222:765, 1983.
3. Klein G, Klein E: Immunol Today 6:208, 1985.
4. Cory S: Adv Cancer Res 47:189, 1986.
5. Fahrlander PD, Marcu KB: In Kahn P, Graf T (eds): "Oncogenes and Growth Control." Heidelberg: Springer-Verlag, 1986, pp 264–270.
6. Kelley S, Siebenlist U: Annu Rev Immunol 4:317, 1986.
7. Land H, Parada LF, Weinberg RA: Science 222:771, 1983.
8. Stewart T, Pattengale P, Leder P: Cell 38:627, 1984.
9. Adams JM, Harris AW, Pinkert CA, Corcoran LM, Alexander WS, Cory S, Palmiter RD, Brinster RL: Nature 318:533, 1985.
10. Leder A, Pattengale PK, Kuo A, Stewart TA, Leder P: Cell 45:485, 1986.
11. Greenberg ME, Ziff EB: Nature 311:438, 1984.
12. Blanchard JM, Piechaczyk M, Dani C, Chambard JC, Franchi A, Pousseger J, Jeanteur P: Nature 317:443, 1985.
13. Dony C, Kessel M, Gruss P: Nature 317:636, 1985.
14. Greenberg ME, Hermanowski AL, Ziff EB: Mol Cell Biol 6:1050, 1986.
15. Dean M, Levine RA, Campisi J: Mol Cell Biol 6:518, 1986.

16. Dean M, Levine RA, Ran W, Kindy MS, Sonenshein GE, Campisi J: J Biol Chem 261:9161, 1986.
17. Nepveu A, Levine RA, Campisi J, Greenberg ME, Ziff EB, Marcu KB: Oncogene 1:243, 1987.
18. Nepveu A, Marcu KB, Skoultchi AI, Lachman HM: Genes Dev 1:938, 1987.
19. Piechaczyk M, Yang JQ, Blanchard JM, Jeanteur P, Marcu KB: Cell 42:589, 1985.
20. Remmers EF, Yang JQ, Marcu KB: EMBO J 5:899, 1986.
21. Bentley DL, Groudine M: Nature 321:702, 1986.
22. Chung J, Sinn E, Reed RR, Leder P: Proc Natl Acad Sci USA 83:7918, 1986.
23. Nepveu A, Marcu KB: EMBO J 5:2859, 1986.
24. Yang JQ, Remmers EF, Marcu KB: EMBO J 5:3553, 1986.
25. Marcu KB, Asselin C, Nepveu A, Weisinger G, Yang JQ: In Alt FW, Harlow E, Ziff EB (ed): "Nuclear Oncogenes." New York: Cold Spring Harbor Laboratory, 1987, pp 165–172.
26. Lipp M, Schilling R, Wiest S, Laux G, Bornkamm GW: Mol Cell Biol 7:1393, 1987.
27. Hay N, Bishop JM, Levens D: Genes Dev 1:659, 1987.
28. Kakkis E, Calame K: Proc Natl Acad Sci USA 84:7031, 1987.
29. Eick D, Bornkamm GW: Nucleic Acids Res 14:8331, 1986.
30. Bauer SR, Piechaczyk M, Marcu KB, Nordan RP, Potter M, Mushinski JF: Curr Top Microbiol Immunol 132:339, 1986.
31. Alt FW, Rosenberg N, Lewis S, Thomas E, Baltimore D: Cell 27:381, 1981.
32. Southern PJ, Berg P: J Mol Appl Genet 2:327, 1982.
33. Feramisco JR, Smart JE, Burridge K, Helfman DM, Thomas GP: J Biol Chem 257:11024, 1982.
34. Yang JQ, Bauer S, Mushinski JF, Marcu KB: EMBO J 4:1441, 1985.
35. Schibler U, Hagenbuchle O, Wellauer PK, Pittet AC: Cell 3:501, 1983.
36. Dignam JD, Lebowitz RM, Roeder RG: Nucleic Acids Res 11:1475, 1983.
37. Sen R, Baltimore D: Cell 46:705, 1986.
38. Schlokat U, Bohmann D, Scholer H, Gruss P: EMBO J 5:3251, 1986.
39. Rosen CA, Sodioski JG, Haseltine WA: Cell 41:813, 1985.
40. Dynan WS, Tjian R: Cell 35:79, 1983.
41. Gidoni D, Dynan WS, Tjian R: Nature 312:409, 1984.
42. Nepveu A, Fahrlander PD, Yang JQ, Marcu KB: Nature 317:440, 1985.
43. Dynan WS, Tjian R: Cell 32:669, 1983.
44. Gidoni D, Kadonaga JT, Barrera-Saldana H, Jakahashi K, Chambon P, Tjian R: Science 230:511, 1985.
45. Jones KA, Yamamoto K, Tjian R: Cell 42:559, 1985.
46. Ishii S, Kadonaga JT, Tjian R, Brady JN, Merlino GT, Pastan I: Science 232:1410, 1986.
47. Nishikura K: Mol Cell Biol 6:4093, 1986.
48. Piechaczyk M, Bonnieu A, Eick D, Remmers E, Yang JQ, Marcu KB, Jeanteur P, and Blanchard JM: Curr Top Microbiol Immunol 132:331, 1986.
49. Jones TR, Cole MD: Mol Cell Biol 7:4513, 1987.
50. Platt T: Annu Rev Biochem 55:339, 1986.
51. Piechaczyk M, Blanchard JM, Marty L, Dani C, Panabieres F, Elsabouty S, Fort P, Jeanteur P: Nucleic Acids Res 12:6951, 1984.

Growth Factors and Their Receptors: Genetic Control and Rational Application, pages 253–258

Structure and Activities of Acidic Fibroblast Growth Factor

K.A. Thomas, J. DiSalvo, S. Ortega, G. Conn, M.-T. Schaeffer, G. Gimenez-Gallego, D. Soderman, T. Mellin, B. Busch, J. Capparella, J. Menke, L. Kelly, and D. Linemeyer

Department of Biochemistry and Molecular Biology, Merck Research Laboratories, Rahway, New Jersey 07065

Acidic fibroblast growth factor (aFGF), found in greatest abundance in the brain, is a potent mitogen for most, if not all, nonterminally differentiated cells of mesodermal and neuroectodermal embryonic origin. The protein has been purified, sequenced, and observed to be homologous to basic FGF and both the int-2 and hst/KS oncogene products. The FGFs are also distantly related to interleukin-1s. A synthetic gene was constructed, expressed, and demonstrated to be active both in culture and in vivo. The mitogenic activity of human aFGF for cells in culture is potentiated from about one to two orders of magnitude by heparin. Much of the heparin dependence appears to be related to Cys residues and their presumed ability to form incorrect intra- and intermolecular disulfide bonds since site-directed mutant forms of aFGF missing some or all of the Cys residues are almost fully active in the absence of heparin. Pure aFGF is angiogenic and accelerates the rate of wound healing in vivo. These activities indicate that aFGF might be of therapeutic utility in man.

Key words: mitogens, protein sequence homology, wound healing, heparin, angiogenesis

Nearly 50 years ago extracts of whole brain were reported to be particularly rich, compared to other organs, in undefined substances that stimulated mitogenesis of fibroblasts in culture. It was not until a few years ago, however, that the mitogens that were likely to have been responsible for this activity were purified to homogeneity. The brain-derived fibroblast growth factors (FGFs), the topic of previous reviews from this laboratory [1,2][1], are now recognized to be composed of at least two separate homologous proteins, designated acidic FGF (aFGF) and basic FGF (bFGF) based on their apparent isoelectric points. These proteins have been identified and named in numerous labs based on their sources, target cells in culture, and avid binding to heparin affinity columns. Only following purification and characterization, especially by par-

[1]Many results cited in these two reviews are not referenced in this article.

Received February 29, 1988.

tial or complete amino acid sequencing, have they been recognized to be identical with one or the other FGF.

PURIFICATION

Neither aFGF nor bFGF are specific to the central nervous system. Total FGF activity, however, appears to be higher in brain than any other organs. Although both FGFs are obtained from a commonly used salt extraction of homogenized brains, aFGF is recovered in greater abundance. Our original purification of aFGF [3] employed conventional chromatographic steps plus reversed-phase HPLC. Soon after these purifications were reported, however, the tight binding of the FGFs for heparin-Sepharose was recognized [4]. The use of this resin increased both the simplicity and efficiency of the purification of both FGFs and is the basis for the alternative nomenclature of heparin-binding growth factor-α (aFGF) and -β (bFGF). Many FGF purification schemes now combine the high selectivity and excellent recovery of heparin-Sepharose and, especially in the case of the acid-stable aFGF, reversed-phase HPLC.

We originally noted that purified bovine brain-derived aFGF chromatographed as a closely-spaced doublet with an apparent mass of about 16–18 kDa on 15% SDS-polyacrylamide electrophoretic gels. The relative amounts of the two bands varied substantially from preparation to preparation, with some products being virtually all one or the other form. The aFGF preparations were equivalently active, however, regardless of the relative amounts of the two bands. Likewise, when the two proteins were separated on shallow gradient reversed-phase HPLC they appeared to be equally active in a Balb/c 3T3 DNA synthesis mitogenesis assay and to have very similar amino acid compositions. We concluded, therefore, that the two bands were microheterogenous forms of the same protein and not two different gene products [3].

PRIMARY STRUCTURE

Aminoterminal sequence analyses of aFGF from a variety of different preparations indicated an excellent correlation between the relative amounts of higher and lower mass forms with the presence or absence of the amino terminal six residues [5]. A longer form of aFGF [6, reported under the nomenclature of endothelial cell growth factor-β], corresponding to the full-length transcript of the gene [7], has an additional 14 residues aminoterminal to the 140-residue form mentioned above. This long microheterogenous form of aFGF is also reported to be equally active so the aminoterminal end of the protein appears not to be crucial for receptor recognition or any step in subsequent mitogenic activity.

We completed the 140-residue amino acid sequence of bovine aFGF by a combination of tryptic, *Staphylococcus V8,* cyanogen bromide and hydroxylamine cleavages coupled with gas-phase aminoterminal and quantitative timed carboxypeptidase C-terminal microsequencing [8]. Based on this amino acid sequence, a synthetic gene was synthesized and fully active recombinant aFGF was expressed in *Escherichia coli* and purified to homogeneity [9]. We subsequently sequenced human brain-derived aFGF [10] and modified the bovine synthetic gene by site-directed mutagenesis at the 11 altered amino acid positions to code for this human sequence. Recombinant human aFGF was similarly expressed and purified.

AMINO ACID SEQUENCE HOMOLOGIES

About the same time that we completed the bovine aFGF sequence, the group at the Salk Institute published the amino acid sequence of bovine pituitary bFGF [11]. The extensive homology, amounting to 55% identity, between the two proteins was immediately obvious [1,8,11]. This similarity is especially interesting given the virtual identity in the spectrum of target cells in culture of these two factors. Radioreceptor binding assays also indicate substantial, if not complete, cross-reactivity between aFGF and bFGF receptor binding, implying identical or very similar receptors for these two homologous protein growth factors. An apparent duplication within the aFGF gene has been inferred from the amino acid sequence making it, in effect, a covalent dimer [12]. Irrespective of the sequence homology between the two FGFs, no clear remnant of the putative gene duplication is retained in bFGF. Either the internal sequence homology within aFGF is artifactual or the two similar halves of bFGF have diverged sufficiently far so that their common ancestry is no longer readily detectable.

Two oncogene products have recently been identified that bear a clear resemblance to the FGFs [2]. The first is the product of the *int-2* gene that is normally expressed only during early embryogenesis but which is induced in adult mice by proximal integration of the mouse mammary tumor virus [13]. The second oncogene product is that of *hst* which originally was identified in a human stomach tumor [14]. Both of the genes code for proteins that are about 34–42% identical with the FGFs with a slightly higher degree of similarity to bFGF compared to aFGF. The *hst* oncogene has also been reported from Kaposi's sarcoma (KS oncogene), a rare cancer observed in higher than normal frequency in AIDs patients [15]. What appears clear from these observations is that other members of the FGF family exist and their inappropriate expression can be associated with the transformed phenotype.

Close comparison between the FGFs and another family of fibroblast mitogens, the interleukin-1s including IL-1α and IL-1β, revealed evidence for a distant common ancestry [1,8]. Although our comparison of the sequences indicated a marginal significance (2 SD) of relatedness by the classical sequence comparison methods developed by Dayhoff and colleagues, the two sets of proteins were similar with a significance of 6 SD from random when analyzed by the quantitative hydropathicity comparison of Sweet and Eisenberg, a method proposed to be able to detect structural relatedness between proteins that have diverged sufficiently so that sequence similarities have degenerated beyond obvious recognition. Based on this analysis, we proposed that the FGFs and IL-1s comprise an extended family of homologous protein growth factors.

HEPARIN EFFECT

The only obvious biochemical difference between the two FGFs appears to be the effect of exogenous heparin on their mitogenic activity in cell culture. Heparin has been reported by various groups to have little effect on either bovine or human bFGF ranging from a slight activation to a small inhibition. These differences might be attributable either to differences in cells, assay conditions or sources of heparin, a somewhat chemically heterogeneous sulphated polysaccharide. Bovine aFGF typically shows up to about a threefold increase in activity in the presence of heparin compared to in its absence. Human aFGF, on the other hand, appears to have a greater dependence on heparin to fully potentiate its mitogenic activity, being only about 1–10% as active in the absence compared to the presence of heparin. The reason for the apparently greater heparin dependence of human compared to bovine aFGF in not certain. Most of the 11 amino acid differences between the two sequences are conservative substitutions with the exception of compensating conversions of Cys to Ser at position 47 and Ser to Cys at site 117 in bovine versus human aFGFs.

Although the mechanism of heparin potentiation has not been established, the complex of aFGF with this glycosaminoglycan might stabilize the active conformation of the protein. One possible means by which heparin could act is by favoring the correct disulfide status of the protein. We have seen by chemical modification of fully active aFGF with iodoacetic acid that the protein does not contain disulfide bonds since all three Cys residues are free. Air oxidation, however, does generate both intra- and intermolecular disulfide bonds, the formation of which is inhibited by heparin, as indicated both by Cys modification and by mobility on nonreduced SDS electrophoretic gels. Conversion of the Cys to Ser residues also indicates that much, if not all, of the heparin dependence of human aFGF is correlated with the presence of these Cys residues. Therefore, one of the functions of heparin appears to be the prevention of incorrect intra- or intermolecular disulfide bond formation that could trap the protein in inactive conformations or states of polymerization. In this model, the difference in the relative heparin dependence between bovine and human aFGFs might be a result of the different locations of one of the three Cys residues and its influence on the rate at which incorrect disulfide bonds might be formed.

Other glycosaminoglycans, including chondroitin-4 and -6 sulphate, dermatan sulphate, and hyaluronic acid show little or no ability to substitute for heparin, a secreted product of mast cells, in potentiating aFGF mitogenic activity (unpublished results in collaboration with S. Meuller and E. Levine). We have also seen that heparan sulphate and the synthetic sulphated polysaccharides dextran sulphate 8000 and pentosan polysulphate are, on the other hand, able to augment the activity of aFGF in culture. The physiological significance of this heparin and heparan potentiation is unknown. Since mast cells are routinely found in damaged or inflamed tissues, heparin might be present in adequate amounts to potentiate aFGF during tissue repair. Furthermore, all basement membranes and the plasma membranes of some cells are known to contain heparan sulphate which has been shown to bind FGFs in tissues. Therefore, under most circumstances, human aFGF might only be fully active on membrane bound heparan or following release of FGF-heparan complexes by heparanases except in the presence of mast cells which would provide a source of free heparin. In both cases, the glycosaminoglycans might be functioning as a targeting mechanism which could partition [16,17] and potentiate aFGF specific mitogenic activity in vivo. While the potentiating effect on human aFGF could be considered a fail-safe mechanism on the partitioning, both aFGF and bFGF would be expected to bind tightly to heparins and heparans.

ACTIVITIES

Although the FGFs are now recognized to be broad spectrum mitogens in culture and very likely in vivo, one of our original motivations for pursuing the biochemistry of aFGF was the mitogenic activity of partially purified brain extracts for vascular endothelial cells. These cells were recognized to be difficult to grow in culture in standard serum-containing media but their growth was substantially improved by addition of partially purified FGFs. Once purified, we soon confirmed that aFGF was indeed mitogenic for these cells, both of large and microvascular origin [5]. The purified aFGF was also found to induce neovascular growth both on the chick egg chorioallantoic membrane and in the hamster cheek pouch. Therefore, the purified growth factor can orchestrate the cascade of events necessary for vascular growth in vivo.

Various pathological responses have been demonstrated or proposed to be correlated with vascular deficiencies including repair of dermal injury. We have tested the ability of aFGF to accelerate dermal wound healing based not only on its angiogenic activity but also on its potential to stimulate other cells of skin to multiply and, perhaps, to migrate into the wounded area from adjacent tissue. The first model we chose was that of a mouse full-thickness punch biopsy to stimulate repair of the epidermal and extensive dermal necrosis encountered in decubitus ulcers, a serious clinical problem, especially in older bed-ridden patients. Pure aFGF, in the presence of an albumin carrier and small amounts of heparin, was applied daily to the wounds. Control wounds were treated with similar solutions devoid only of the aFGF. Daily wound surface area measurements were made. About 500 ng of aFGF per day was sufficient to generate an approximately twofold acceleration in the rate of wound closure within only 4 days of treatment.

A second healing model was developed to monitor breaking strength of controled full-thickness lacerations. Thicker-skinned rats were used for these studies so that the wounds could be more uniformly closed in a controlled fashion such that the surface on each side of the laceration could be positioned face-to-face. Wounds were treated daily with 2 μg of aFGF and excised as a function of time, and tensile, or break, strength was measured in aFGF-treated versus control incisions. Following 10 days of treatment, the average tensile strength of the treated wounds was about double that of the control incisions. At this same time, an increased cellularity and amount of connective tissue was observed in the wounds that received aFGF.

These in vivo results clearly demonstrate that wound healing is not optimal even in young healthy animals. The mechanisms of these effects are, nevertheless, not certain since aFGF has so many potential target cells in vivo. What is clear, however, is that pure aFGF does demonstrate efficacy in promoting acceleration of wound healing in these models and, therefore, is a reasonable candidate for treating dermal wounds in man.

REFERENCES

1. Thomas KA, Gimenez-Gallego G: TIBS 11:1–4, 1986.
2. Thomas KA: Fed Proc 1:434–440, 1987.
3. Thomas KA, Rios-Candelore M, Fitzpatrick S: Proc Natl Acad Sci USA 81:357–361, 1984.
4. Shing Y, Folkman J, Sullivan R, Butterfield C, Murray J, Klagsbrun M: Science 223:1296–1299, 1984.
5. Thomas KA, Rios-Candelore M, Gimenez-Gallego G, DiSalvo J, Bennett C, Rodkey J, Fitzpatrick S: Proc Natl Acad Sci USA 82:6409–6413, 1985.
6. Burgess WH, Mehlman T, Marshak D, Fraser BA, Maciag T: Proc Natl Acad Sci USA 83:7216–7220, 1986.
7. Jaye M, Howk R, Burgess W, Ricca GA, Chiu IM, Ravera MW, O'Brien SJ, Modi WS, Maciag T, Drohan WN: Science 223:541–545, 1986.

8. Gimenez-Gallego G, Rodkey J, Bennett C, Rios-Candelore M, DiSalvo J, Thomas KA: Science 230:1385–1388, 1985.
9. Linemeyer DL, Kelly LJ, Menke JG, Gimenez-Gallego G, DiSalvo J, Thomas KA: Biotechnology 5:960–965, 1987.
10. Gimenez-Gallego G, Conn G, Hatcher VB, Thomas KA: Biochem Biophys Res Commun 138:611–617, 1986.
11. Esch F, Baird A, Ling N, Ueno N, Hill F, Denoroy L, Klepper R, Gospodarowicz D, Bohlen P, Guillemin R: Proc Natl Acad Sci USA 82:6507–6511, 1985.
12. Strydom DJ, Harper JW, Lobb RR: Biochemistry 25:945–951, 1986.
13. Dickson C, Peters G: Nature 326:833, 1987.
14. Yoshida T, Miyagawa K, Odagiri H, Sakamoto H, Little PFR, Terada M, Sugimura T: Proc Natl Acad Sci USA 84:7305–7309, 1987.
15. Bovi PD, Curatola AM, Kern FG, Greco A, Ittmann M, Basilico C: Cell 50:729–737, 1987.
16. Baird A, Ling N: Biochem Biophys Res Commun 142:428–435, 1987.
17. Vlodavsky I, Folkman J, Sullivan R, Fridman R, Ishai-Michaeli R, Sasse J, Klagsbrun M: Proc Natl Acad Sci USA 84:2292–2296, 1987.

Growth Factors and Their Receptors: Genetic Control and Rational Application, pages 259–268

Clinical Applications of Hematopoietic Growth Factors

David G. Nathan and Colin A. Sieff

Department of Pediatrics, Harvard Medical School and Department of Medicine and Division of Hematology and Pediatric Oncology, Children's Hospital and Dana Farber Cancer Institute, Boston, Massachusetts 02115.

Human recombinant granulocyte macrophage colony stimulating factor (rhGM-CSF) has been investigated in normal, pancytopenic, and irradiated simian models and shown to influence leukocyte, platelet, and erythrocyte production. Studies of rhGM-CSF and rh interleukin 3 (rhIL-3) in vitro in serum-free cultures demonstrate that the human hematopoietic growth factors clearly interact. Neither IL-3 nor GM-CSF act independently but instead require interaction with erythropoietin, G-CSF, or M-CSF to carry out their functions. Pretreatment of simians in vivo with IL-3 markedly increases the sensitivity of hematopoietic progenitors to GM-CSF.

Key words: IL-3, GM-CSF, hematopoiesis

The various marrow failure syndromes have occupied the attention of hematologists since the days of Ehrlich, but it was not until Erslev's confirmation and extension of the classic findings of Carnot and deFlandre [1] that hope for the rational clinical application of hematopoietic growth factors (HGF's) in the refractory cytopenias began to emerge. Anticipation was heightened by Metcalf's observation that the urine of mice with pertubations of granulopoiesis contains a colony stimulating activity (CSA) that would, when injected intraperitoneally, stimulate granulocyte production in normal mice [2]. Frustrating further studies was the enormous difficulty encountered when attempts were made to purify the activity for preclinical use. Innumerable experiments performed in the 25 year period between the late 1950s and early 1980s involved injection of serum, urine, or partially purified proteins containing CSA. The proteins were invariably associated with endotoxin or other physiologically active trace contaminants. It was therefore impossible to gather reproducible and reliable information about CSA itself. Conflicts over interpretation of data ensued. A confusing nomenclature for the hematopoietic growth factors emerged based on poorly defined and less than reproducible activities in vivo and in vitro. The cellular sources of the activities became a battleground of in vitro experiments that were occasionally vituperative and often simply mysterious.

Received March 7, 1988.

Despite this welter of conflicting experiments certain concepts emerged that proved to be both useful and important [3]. There are several HGF's that influence the hematopoietic system, and some of them work synergistically. In vitro experiments strongly suggest that monocytes and T cells as well as the stromal cells of marrow including fibroblasts and endothelial cells interact to produce the HGF's. Some of these factors participate in the proliferation and differentiation of multilineage as well as single-lineage progenitors and are known as multipoietins. Interleukin 3 (IL-3) and granulocyte macrophage colony stimulating factor are examples. Others, such as G-CSF, M-CSF, and erythropoietin (Ep), collaborate with multipoietins to induce progenitor differentiation and can directly induce differentiation of subsets of single-lineage progentors without the synergistic function of a multipoietin. The cellular sites of G-CSF, M-CSF, and GM-CSF are widely disseminated. The signal from the monocyte to the T cell, endothelial cell, and fibroblast to produce their growth factors is mediated by a monkine that has been identified as IL-1 [4,5].

The T cell is itself a potent producer of HGF's and was particularly identified as a producer of what Iscove called burst promoting activity (BPA) [6,7]. The latter is an activity defined in vitro that is required to induce the formation of erythroid colonies in addition to erythropoietin.

A subset of renal cells, as well as fetal liver cells produce erythropoietin in response to hypoxia [8–11]. The regulation of erythropoietin production is mediated through anemia and hypoxia, but the serum levels of the hormone are very variable. The levels in serum and the urinary excretion are much higher at any hemoglobin level in conditions characterized by erythroid hypoplasia rather than hyperplasia, but turnover studies do not suggest that the hormone is utilized by marrow cells [12]. The regulation of the leukocyte inducing HGF's (GM-CSF, G-CSF, and M-CSF) is not understood in any detail, but we know that stimulation of macrophages by endotoxin induces IL-1 release and is followed by enhanced urinary excretion of granulocyte and macrophage colony stimulating activity and granulocyte production in vivo [13]. Unlike erythropoietin and anemia, serum levels of GM-CSF correlate poorly with the degree of neutropenia. This could be due to the very low correlation of circulating phagocytes with the much larger phagocyte pool or to storage of the factors on the surface of the stromal cells [14].

Much of the information briefly outlined above had already been acquired when Clark and his co-workers successfully adapted an expression vector system to produce and subsequently to purify human recombinant granulocyte macrophage colony stimulating factor (rhGM-CSF) from the Mo-1 human T cell line that had been derived by David Golde and his colleagues [13,15]. It remained for Sieff and Emerson and co-workers [16,17] to detect that human GM-CSF is a multipoietin with BPA as well as megakaryocyte and leukocyte inducing activity in vitro. Since then, several other human HGF's have been identified and purified by similar techniques. Table I lists the presently known HGF's and their functions and also provides a brief description of their genetic structures. Figure 1 reviews the hierarchy of the progenitors with which they interact. The figure emphasizes that the multipoietic HGF's (GM-CSF and IL-3) interact with the single-lineage HGF's to induce differentiation of all classes of progenitors from the least mature (multilineage) to the most mature (single lineage). The single-lineage HGF's (M-CSF, G-CSG, and erythropoietin) can alone induce differentiation of a subset of mature (single, lineage) progenitors, including CFU-M, CFU-G, and BFU-F and most CFU-E [18]. The process by which immature (multi-

TABLE I. The Hematopoietic Growth Factors*

			Protein		Biological activity	
Factor	Chromosome	RNA (kb)	Core (kD)	Glycosylated (kD)	Progenitor cells	Mature cells
IL-3	5q 23–31	1.0	14–15	20–26	All CFC	eo
GM-CSF	5q 23–31	1.0	14–15	14–35	All CFC	n,m,eo
G-CSF	17q 11.2–21	1.6	18–19	19–20	CFU-G	n
M-CSF	5q 33.1	4.0	26	35–45 (×2)	CFU-M	m
		1.6	16	20–25 (×2)		
ep	7q 11–22	1.6	18	34–39	BFU-E CFU-E	

*kb = kilobases; kD = kilodaltons; eo = eosinophils; m = monocytes; n = polymorphonuclear neutrophils; CFC = colony forming cells.

lineage) stem cells or progenitors mature to single-lineage progenitors is unknown and has been thought to be entirely stochastic [19]. Recent evidence, however, suggests that the induction of cycling of stem cells may be positively influenced by the combination of IL-6 and IL-3 [19]. It is important to emphasize that the majority of the mature single-lineage progenitors are not induced to form colonies of mature precursors in the sole presence of the appropriate lineage HGF. They require the additional presence of a multipoietin (GM-CSF or IL-3) to complete the differentiation process [18]. Escape from the necessity for a "multipoietin" in addition to a "unipoietin" to achieve precursor colony formation is part of the process of progenitor cell maturation. Thus, subsets of single-lineage progenitors exist that respond to single-lineage HGF's such as erythropoetin without the additional requirement for GM-CSF or IL-3 [18,20].

IL-3 is a more potent BPA than GM-CSF. At equal doses it induces larger and more numerous erythroid colonies in the presence of erythropoietin than does GM-CSF [21]. But GM-CSF is far more ubiquitous than IL-3. The latter gene is transcribed and translated only in stimulated T cells [22], whereas a broad array of hematopoietic and mesenchymal cells can be induced to express and translate the GM-CSF gene. For a review, see Nathan and Sieff [3]. This is surprising because both genes are found in close association (within 10 kb) on chromosome 5 and share similar 5′ promotor regions and a 3′ A/T-rich region associated with mRNA instability [23].

Finally, recent studies have emphasized the important role of monocyte-derived IL-1 in hematopoiesis. Initially identified as hematopoietin 1 or synergistic factor by Stanley and his colleagues [24], it is now apparent that IL-1 has two important HGF functions. The hormone induces the formation of GM-CSF in certain mesenchymal cells, and in some fashion not currently understood, increases the sensitivity of progenitor cell surfaces to HGF's such as M-CSF and IL-3 [25,26].

IN VIVO EFFECTS

As erythropoietin, G-CSF, and GM-CSF have become available for clinical use, attention has been directed toward their application in various disorders of the hematopoietic system. This might have been expected because correction or amelioration of marrow failure syndromes by administration of hematopoetic growth factors has been

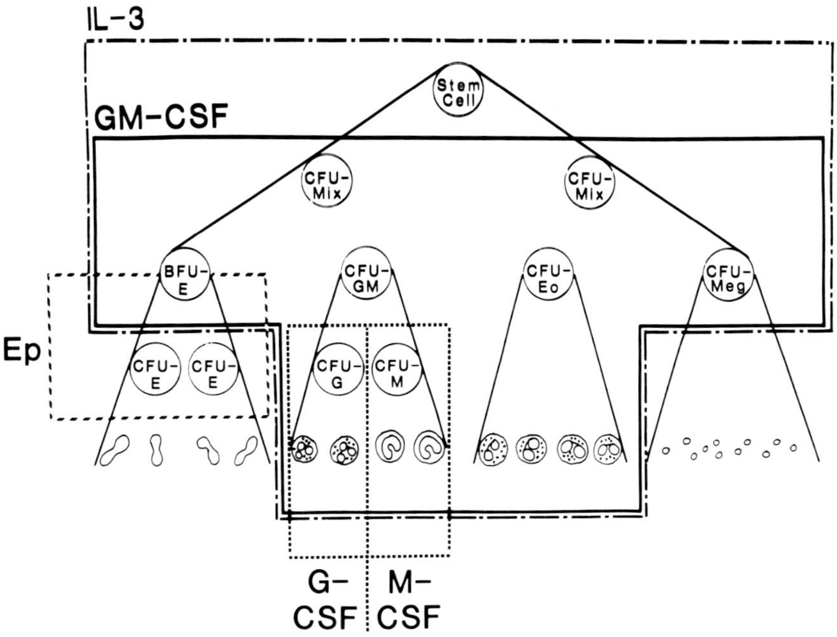

Fig. 1. Lineage diagram showing the overlapping spectra of biological activities of the human hematopoietic factors. Although murine IL-3 (multi-CSF) acts on pluripotent stem cell, this has not been demonstrated for human IL-3. IL-3 and GM-CSF act on similar target cells, while Ep, G-CSF, and M-CSF act on more mature progenitor cells of their respective cell lineages.

and continues to be the major practical goal of research in hematopoiesis. The goal could not be achieved, however, until recombinant DNA technology provided sufficient amounts of the hormones to permit interpretable investigations. It is for this reason and others that the work of Clark and his colleagues has received so much attention in hematology.

The discovery, cloning, and expression of the gene for murine multi-CSF or IL-3 presented the first opportunity to evaluate hematopoietic growth factors in an unambiguous fashion [27,28]. Sublethally irradiated mice were infused for 7 days with recombinant multi-CSF or with control protein [29]. The spleens of the multi-CSF-treated marrow recipients were much larger than those of the controls, were more cellular, and contained more progenitors. The increase in progenitor cells affected erythroid and myeloid lineages. In contrast, bone marrow cellularity was unaffected and progenitor content was reduced. Metcalf et al. [30] injected mice with purified bacterially synthesized IL-3 by the intraperitoneal route and obtained similar results. In addition, tenfold increases in blood eosinophil and 2–3-fold increases in neutrophil and moncyte levels were observed. The intraperitoneal injections also resulted in 6–15-fold increase in peritoneal phagocytes with an increase in macrophage phagocytic activity.

These experiments clearly demonstrate that murine IL-3 influences the replication and growth potential of primitive hemopoietic progenitors, and strongly suggest that whatever effect such hormones have on blood counts are largely related to their influences on progenitor function and not simply to their effects on peripheral blood cell kinetics. They also suggest that the function of mature cells can be altered in vivo, an effect that would be expected to decrease rather than increase numbers of circulating phagocytes.

The first indications that the human hematopoietic growth factor GM-CSF could broadly stimulate hematopoiesis in vivo resulted from the studies of Donahue et al. [31]. These workers infused cynmologous macaques with COS cell produced GM-CSF and markedly stimulated hematopoiesis in the recipients. In preliminary studies, Donahue and co-workers demonstrated that human recombinant GM-CSF induced colony formation by a broad array of simian progenitor cells. They showed that the hormone stimulated simian BFU-E, CFU-GM, and CFU-GEM to form progenitor derived colonies at nanomolar concentrations. They then produced a ^{35}S-labeled sample of GM-CSF by incubation of producer COS cells with ^{35}S-methionine. This material was purified, injected into simian recipients, and the rate of disappearance of radioactivity from the blood determined. The disappearance curve was complex, suggesting a multicompartment turnover model, but the overall initial half-time of 15–20 min clearly demonstrated that infusion of the hormone at a rate sufficient to maintain a functional blood level could be achieved. The results of such infusions into normal *Macaca fascicularis* were striking. Immediate and large increments in all classes of leukocytes including eosinophils and lymphocytes were observed during the hormone infusion. The response was evident within 24 h, suggesting that the effect must be due, in part, to the transfer of leukocytes from a storage pool to the blood. However, the eosinophilia suggested an effect on progenitor differentiation because the fraction of eosinophils in the storage pool was much lower than the fraction observed in the blood following GM-CSF administration. Modest reticulocytosis was also observed, and the rate of emergence was entirely compatible with an effect of the hormone on erythroid progenitors. When the hormone treatment was terminated, the blood counts rapidly fell toward normal. A particularly striking reticulocytosis was observed in an animal with an acquired type D retroviral infection, secondary pancytopenia, and an elevated erythropoietin level. This "preclinical" trial in a severely ill monkey encouraged the conclusion that GM-CSF might play an important therapeutic role in various cytopenias such as those observed in viral infections, including AIDS, or in autologous or allogeneic marrow transplantation.

To investigate the potential role of human recombinant GM-CSF in marrow transplantation, Nienhuis and co-workers [32] infused autologous marrow into rhesus monkeys (*Macaca mulatta*) 2 h after the animals had received 1,200 rads of total body irradiation at 12 rads/min. In this model, a granulocyte nadir is achieved in about 5 days, and recovery of granulocytes to a total count of 1,000 mm^3 does not occur until day 17 or later. In the experiments designed by Nienhuis et al. human recombinant GM-CSF was administered at a daily dose of 50 units/min/kg continuously from 10 to 19 days prior to radiation and/or from 9 to 17 days beginning 2 or 3 days after irradiation. Both dosage schedules produced the same final results, though the blood counts of the animals that received the hormone prior to radiation were much higher at the onset of radiation than those who received the hormone only after radiation. In five separate studies, granulocytes recovered to a level greater than 1,000/mm^3 in 9 days rather than the minimum of 17 days observed in two untreated controls. In four of the five experiments, platelets recovered more rapidly as well.

Human G-CSF has also undergone simian preclinical trials. Welte and colleagues [33] treated cynomolgous monkeys with two daily subcutaneous injections of purified G-CSF for 14–28 days. A dose-related increase in polymorphonuclear neutrophils was observed, the plateau being reached after 1 week of treatment. At the intermediate dose of 10 μg/kg/day, total white blood cell counts of 40,000–50,000/μl were observed. Neutrophil function was also enhanced. Encouraging results were also achieved in two cyclophosphamide treated animals that received G-CSF either from 6 days before until 21 days after the cyclophosphamide treatment, or for 14 days from day 3 after cyclophosphamide. In both monkeys, the neutrophil count increased dramatically by day 6–7 after cyclophosphamide, reaching levels of 50,000/μl by the tenth day. The control animal remained pancytopenic for 3–4 weeks after treatment.

A phase I/II trial of GM-CSF in 16 patients with acquired immunodeficiency syndrome (AIDS) and leukopenia has now been completed [34]. Forty-eight hours after a test bolus dose, patients received a 14 day continuous intravenous infusion of GM-CSF at doses that ranged from 1.3×10^3 to 2×10^4 U/kg/day (0.5–8 μg/kg day). Dose-dependent increases in circulating neutrophils, monocytes, and eosinophils were observed. The peak leukocyte count ranged from 4,500/μl at the lowest, to 48,000/μl at the highest, doses. No serious toxic effects occurred. Low-grade fever, myalgia, and flushing were noted in some patients and mild phlebitis was associated with administration of the drug through a peripheral vein. No phlebitis or clinical thrombosis occurred when the drug was administered by central venous catheter. Preliminary results from a phase I/II study of recombinant G-CSF in cancer patients showed a dose-related increase in neutrophils when the factor was given before the first course of chemotherapy [35]. When the G-CSF was administered after the first course of chemotherapy, a threefold higher white blood cell count was observed on day 14 in patients who received G-CSF compared with patients not receiving the factor; furthermore, the neutrophil count did not fall below $1,000/m^3$ in the G-CSF-treated group, and more of them were able to receive the next course of chemotherapy on schedule.

Human trials of recombinant erythropoietin have clearly documented that it is effective in reducing or eliminating the red cell transfusion requirement of patients with end-stage renal disease [36,37]. A dose-related rise in hematocrit was observed. An increase in blood pressure in four of 25 patients, and increases in serum creatinine and potassium levels in most patients were the only significant toxic effects noted. In addition to the partial or complete correction of anemia, one group of investigators has reported an improvement in bleeding time of uremic patients in whom hematocrit elevations have been observed following treatment with the hormone [38].

Erythropoietin has also been administered to baboons, in which it induces the production of erythrocytes enriched with hemoglobin F [39]. Thus, plans are being readied to administer erythropoietin to patients with sickle cell anemia with the hope that an observed increase in Hb F will reduce the hemolytic rate and perhaps alleviate the symptoms of sickling. In this writer's view, it will be necessary to maintain the hematocrits of such patients at their usual low levels since increased viscosity due to an elevation of hematocrit will tend to cancel any benefits of an increase in the hemoglobin F content of individual sickle cells (assuming that the Hb F contents achieved reduce the rate and extent of sickling in individual erythrocytes) [40].

rhGM-CSF irregularly induces erythropoiesis and thrombopoiesis. Though both Donahue and Nienhuis and their co-workers noted salutory effects of the hormone on both reticulocyte and thrombocyte production, the effects were minimal in normal animals. In contrast, they were observed readily in cytopenic animals whether the cytopenia was induced by irradiation or a type D viral infection. The recent studies of Mayer and co-workers are of interest in this regard [41]. They found that rhGM-CSF was considerably more effective as a stimulant of phagocyte production in rhesus monkeys when it is administered subcutaneously rather than intravenously, but the hormone had little or no effect on reticulocytes or thrombocytes in normal animals. They did note, however, that the hormone-induced granulocytes of such treated animals were activated and possessed enhanced bactericidal activity. This activation may or may not contribute to potential toxicity such as pulmonary sequestration and acute respiratory distress syndrome. Indeed, some workers are reporting in unpublished discussions at conferences that monkeys exposed to high-dose chemotherapy to induce cytopenia may recover rapidly after hormone administration only to suffer extensive gut necrosis, and two patients in Europe who received the hormone to combat chemotherapy-induced cytopenia have actually succumbed to gut necrosis at doses of chemotherapy at which this complication had not been previously observed. Obviously, we do not yet know the range of toxicity of GM- or G-CSF. The activation and increased adhesiveness of neutrophils and monocytes that follows treatment with the hormones may cause untoward accumulation of phagocytes in the microvasculature of lung and gut. This complication may limit the utility of the hormone in chemotherapy or radiation-treated patients in whom lung and gut damage are common. Only careful clinical trials will help us to understand the limits of these problems.

One of the most common causes of bone marrow failure is chemotherapy of cancer, particularly the therapy of acute myelogenous leukemia (AML). Hopes for the use of recombinant G- or GM-CSF in the cytopenia of treated patients with acute myelogenous leukemia were reduced when Griffin and his co-workers observed that AML colony forming cells are induced to form colonies of blasts by both hormones [42]. Delwel and co-workers have confirmed the findings with IL-3 [43]. Some leukemic cells actually produce GM-CSF [44]. Miyauchi and co-workers have presented data to suggest that the HGF's may have different selective effects on normal blood cell proliferation and differentiation in different patients [45]. Some patients might be expected to exhibit improvement due to enhanced blast cell differentiation as a result of treatment with a particular combination of HGF's. Others would be expected to exhibit relapse due to excessive blast cell proliferation. Despite these uncertainties, Yuo and co-workers [46] and Vadhan-Raj and colleagues [47] have treated patients with various myelodysplastic syndromes with varying doses of G-CSF and GM-CSF, respectively. The results of the Houston group [47] with GM-CSF have been of particular interest. Short-term (8–32 week) improvements in the pancytopenia of myelodysplasia have been observed in the majority of patients and no episode of blast crisis has occurred. This unexpected result of a high-risk experiment will be followed with great interest.

The most recent preclinical and clinical trials of HGF's include several experiences with IL-3, GM-CSF, and G-CSF. All of the trials are preliminary and of a phase I/phase II type. Vadhan-Raj and co-workers treated 25 patients with decreased bone marrow function due to malignancy or other causes with daily intravenous infusions of RH GM-CSF for 2 weeks and repeated the treatment after a 2-week rest period [48]. The tested dose range was from 15 to 500 $\mu g/m^2/day$. Increments in the granulocyte counts were regularly observed, and increases in platelet counts and reticulocytes were also observed in about half of the patients. Three patients with smouldering acute myelogenous leukemia were treated and exhibited increased blast cells in marrow and blood. It is clearly dangerous to treat AML with this drug. Its application in the setting of AML needs extreme justification.

Brandt and co-workers [49] have reported on the effect of recombinant rhGM-CSF on hematopoietic reconstitution following high-dose chemotherapy and autologous bone marrow transplantation. They treated 19 patients with either metastatic breast cancer or melanoma with an ablative course of chemotherapy followed 3 days later by autologous marrow cells. rhGM-CSF was then given by continuous intravenous infusion for 14 days beginning 3 h after bone marrow infusion. The doses ranged from 2 to 32 μg/kg/day. Recovery of blood counts was compared to 29 age-, diagnosis-, and treatment-matched historical controls. White blood cell recovery was accelerated in the rhGM-CSF-treated group. No effects on platelet or reticulocyte recovery were demonstrated. The white count recovery was dose related and little toxicity was observed between 2 and 16 μg/kg/day.

Morstyn and colleagues [50] have administered G-CSF to patients with advanced cancer receiving melphalan. Twelve patients were treated in groups of three, each of which received from either 0.3, 1, 3, or 10 μg/kg of G-CSF intravenously on days 1–5 and days 10–18 of a protocol in which a cytopenic dose of melphalan was given on day 9. Serum levels of G-CSF fell in a biphasic manner with a "half-life" of 5–10 and 100 min, respectively. No comparison was offered between treated and untreated patients, but the neutropenia after melphalan was partially abolished at doses as low as 1 and 3 μg per kilogram, and the subcutaneous route of administration produced higher neutrophil levels than did the intravenous route.

Preclinical studies of G-CSF and GM-CSF have been carried out in primates subjected to repeated cycles of cyclophosphamide (CPM) (60 mg/kg per day × 2 every 28 days for four cycles). Two animals were pretreated with either rhG-or rhGM-CSF at 10 μg/kg/day subcutaneously prior to CPM and daily thereafter through subsequent cycles [51]. Two other animals received rhG-CSF or rhGM-CSF, 10 μg/kg/day subcutaneously beginning 72 h following each dose of cyclophosphamide and for 12 days thereafter. One animal received only cyclophosphamide. The results suggested that either continuous treatment with rhGM-CSF or postchemotherapy with rhG-CSF is the most effective means of administration of these agents to ameliorate CPM-induced neutropenia.

In an important study [52], Donahue et al. have shown that primates treated with the combination of 20 μg/kg/day of rhIL-3 from day 1 through 7, and 2 μg/kg/day of rhGM-CSF from day 8 exhibit prompt and marked changes in all classes of phagocytes, lymphocytes, reticulocytes, and platelets. That dose of GM-CSF has little or no effect when given alone, and that dose of IL-3 alone only variably and slightly stimulates reticulocytes and platelet counts or phagocytic counts in the primate species that they studied. When the GM-CSF preceded the IL-3 dose, there was no effect. The observation suggests that IL-3 is capable of acting on an immature population of progenitors which in turn can be stimulated to proliferate and terminally differentiate in the presence of a second low dose of hematopoietic growth factor such as GM-CSF.

The Donahue studies suggest that combination therapy may be very important when hematopoietic growth factors are applied. The recent report of Sieff et al. [18] has demonstrated that neither IL-3 nor GM-CSF (to a lesser extent) have any effect on the appearance of progenitor derived colonies without co-treatment with either G-CSF, M-CSF, or erythropoietin. Conversely, neither G-CSF nor M-CSF nor erythropoietin induces the differentiation of all committed progenitors to form colonies. Only a subset of progenitors exhibits such complete sensitivity to the single-lineage HGF. Furthermore, Il-3 and GM-CSF act synergistically in vitro. Thus, it would appear that future clinical trials will certainly feature combinations of HGF's in an attempt to inducc correction of either acquired or congenital cytopenia.

CONCLUSIONS

The advent of recombinant growth factors has led to a startling revolution in the study of hematopoietic mechanisms. In the 2 years since the genes for these hormones have been successfully cloned and purified, marked advances have taken place in our understanding of hematopoietic mechanisms and in their therapeutic potential. Much clinical investigation needs to be done to understand the limits and the promise of these hormones, but this exciting new area is likely to lead to significant gains for patients with persistent cytopenia and its morbid consequences.

ACKNOWLEDGMENTS

Some of the work presented here was performed by our collaborators Robert Donahue and Steven Clark of Genetics Institute and by Bernard Mathey-Prevot and Charlotte Niemeyer of the Dana Farber Cancer Institute and The Children's Hospital.

REFERENCES

1. Erslev A: Blood 8:349–357, 1953.
2. Metcalf D, Stanley ER: Br J Haematol 21:481–492, 1971.
3. Nathan DG, Sieff CA: In Brown EB (ed): "Progress in Hematology," Vol XV. Orlando: Grune & Stratton, 1987, pp 1–18.
4. Adamson JW: J Clin Invest 81:92–97, 1988.
5. Sieff CA, Tsai S, Faller DV: J Clin Invest 79:48–51, 1987.
6. Iscove NN: In Golde DW, Cline MJ, Metcalf D, Fox CF (eds): "ICN-UCLA Symposium on Hematopoietic Cell Differentiation." Molecular and Cell Biology. New York: Academic Press, 1978, vol 10, pp 37–82.
7. Nathan DG, Chell L, Hillman DG, Clarke B, Breard J, Merler E, Housman DE: J Exp Med 147:324–339, 1978.
8. Koury ST, Bondurant MC, Koury MJ: Blood 70:558, 1987.
9. Kurtz A, Jelkmann W, Pful A, Malmstrom K, Bauer C: Endocrinology 118:567–572, 1986.
10. Goldberg MA, Glass GA, Cunningham JM, Bunn HF: Proc Natl Acad Sci USA 84:7972–7926, 1987.
11. Spivak J: Int J Cell Cloning 4:139–166, 1986.
12. Rosse WF, Waldmann TA: J Clin Invest 43:7, 1964.
13. Wong GG, Witek JS, Temple PA, Wilkens KM, Leary AC, Luxenberg DP, Jones SS, Brown EL, Kay RM, Orr EC, Shoemaker C, Golde DW, Kaufman RJ, Hewick RM, Wang EA, Clark SC: Science 228:810–815, 1985.
14. Ford AM, Watt SM, Molgaard HV, Greaves MF: Blood 70:192a, 1987.
15. Golde DW, Bersch N, Quan SG, Lusis AG: Proc Natl Acad Sci USA 77:593, 1980.
16. Sieff CA, Emerson SG, Donahue RE, Nathan DG, Wang EA, Wong GG, Clark SC: Science 230:1171–1173, 1985.
17. Emerson SG, Sieff CA, Wang EA, Wong GG, Clark SC, Nathan DG: J Clin Invest 76:1286–1290, 1985.
18. Sieff CA, Ekern S, Nathan DG: Clin Res 1988. Submitted.
19. Ikebuchi K, Wong GG, Clark SC, Ihle JN, Hirai Y, Ogawa M: Blood 70(Suppl 1):173a, abstract 549, 1987.

20. Sieff CA, Emerson SG, Mufson A, Gesner TG, Nathan DG: J Clin Invest 77:74–81, 1986.
21. Sieff CA, Niemeyer CM, Nathan DG, Ekern SC, Bieber FR, Wong GG, Clark SC: J Clin Invest 80:818–823, 1987.
22. Niemeyer CM, Sieff CA, Mathey-Prevot B, Bierer BE, Clark S, Nathan DG: Blood 70:182a, 1987.
23. Shaw G, Kamen R: Cell 46:659–667, 1986.
24. Jubinsky PT, Stanley ER: Proc Natl Acad Sci USA 82:2764–2768, 1985.
25. Bartelmeg SH, Stanley ER: J Cell Physiol 122:370–378, 1985.
26. McNiece IK, Bradley TR, Kriegler AB, Hodgson GS: Exp Hematol 14:856–860, 1986.
27. Fung MC, Hapel AJ, Ymer S, Cohen DR, Johnson RM, Campbell HD, Young IG: Nature 307:233–236, 1984.
28. Yokota T, Lee T, Rennick D, Hall C, Arai N, Mosmann T, Nabel G, Cantor H, Ara K: Proc Natl Acad Sci USA 81:1070–1074, 1984.
29. Kinder J, Thorens B, De Kossodo S, Allet B, Eliason JF, Thatcher D, Farber N, Vassalli P: Proc Natl Acad Sci USA 83:1001–1005, 1986.
30. Metcalf D, Begley CG, Johnson GR, Nicola NA, Lopez AF, Williamson DJ: Blood 68:46–57, 1986.
31. Donahue RE, Wang EA, Stone DK, Kamen R, Wong GG, Sehgal PK, Nathan DG, Clark SC: Nature 321:872–875, 1986a.
32. Nienhuis AW, Donahue R, Karlsson S, Clark SC, Agricola B, Antinoff N, Pierce JE, Turner P, Anderson WF, Nathan DG: J Clin Invest 80:573–577, 1987.
33. Welte K, Bonilla MA, Gillio AP, Boone TP, Potter GK, Gabrilove JL, Moore MA, O'Reilly RJ, Souza LM: J Exp Med 165:941–948, 1987.
34. Groopman JE, Mitsuyasu T, DeLeo MJ, Oette DH, Golde DW: N Engl J Med 317:593–598, 1987.
35. Gabrilove J, Jahnbowski A, Fain K, Sher H, Grous J, Sternberg C, Yagoda A, Clarkson B, Moore MAS, Bonilla MA, Oeffgen HF, Alton K, Downing M, Welte K, Souza LM: Blood 70(Suppl 1):135a, 1987.
36. Winearls CG, Oliver DO, Pippard MJ, Reid C, Downing MR, Cotes PM: Lancet ii:1175–1178, 1986.
37. Eschbach JW, Egrie JC, Downing MR, Browne JK, Adamson JW: N Engl J Med 316:73–78, 1987.
38. Moia M, Mannucci PM, Vizzotto L, Casati S, Cattaneo M, Ponticelli C: Lancet ii:1227–1229, 1987.
39. Al-Khatti A, Veith RW, Papayannopoulou T, Fritsch EF, Goldwasser E, Stamatoyannopoulos G: N Engl J Med 317:415–420, 1987.
40. Naguchi TC, Rodgers GP, Sergeant G, Schechter AN: N Engl J Med 318:96–99, 1987.
41. Mayer P, Lam C, Obenaus H, Liehl E, Besemer J: Blood 70:206–213, 1987.
42. Griffin JD, Young D, Herrmann F, Wiper D, Wagner K, Sabbath KD: Blood 67:1448, 1986.
43. Delwel R, Dorssers L, Touw I, Wagemaker G, Löwenberg B: Blood 70:333–336, 1987.
44. Young DC, Wagner K, Griffin JD: J Clin Invest 79:100–106, 1987.
45. Miyauchi J, Kelleher CA, Yang YC, Wong GG, Clark SC, Minden MD, Minkin S, McCulloch EA: Blood 70:657–663, 1987.
46. Yuo A, Kitagawa S, Okabe T, Urabe A, Komatsu Y, Itoh S, Takaku F: Blood 70:404–411, 1987.
47. Vadhan-Raj S, Keating M, LeMaistre A, Hittelman WN, McCredie K, Trujillo J, Broxmeyer HE, Henney C, Gutterman JU: N Engl J Med 317:1545–1552.
48. Vadhan-Raj S, LeMaistre A, Keating M, Walters R, Kantarjian H, Gutterman JU: Blood 70(Suppl 1):144a, 1987.
49. Brandt SJ, Kurtzberg J, Atwater SK, Borowtiz MJ, Jones RB, Shpall EJ, Gilbert CJ, Bast RC, Jr, Oette DH, Peters WP: Blood 70(Suppl 1):131a, 1987.
50. Morstyn G, Duhrsen U, Campbell L, Villeval J, Kannourakis G, Boyd A, Keech J, Abbott M, Nicola NA, Boyd J, Green M, Sheridan W, Cebon J, Thomas RJ, Metcalf D, Fox R: Blood 70(Suppl 1):140a 415, 1987.
51. Bonilla MA, Gillio AP, Potter GK, O'Reilly RJ, Souza LM, Welte K: Blood 70(Suppl 1):130a, 1987.
52. Donahue RE, Seehra J, Norton C, Turner K, Metzger M, Rock B, Carbone S, Seghal R, Yang YC, Clark SC: Blood 70(Suppl 1):133a, 1987.

Growth Factors and Their Receptors: Genetic Control and Rational Application, pages 269–282

Growth, Differentiation, and Neoplastic Transformation of Human Bronchial Epithelial Cells

Curtis C. Harris, Brenda Gerwin, Yang Ke, Tohru Masui, Masao Miyashita, Andrea Pfeifer, Roger Reddel, Vincent L. Wilson, and John F. Lechner

Laboratory of Human Carcinogenesis, National Cancer Institute, Bethesda, Maryland 20892

We are investigating the molecular and cellular mechanisms controlling growth and differentiation of human bronchial epithelial cells and the dysregulation of these controls during the multistage process of carcinogenesis. Our results and those of others are consistent with the hypothesis that the preneoplastic and neoplastic bronchial epithelial cells have a selective clonal expansion advantage because of diminished responsiveness to negative growth factors and/or autocrine production of positive growth factors. Aberrations in regulation of the multistage pathway leading to terminal squamous differentiation of the bronchial epithelial cells are also observed in neoplastic cells. The role of activated proto-oncogenes in controlling proliferation and squamous differentiation is being studied by transfection of expression vectors containing *ras, myc,* and *raf* in the sense or antisense orientation singly or in combination into normal, SV-40 "immortalized," or neoplastic bronchial epithelial cells. Activated *ras* and *myc* proto-oncogenes cause abnormalities in growth and squamous differentiation pathways of human bronchial epithelial cells.

Key words: transforming growth factor type beta, cAMP, myc, fos, ras, raf, 12-O-tetradecanoylphorbol-13-acetate, DNA methylation, oxidants

Normal human cells in vitro appear to retain many normal phenotypic properties, remain diploid, eventually undergo senescence, and rarely, if ever, "spontaneously" transform to malignant cells. Retained properties may include synthesis of classes of proteins associated with specific cell types such as collagens, keratins, or melanin; responsiveness to hormones; and antigenic specificity. In addition, human cells with abnormal phenotypes such as either enzymatic deficits or malignant properties frequently maintain these phenotypes in vitro. Normal human cells cultured in vitro have thus proven to be extremely useful to scientists studying the molecular and biochemical aspects of human carcinogenesis. Such studies have been facilitated by

Received March 24, 1988.

TABLE I. Differentiation Potential of Cultured Human Bronchial Epithelial Cells

A. In vitro	B. In vivo
1. Fibronectin, vitrogen, albumin substrate	1. De-epithelialized tracheal xenografts
a. Squamous	a. Mucous
2. Fibroblastic feeder cells	b. Ciliated
a. Squamous	c. Squamous
b. Ciliated	
3. Collagen gel substrate	
a. Squamous	
b. Mucous	

the recent development of improved methods for culturing normal human epithelial tissues and cells [reviewed in 1]. Chemically defined media have been developed for culturing many of these tissues and cells from normal organs, including those with a high rate of cancer in humans. Serum-free media have several advantages in studies of cultured human cells, including a) less experimental variability compared to serum-containing media; b) selective growth conditions for normal cells of different histological types (e.g., epithelial versus fibroblastic) or for normal versus malignant cells; c) ease of identification of growth factors, inhibitors of growth, and inducers of differentiation; and d) ease of isolating and analyzing secreted cellular products. Advances in cell biology, including the delineation of biochemical and morphological markers of specific cell types, have also facilitated the identification of cells in vitro (including keratins as markers for epithelial cells and collagen types I and III for identifying fibroblasts). These advantages have created experimental approaches to answering critical questions in human cell carcinogenesis [1,2]. For example, we have been investigating the molecular mechanisms controlling growth and differentiation of human lung cells and the dysregulation of these controls during carcinogenesis.

In this brief review, we will emphasize the balance between growth and differentiation pathways of normal versus neoplastic human bronchial epithelial cells in vitro. The normal cells are responsive to certain growth factors and have a pluripotent differentiation potential (Table I). Their proclivity to undergo mucous, ciliated, or squamous differentiation is dependent on their microenvironment. Aberrations in these differentiated phenotypes commonly occur during carcinogenesis.

IN VITRO STUDIES OF MULTISTAGE CARCINOGENESIS

The balance between growth and terminal differentiation is strictly controlled in normal epithelial cells. Furthermore, carcinogenesis studies using murine epidermal cells suggest that defects in differentiation occur during tumor initiation and that selective clonal expansion of these initiated cells occurs during tumor promotion [3]. Studies using human epithelial cells are producing results supporting this hypothetical sequence of aberrations in control of growth and differentiation.

Human epithelial cells respond to several chemical classes of growth factors. For example, epidermal growth factor (EGF), insulin, and hydrocortisone are mitogenic for most types of human epithelial cells [4]. EGF is apparently a universal growth factor and binds to high-affinity membrane receptors. At high concentrations, insulin presumably exerts its mitogenic effects by binding to the membrane receptors for insulin-like growth factors.

Many epithelial cell growth factors, such as the mitogenic factors found in pituitary extracts, have not as yet been identified. The growth of epithelial cells cultured in serum-containing media is enhanced by agents that elevate intracellular levels of cyclic AMP [5]. Epinephrine, cholera toxin, and other cyclic AMP-elevating agents are not directly mitogenic in many types of epithelial cells but act indirectly by negating the growth-inhibitory effects of serum [6,7]. Epithelial and nonepithelial cells may respond differently to growth factors. For example, platelet-derived growth factor enhances the growth of fibroblasts and mesothelial cells, but bronchial epithelial cells are generally unresponsive.

The concept of autocrine production of growth factors has been proposed to explain the uncontrolled growth of some neoplastic cells [8]. "Ectopic" hormones produced by carcinomas are candidates for autocrine growth factors. Gastrin-releasing peptide (the mammalian equivalent of bombesin) is secreted by most small cell carcinomas of the lung [9], and intracellular human chorionic gonadotropin is detected in many non-small-cell carcinomas of the lung [10]. A monoclonal antibody to bombesin blocks the binding of the hormone to cellular receptors and inhibits clonal growth of small cell carcinomas in vitro and their growth as xenografts in vivo [11]. Both of these hormones enhance the growth of normal human bronchial epithelial (NHBE) cells in vitro by binding to specific membrane receptors [12,13].

Normal epithelial cells may differ from preneoplastic cells in their response to tumor promotors. For example, 12-O-tetradecanoylphorbol-13-acetate (TPA) inhibits the growth of normal human colonic epithelial cells and is mitogenic in cultures of epithelial cells from adenomatous polyps [14]. Therefore, an imbalance between the pathways of growth and differentiation (Fig. 1) could provide a selective clonal expansion advantage for preneoplastic and neoplastic human cells, whereas normal epithelial cells respond by terminally differentiating.

Other hypotheses of selective clonal expansion can be envisioned, including defective control of cellular growth leading to differential responses to exogenous as well as endogenous toxic agents (Table II). As noted above, autocrine production of growth factors could provide a selective clonal expansion advantage for preneoplastic and neoplastic cells. A heightened sensitivity to nominal concentrations of systemic and/or locally elaborated mitogens could also provide a selective expansion advantage. This increased sensitivity could arise from an increase in a) the number of receptors, b) their binding affinity for growth factors, and/or c) the efficiency of signal transduction. Although squamous cell carcinomas tend to have an increased number of EGF receptors per cell, the functional significance of these cell membrane changes is unknown; in many carcinomas the receptor number is either unchanged or decreased [15]. Constitutive expression of an autocrine-secreted growth factor appears to be insufficient for driving an otherwise normal cell to the neoplastic state. For example, autocrine secretion of platelet-derived growth factor B does not cause neoplastic transformation of human fibroblasts [16]. The dual abnormalities of autocrine secretion of either EGF or colony-stimulatory factor 1 and constitutive expression of their respective membrane receptors will transform NIH 3T3 cell lines to malignancy [17–19]. Such results add further support to the concept that multiple genetic defects are required in carcinogenesis.

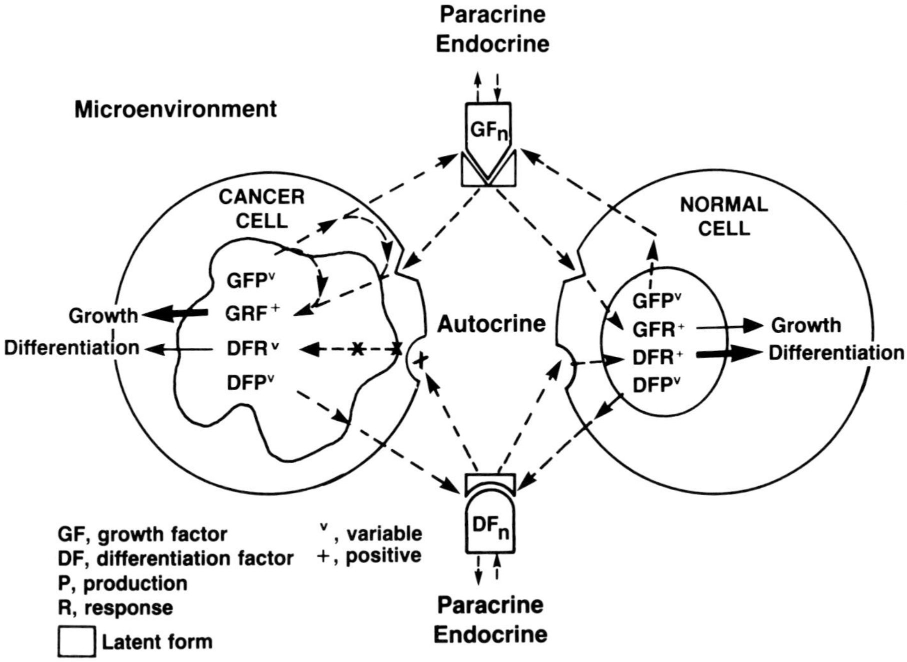

Fig. 1. Secretion of and responsiveness to growth and differentiation factors are tightly regulated in normal cells. Multiple defects that cause an imbalance in the growth and differentiation pathways are observed in cancer cells.

Stimulation of mitogenesis by a constitutively switched-on receptor that does not require binding by the ligand is another hypothetical possibility and has been proposed for the action of a truncated version of the EGF receptor, the oncogene *erb-B* product [20]. This and other connections between oncogenes and mediators in the cellular growth pathways is the subject of a recent review [21].

Human bronchial epithelial cells readily differentiate to squamous cells in vitro, and the recent success in culturing these cells is in part due to empirical observation of the conditions that inhibit terminal cellular differentiation. Cyclic AMP inducers act by neutralizing the growth-inhibitory and differentiating effects of TGF-β_1 found in serum [7]. Although TGF-β_1 increases DNA synthesis in human fibroblastic (Masui, unpublished observation) and mesothelial cells [22], it inhibits growth of epidermal and bronchial epithelial cells [7,23–27]. TGF-β_1 binds to high-affinity membrane receptors and induces several markers of squamous differentiation in bronchial epithelial cells grown at clonal density, including a) inhibition of clonal cell growth, b) irreversible inhibition of DNA synthesis, c) an increase in extracellular plasminogen activator inhibitor, and d) an increase in cellular surface area [6]. Similar results have been observed in rabbit tracheal epithelial cells [25]. In contrast, growth inhibition by TGF-β_1 is reversible in foreskin human epidermal cells [24]. Differences in the culture media used, cell type differences, degree of cellular confluence, and/or the age of the donors may explain these apparently divergent results. For example, the age of the

TABLE II. Possible Selective Clonal Expansion Advantages of Preneoplastic and Neoplastic Cells

A. Defect in control of differentiation, e.g., resistance to induction of terminal differentiation by endogenous and exogenous factors
B. Defect in control of growth
   1. Autocrine production of growth factors
   2. Increased sensitivity to growth factors
   3. Decreased sensitivity to inhibitors of growth
C. Differential response to cytotoxic agents
   1. Inhibition of viral cytopathological response
   2. Resistance to damage by electrophiles
   3. Resistance to oxidative stress
D. Other
   1. Escape from intercellular control mediated by cell-cell communication
   2. Increased capacity to repair DNA damage

donor has been shown to modify the response of epithelial cells to TPA-induced growth arrest and squamous differentiation [28] and subpopulations of less responsive cells may be present in cultures of NHBE cells [23].

Serum induces terminal squamous differentiation of NHBE cells [29]. Anti-TGF-β_1 antibody neutralizes the inhibition of DNA synthesis by either TGF-β_1 or serum in a dose-dependent fashion [7]. These findings indicate that TGF-β_1 is critical for growth inhibition by serum and its induction of squamous differentiation in NHBE cells. Serum is a pathological fluid, and TGF-β most probably plays a role in the repair of normal tissues following wounding [30]. Other endogenous molecules have been found to affect differentiation pathways in epithelial cells. 10,25-Dihydroxyvitamin D_3 induces squamous differentiation of epidermal cells and has been proposed to be important in the maturation of skin [31]. Diethylstilbestrol also inhibits growth and induces the formation of cross-linked envelopes in human cervical epithelial cells [32]. Finally, because normal human bronchial epithelial cells undergo squamous differentiation after confluence in vitro, one may speculate that they secrete an inducer(s) of differentiation that reaches an effective concentration in this cell-packed microenvironment and/or that cell-cell contact is involved. Recent advances in molecular biology and in microisolation and microsequencing of macromolecules now make the search for such inducers feasible.

Exogenous agents can also induce squamous differentiation of epithelial cells [13]. These agents appear to mediate their effects by activating protein kinase C and/or by increasing the concentration of free cytosolic calcium ion. Their effects may result from direct interaction between the agent and the cell (e.g., TPA activation of protein kinase C) and/or indirect mechanisms mediated by membrane lipid peroxidation and generation of active oxygen species that modify mitochondrial membranes, leading to release of calcium ions into the cytosol [33]. In addition, aldehydes, including those found in tobacco smoke (i.e., formaldehyde, acrolein, and acetaldehyde), peroxides, and the neutral fraction of tobacco smoke condensate, produce many of the same changes [34]. It is noteworthy that growth inhibition, produced by either depleting the culture medium of mitogenic factors or exposing cells to cytotoxic agents, e.g., ethionine, does not necessarily induce terminal squamous differentiation of human epithelial cells [7] (unpublished results). Thus, induction of squamous differentiation

TABLE III. TGF-β_1 and Epinephrine Effects on Cellular Cyclic AMP and on Growth of NHBE Cells†

Agent	Cyclic AMP (pmol/10^6 cells)	Growth (%)
Control	2.9	100
+ TGF-β	2.0	21***
+ Epinephrine	50*	200*
+ Epinephrine + TGF-β	52*	74**

†The cells were exposed to test media for 4 h, and cyclic AMP concentration in 10% TCA extract was assayed by radioimmunoassay. Concentrations of TGF-β at 1.2 pM and epinephrine at 1.6 μM were used. Growth of NHBE cells was examined in triplicate with [^{3}H]-thymidine incorporation assay. Significantly different from control, Student's t-test: *$P < .05$; **$P < .01$; ***$P < .005$.

is not simply a response to nonspecific growth inhibition and/or cytotoxicity. The data are consistent with the model proposed by Scott and Maercklein [35] that implicates a distinct arrest point in the cell cycle for the differentiation pathway.

MECHANISMS OF SQUAMOUS DIFFERENTIATION OF HUMAN BRONCHIAL EPITHELIAL CELLS

Cyclic AMP and Inositol Phospholipid Second Messengers

Growth or differentiation factors bind to specific membrane receptors and trigger the release of second messengers that lead to modulation of gene expression and ultimately control cellular phenotype and behavior. The signal transduction system using cyclic AMP as the second messenger has been the most extensively studied [36]. More recently, second messengers generated by another signalling pathway that utilizes inositol phospholipids have been identified as critical signals for growth and differentiation factors in mammalian cells [37].

In some types of epithelial cells an increase in the cyclic AMP level is correlated with growth stimulation [5]. Although elevation of cyclic AMP levels by epinephrine alone does not enhance growth of NHBE cells, cell growth is enhanced by epinephrine or other cyclic AMP enhancers (3-isobutyl-1-methyl-xanthine, dibutyryl cyclic AMP, and cholera toxin) in the presence of both EGF and bovine pituitary extract [6]. Moreover, cholera toxin antagonizes the growth inhibitory effect of serum on NHBE cells [38].

To investigate whether the effect of TGF-β_1 might be modulated by the cyclic AMP system, the effect of epinephrine was examined [7]. Epinephrine stimulates growth of NHBE cells in the absence of TGF-β_1, and neutralizes the inhibitory effect of TGF-β_1 on DNA synthesis (Table III). NHBE cells do not become squamous in the presence of epinephrine in TGF-β_1-supplemented medium. On the other hand, small amounts of TGF-β_1 (0.12 pM) that only weakly inhibit DNA synthesis of NHBE cells significantly inhibit the growth enhancement caused by epinephrine. TGF-β_1 did not cause changes in cyclic AMP levels in the cells, whether or not epinephrine was present. Therefore, epinephrine and TGF-β_1 appear to affect different intracellular pathways that control growth and differentiation processes of NHBE cells. Interestingly, we found that the differentiation-inducing effect of TPA is also partially antagonized by epinephrine (Masui et al., unpublished results). This observation supports the suggestion that TGF-β_1 and TPA may share common steps and that epinephrine may cause certain effects antagonistic to differentiation induction.

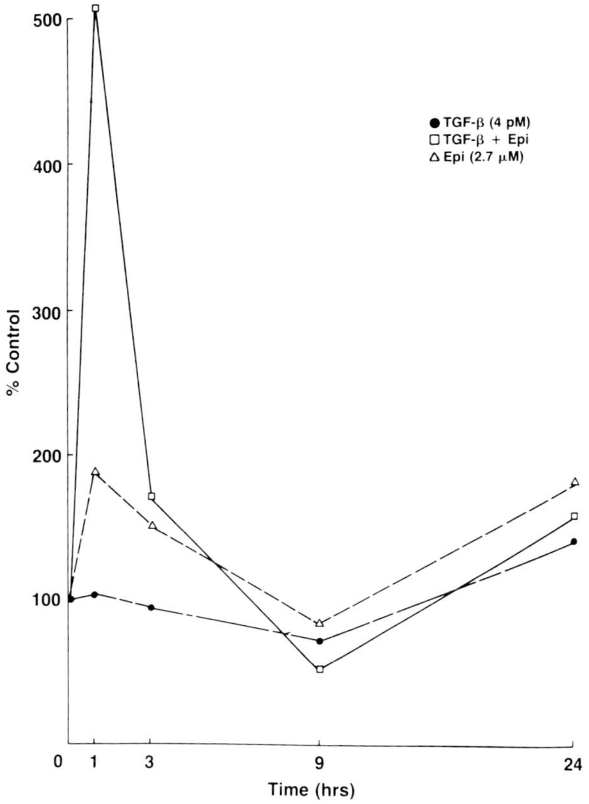

Fig. 2. Effects of TGF-β_1 and/or epinephrine on steady-state levels of c-fos mRNA in NHBE cells.

We have recently investigated the effects of TGF-β_1 and/or epinephrine on the mRNA steady-state levels of the proto-oncogenes c-myc, c-fos, and c-Ha-ras in NHBE cells [39]. Specific mRNA was detected by Northern blotting and normalized by probing the constitutive enzyme glyceraldehyde-3-phosphate-dehydrogenase. The level of c-myc mRNA was reduced by TGF-β_1 up to 40% at 1 h and returned to the control level after 24 h. Epinephrine increased the expression of c-myc about 2–3-fold and neutralized the reduction of c-myc mRNA by TGF-β_1. An increase in the level of c-fos mRNA was induced at 1 h by TGF-β_1 (1.5-fold) or epinephrine (2–3-fold). The level of c-fos mRNA returned to the control level after 9 h. TGF-β_1 and epinephrine in combination acted synergistically to increase c-fos mRNA levels (Fig. 2). Elevation of intracellular cAMP by epinephrine stimulates the positive DNA cis-acting regulatory element found in the 5′ untranslated portion of c-fos that is responsive to cAMP [40]. The synergism of epinephrine with TGF-β_1 (Fig. 2) suggests that it may also act via as an as yet unknown second messenger and activate a positive cis-acting DNA regulatory element of c-fos. The level of c-H-ras mRNA was not affected by either TGF-β_1 or epinephrine. These data suggest that in NHBE cells a transient alteration in c-myc mRNA levels reflects or is involved in a signal in the proliferative pathway. On the other hand the levels of c-fos and c-Ha-ras mRNAs were uncoupled from the processes of growth or squamous differentiation of NHBE cells.

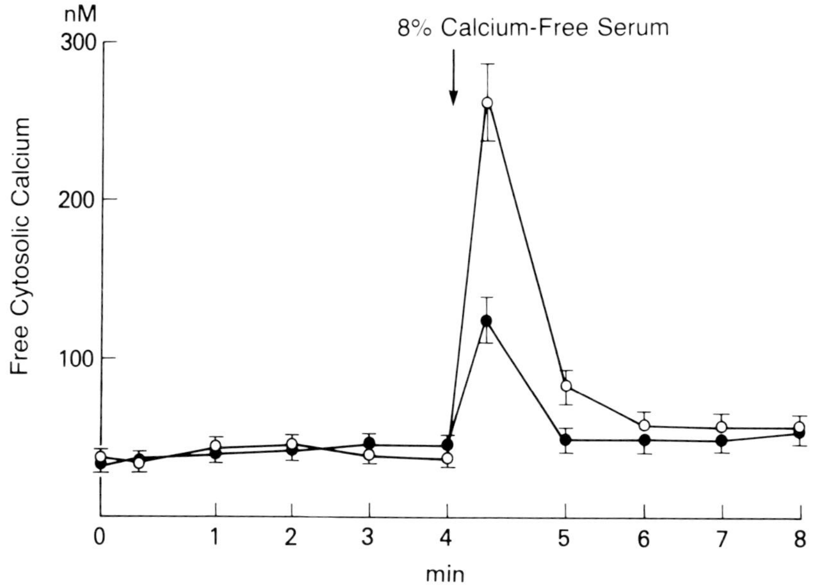

Fig. 3. Attenuation of serum-induced $[Ca^{2+}]_i$ by TPA pretreatment of BEAS-2B cells.

Changes in free cytosolic calcium ion concentration ($[Ca^{2+}]_i$) have been observed in mammalian cells during stimulation of either growth or differentiation. For example, a decrease in $[Ca^{2+}]_i$ is associated with DMSO-induced differentiation of Friend erythroleukemia [41,42]. However, since Ca^{2+}-chelating agents or agents that block extracellular Ca^{2+} uptake inhibit DMSO-induced differentiation of erythroleukemia cells [43], the role of $[Ca^{2+}]_i$ in these processes is presently unclear and may vary in different cell types. In HL-60 human promyelocytic leukemia cells, inhibition by verapamil of $[Ca^{2+}]_i$ elevation enhances differentiation induced by 1α25-hydroxyvitamin D_3, retinoic acid, or DMSO, but not by TPA [44]. Calcium ionophore A23187 also potentiates monocytic differentiation of HL-60 cells induced by TPA [45], but causes Ca^{2+} influx in lymphocytes and promotes proliferation [46].

A high concentration of extracellular calcium induces terminal squamous differentiation in mouse epidermal keratinocytes [47–49]. However, when NHBE cells are incubated at clonal density, a high concentration of extracellular calcium ion has no effect on growth or squamous differentiation, but it does cause squamous differentiation in concert with a small amount of serum [50]. When NHBE cells reach culture confluence, they become sensitive to the squamous differentiation effects of high concentrations of extracellular calcium ion (unpublished results). In both NHBE cells and adenovirus 12-SV-40 hybrid virus transformed NHBE cells (BEAS-2B cells), ionomycin rapidly elevates $[Ca^{2+}]_i$ [51]. This rise in $[Ca^{2+}]_i$ may accelerate the process of cross-linked envelope formation during squamous differentiation because the protein cross-linking reaction occurs by a Ca^{2+}-dependent enzyme, transglutaminase [52]. Serum causes growth inhibition of human epidermal keratinocytes and terminal squamous differentiation of NHBE cells [29,38] and BEAS-2B grown at low cell density [53]. When compared to NHBE cells, BEAS-2B cells are about 100-fold less sensitive to TPA induction of differentiation even though the number and affinity of their TPA receptors are equivalent to NHBE cells [53]. Our recent studies indicate that calcium-free fetal bovine serum (FBS) increased $[Ca^{2+}]_i$ rapidly and transiently while inhibiting cell growth and causing the same morphological changes as those caused by FBS. Although TGF-β_1 is an important factor in the induction of squamous differentiation by serum [7], TGF-β_1 did not alter $[Ca^{2+}]_i$. These results suggest that the presence of another factor(s) in serum is responsible for changes in $[Ca^{2+}]_i$.

Certain factors in serum bind to specific cellular receptors and this receptor-agonist coupling activates phospholipase C and which generates two interacting second messengers. One receptor-mediated signal is 1,4,5-inositol trisphosphate, which increases $[Ca^{2+}]_i$, and another is diacylglycerol, which activates protein kinase C [37]. Activation of protein kinase C by either endogenous diacylglycerol or exogenous TPA is linked to extrusion of free-cytosolic Ca^{2+} to the endoplasmic reticulum, mitochondria, or extracellular space by activated Ca^{2+}-dependent-ATPase [54–56]. Although TPA is an exogenous inducer of squamous differentiation in NHBE cells [50], it did not alter $[Ca^{2+}]_i$ in either NHBE cells or BEAS-2B cells (Fig. 3). Furthermore, TPA pretreatment inhibited the generation of a calcium signal caused by calcium-free FBS. Studies in several cell types have shown that both GTPγS stimulation of phospholipase C activity and 1,4,5-inositol trisphosphate production are attenuated by TPA [reviewed in 37]. One possible mechanism for these effects would involve phosphorylation of the α-subunit of G protein by protein kinase C. These data and the results reported here suggest that activation of protein kinase C may be a negative regulator of calcium homeostasis in human bronchial epithelial cells.

The relationship between $[Ca^{2+}]_i$ and squamous differentiation has been investigated utilizing cell lines subcloned from BEAS-2B cells. One cell line (S_6) retains the parental sensitivity to serum-induction of squamous differentiation while the other (R_1) is resistant to these effects of serum. The S_6 cell line produced a pattern of $[Ca^{2+}]_i$ change similar to that found in BEAS-2B cells when exposed to calcium-free FBS. The R_1 serum-resistant subclone of BEAS-2B cells generated a lesser increase in $[Ca^{2+}]_i$ in response to calcium-free FBS. Thus, in contrast to observations with TGF-β_1 and TPA, elevation of $[Ca^{2+}]_i$ is associated with serum-inducted squamous differentiation of NHBE cells and BEAS-2B cells, suggesting that serum factors other than TGF-β_1 mediate changes in $[Ca^{2+}]_i$. We speculate that certain serum factors may activate phospholipase C, leading to the formation of 1,4,5-inositol trisphosphate and increased $[Ca^{2+}]_i$. It is interesting that phospholipase C added to the culture media of murine or human normal keratinocytes causes terminal squamous differentiation [57,58].

In summary, an increase in $[Ca^{2+}]_i$ was observed during the process of serum-induced squamous differentiation in either NHBE cells or BEAS-2B cells, and this increase was minimal in the R_1 subclone of BEAS-2B cells. However, a calcium signal was not generated during squamous differentiation induced by either TGF-β_1 or TPA. Therefore, early changes in $[Ca^{2+}]_i$ appear not to be necessary to initiate the process of squamous differentiation in human bronchial epithelial cells.

DNA Methylation

Epigenetic changes in gene expression can also be mediated by methylation of cytidine. The nucleotide analog 5-azacytidine (5AzaC), induces differentiation in a

TABLE IV. A Comparison of the Abilities of Various Agents to Induce Squamous Differentiation and Alter the Genomic 5-Methyldeoxycytidine Content of NHBE Cells

Agent	Concentration	Population doubling per day[a]	DNA synthesis[b]	Morphology	Cross-linked envelopes[c]	% 5mdC[d]
None	—	0.62	100	Normal	100	3.09 ± 0.07 (28)
5-Azacytidine	5 M	0.16	12 ± 2	Squamous	281	2.19 ± 0.04 ($P < 0.001$) (3)
12-O-tetradecanoyl-phorbol-13-acetate	10 nM	0.09	25 ± 2	Squamous	483	2.83 ± 0.08 ($P < 0.001$) (4)
Fetal calf serum	8%	0.04	64 ± 3	Squamous	325	2.59 ± 0.13 ($P < 0.001$) (4)
Transforming growth factor-beta$_1$	1.2 pM	0.10	41 ± 2	Squamous	224	2.74 ± 0.04 ($P < 0.001$) (4)

[a]Values for population doubling per day represent the average of at least three determinations.
[b]Total DNA synthesis was monitored by <^{3}Hβthymidine incorporation and expressed as percent of controls. Values represent the mean ± SD of 4 separate experimental samples.
[c]The No. of cells producing cross-linked envelopes in the presence of calcium ionophore A23187, expressed as percent of control. Values represent the average of two separate experiments. The percentage of cells producing cross-linked envelopes was 2% and 39% of the respective controls for the two separate experiments.
[d]Values represent the total percent of deoxycytidine residues methylated and are reported as the mean ± SD of (N) samples from at least two separate experiments.

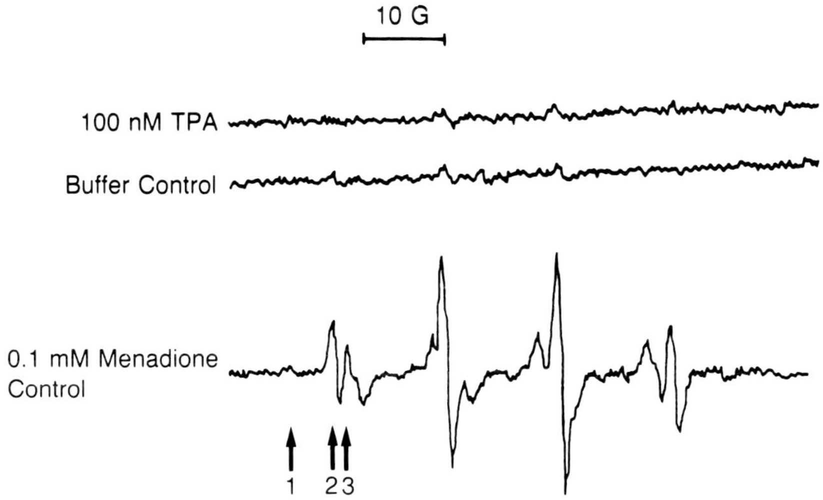

Fig. 4. Electron paramagnetic resonance in NHBE cells exposed to menadione or TPA. 2,2.5-trimethyl-1-pyrrolidinyloxyl (peak 1), DMPO-OH (peak 2), and DMPO-OOH (peak 3). Microwave power was 20 mW, modulation frequency was 100 kHz with an amplitude of 0.63 G, sweep time was 12.5 G/min, and the receiver gain was 6.3×10^2 with a response time of 1.0 s [74].

number of model systems [see reviews: 59,60]. Treatment of dividing murine C3H/10T½cells with 5AzaC induced the formation of adipocytes, chondrocytes, and myotubules [61,62]. Davis et al. [63] have recently used subtraction cDNA hybridization methods to isolate a gene (MyoD) from 5AzaC-treated C3H/10T½ cells that is related to c-myc and converts C3H/10T½ cells to stable myoblasts. Erythroid differentiation has been initiated in Friend leukemia cells [64] and in HL60 cells [65,66] by 5AzaC.

The induction of differentiation by 5AzaC has been correlated with the ability of this agent to induce decreases in levels of 5-methyldeoxycytidine (5mdC) in DNA in these model systems [64,66–68]. As the prototype inhibitor of 5mdC formation, DNA which has incorporated 5AzaC binds to and blocks the action of the DNA cytidine methyltransferase enzyme responsible for the maintenance of the 5mdC patterns in newly synthesized DNA [69,70]. The first level of control for the expression of some mammalian genes appears to be selective DNA methylation sites [71,72], so that changes in DNA 5mdC patterns might be expected to occur during differentiation. The requirement that global changes in the DNA 5mdC content occur as an integral component of the differentiation process, however, remains controversial.

We have recently investigated the effects of a variety of squamous differentiation-inducing agents on the genomic level of 5mdC [73]. 5-Azacytidine, TPA, or TGF-β_1 significantly diminish the genomic 5mdC content of NHBE cells concomitantly with the induction of squamous differentiation (Table IV). These effects of TPA or TGF-β_1 occurred at concentrations of 1 nM and 1.2 pM, respectively, which are approximately one-tenth of kd values of their specific receptors. Therefore, the decreases in genomic 5mdC induced by these agents are probably mediated by receptor-ligand interactions.

TABLE V. Strategy for Studying Neoplastic Transformation of Human Bronchial Epithelial Cells by Activated Proto-Oncogenes

A. Select proto-oncogenes activated in human lung cancer
B. Use progenitor epithelial cells of bronchogenic carcinoma
C. Use high-frequency gene transfection methods
D. Use conditions to select preneoplastic and neoplastic cells
E. Investigate dysregulation in molecular controls of growth and terminal differentiation
F. Determine tumorigenic potential in athymic nude mice

TABLE VI. Multistage Transformation of Human Bronchial Epithelial Cells by Transfected vHa-ras

Progressive phenotypic and genotypic changes
1. Decreased response to inducers of squamous differentiation
2. Increased response to serum mitogens
3. Increased frequency of chromosomal aberrations and aneuploidy
4. Increased cell population doublings
5. Cell "crisis"
6. Continuous cell line
7. Tumorigenicity
8. Tumor cells have increased ras p21 expression

Oxygen Radicals

As noted above, the tumor promoter TPA inhibits growth and induces terminal squamous differentiation of NHBE cells when added to the culture medium [50]. We have investigated the possibility of free oxygen radicals being involved as intermediates in this process [74]. Electron paramagnetic resonance measurements using the spin trapping agent 5,5-dimethyl-1-pyrroline-1-oxide failed to detect free oxygen radicals in NHBE cells exposed to TPA, although radicals were detected in bronchial epithelial cells after a nontoxic exposure to menadione, which did not induce squamous differentiation, and in human neutrophils after exposure to TPA (Fig. 4). Addition to the culture medium of free radical scavengers, i.e., superoxide dismutase, reduced glutathione, N-acetylcysteine, D-α-tocopherol, and copper$^{(II)}$ (3,5-diisopropyl salicylic acid)$_2$ (CuDIPS) did not affect the dose-dependent growth inhibition of TPA on the bronchial epithelial cells. Moreover, exposure of the bronchial epithelial cells to TPA did not result in increased DNA single strand breaks measured by the alkaline elution methodology, as would be expected with a free-radical mediated mechanism. Thus, our results argue against the importance of free oxygen radicals in the inhibition of growth and the induction of squamous differentiation by TPA in NHBE cells.

NEOPLASTIC TRANSFORMATION OF HUMAN BRONCHIAL EPITHELIAL CELLS

The strategy that we have formulated for investigating the neoplastic transformation of NHBE cells is shown in Table V. Four families of proto-oncogenes, *ras, raf, fur,* and *myc* have so far been associated with human lung cancer. Human bronchial epithelial cells in vitro are being used to investigate the role of these specific oncogenes and growth regulatory genes in carcinogenesis and tumor progression. Using the protoplast fusion method for high frequency gene transfection, the v-Ha-*ras* oncogene initiates a cascade of events in the normal human bronchial cells leading to their apparent immortality, decreased responsiveness to inducers of squamous differentiation, aneuploidy, and tumorigenicity with metastasis in athymic nude mice [75] (Table VI). Transfer of the SV-40 T antigen gene leads to bronchial epithelial cell lines that are nontumorigenic at early passage [76], e.g., BEAS-2B cells, that have a nearly normal pathway of terminal squamous differentiation and can be readily transformed to malignant cells by Ha-*ras* [77] or Ki-*ras* [78] or the combination of transfected c-*myc* and c-*raf*-1 [79]. These and other results indicate that activated proto-oncogenes dysregulate pathways of growth and differentiation of human bronchial epithelial cells and play an important role in human lung carcinogenesis. These studies suggest the possibility of reversal of the malignant phenotype by use of antisense mRNA from these proto-oncogenes.

REFERENCES

1. Harris CC: Cancer Res 47:1, 1987.
2. Brash DE, Reddel RR, Quanrud M, Ke Y, Farrell MP, Harris CC: Mol Cell Biol 7:2031, 1987.
3. Yuspa SH: In Pullman B (ed): "Interrelationship Among Aging, Cancer and Differentiation." New York: D. Reidel Publishing, 1985, pp 67–81.
4. Barnes D, Sato G: Anal Biochem 102:255, 1980.
5. Green H: Cell 15:801, 1978.
6. Willey JC, LaVeck MA, McClendon IA, Lechner JF: J Cell Physiol 124:207, 1985.
7. Masui T, Wakefield LM, Lechner JF, LaVeck MA, Sporn MB, Harris CC: Proc Natl Acad Sci USA 83:2438, 1986.
8. Sporn MB, Roberts AB: Nature 313:747, 1985.
9. Moody TW, Pert CB, Gazdar AF, Carney DN, Minna JD: Science 214:1246, 1981.
10. Trump BF, Wilson T, Harris CC: In Ishikawa S, Hayata Y, Suematsu K (eds): "Lung Cancer 1982." Amsterdam: Excerpta Medica, 1982, pp 101–124.
11. Cuttitta F, Carney DN, Mulshine J, Moody TW, Fedorko J, Fischler A, Minna JD: Nature 316:823, 1985.
12. Willey JC, Lechner JF, Harris CC: Exp Cell Res 153:245, 1984.
13. Harris CC, Yoakum GH, Lechner JF, Willey JC, Masui T, Gerwin B, Schlegel S, Mark G: In Harris CC (ed): "Biochemical and Molecular Epidemiology of Cancer." New York: Alan R. Liss, 1986, pp 213–226.
14. Friedman EA: Cancer Res 41:4588, 1981.
15. Stoscheck CM, King LE, Jr: Cancer Res 46:1030, 1986.
16. Bywater M, Mark G, Heldin C-H, Westermark B, Betsholtz C: J Cell Biochem 12A:162, 1988.
17. Velu TJ, Beguinot L, Vass WC, Willingham MC, Merlino GT, Pastan I, Lowy DR: Science 238:1408, 1987.
18. DiFiore PP, Pierce JH, Fleming TP, Hazan R, Ullrich A, King CR, Schlessinger J, Aaronson SA: Cell 51:1063, 1987.
19. Roussel MF, Dull TJ, Rettenmier CW, Ralph P, Ullrich A, Scherr CJ: Nature 325:549, 1987.
20. Downward J, Yarden Y, Mayes E, Sorace JG, Totty N, Stockwell P, Ullrich A, Schlessinger J, Waterfield MD: Nature 307:521, 1984.
21. Goustin AS, Leaf EB, Shipley GD, Moses HL: Cancer Res 46:1015, 1986.
22. Gabrielson EW, Lechner JF, Gerwin BI, Sporn MB, Wakefield LM, Roberts AB, Harris CC: Proc Am Assoc Cancer Res 27:214, 1986.
23. Hesterberg TW, Maness SC, Iglehart JD, Sanchez JH, Boreiko CJ: Carcinogenesis 8:1511, 1987.
24. Shipley GD, Pittelkow MR, Wille JJ, Scott RE, Moses HL: Cancer Res 46:2068, 1986.
25. Jetton AM, Shirley JE, Stoner G: Exp Cell Res 167:539, 1986.
26. Pittelkow MR, Wille JJ, Scott RE: J Invest Dermatol 186:410, 1986.
27. Bertolero F, Kaigh ME, Camalier RF, Saffiotti U: In Vitro Cell Dev Biol 22:423, 1986.
28. Parkinson EK, Al-Yaman FM, Appleby MW: Carcinogenesis 8:907, 1987.
29. Lechner JF, McClendon IA, LaVeck MA, Shamsuddin AM, Harris CC: Cancer Res 43:5915, 1983.

30. Sporn MB, Roberts AB, Shull JH, Smith JM, Ward JM, Sodek J: Science 219:1329, 1983.
31. Hosomi J, Hosoi J, Abe E, Suda T, Kuroki T: Endocrinology 113:1950, 1983.
32. Stanley MA, Crowcroft NS, Quigley JP, Parkinson EK: Carcinogenesis 16:1011, 1985.
33. Cerutti PA: Science 227:375, 1985.
34. Saladino AJ, Willey JC, Lechner JF, Grafstrom RC, Harris CC: Cancer Res 45:2522, 1985.
35. Scott RE, Maercklein PB: Proc Natl Acad Sci USA 82:2995, 1985.
36. Gilman AG: Cell 36:577, 1984.
37. Berridge MJ: Annu Rev Biochem 56:156, 1987.
38. Lechner JF, Haugen A, McClendon IA, Shamsuddin AM: Differentiation 25:229, 1984.
39. Mausi T, Gerwin BI, Lechner JF, Harris CC: J Cell Biochem [Suppl] 12A:125, 1988.
40. Verma IM, Sassone-Corsi P: Cell 51:513, 1987.
41. Faletto DL, Macara IG: J Biol Chem 260:4884, 1985.
42. Faletto DL, Macara IG: Fed Proc 5:1954, 1985.
43. Supino R, Gibelli N, Galatulas I, Zunino F: Cell Biol Int Rep 9:1059, 1985.
44. Okazaki T, Mochizuki T, Tashima M, Sawada H, Uchino H: Cancer Res 46:6059, 1986.
45. Tyers M, Harley CB: Fed Proc 5:1594, 1985.
46. Lichtman AH, Segel GB, Lichtman MA: Blood 61:413, 1983.
47. Hennings H, Michael D, Cheng C, Steinert P, Holbrook K, Yuspa SH: Cell 19:245, 1980.
48. Hennings H, Holbrook KA, Yuspa SH: J Cell Physiol 116:265, 1983.
49. Hennings H, Holbrook KA, Yuspa SH: J Invest Dermatol 81:50s, 1983.
50. Willey JC, Moser CE Jr, Lechner JC, Harris CC: Cancer Res 44:5124, 1984.
51. Miyashita M, Smith MW, Willey JC, Lechner JF, Trump BF, Harris CC: Proc Am Assoc Cancer Res 29:166, 1988.
52. Rice RH, Green H: J Cell Biol 76:705, 1978.
53. Ke Y, Reddel RR, Gerwin BI, Miyashita M, McMenamin M, Lechner JF, Harris CC: Differentiation 38:60, 1988.
54. Drummond AH: Nature 315:752, 1985.
55. Lagast H, Pozzan T, Waldvogel FA, Lew PD: J Clin Invest 73:878, 1984.
56. Rink TJ, Sanchez A, Hallam TJ: Nature 305:317, 1983.
57. Jeng AY, Lichti U, Strickland JE, Blumberg PM: Cancer Res 45:5714, 1985.
58. Parkinson EK: Carcinogenesis 8:857, 1987.
59. Holliday R: Science 238:163, 1987.
60. Jones PA: Cancer Res 46:461, 1986.
61. Constantinides PG, Jones PA, Gevers W: Nature 267:364, 1977.
62. Taylor SM, Jones PA: Cell 17:771, 1979.
63. Davis RL, Weintraub H, Lassar AB: Cell 51:9877, 1987.
64. Creusot F, Acs G, Christman JK: J Biol Chem 257:2041, 1982.
65. Bodner AJ, Ting RC, Gallo RC: JNCI 67:1025, 1981.
66. Christman JK, Mendelsohn N, Herzog D, Schneiderman N: Cancer Res 43:763, 1983.
67. Jones PA, Taylor SM: Cell 20:85, 1980.
68. Taylor SM, Jones PA: J Mol Biol 162:679, 1982.
69. Santi DV, Norment A, Garrett CE: Proc Natl Acad Sci USA 81:6993, 1984.
70. Christman JK, Schneiderman N, Acs G: J Biol Chem 260:4059, 1985.
71. Ley TL, Chiang YL, Haidaris D, Anagnou NP, Wilson VL, Anderson WF: Proc Natl Acad Sci USA 81:6618, 1984.
72. Korba BE, Wilson VL, Yoakum GH: Science 228:1103, 1985.
73. Wilson VL, Masui T, Smith RA, Harris CC: Carcinogenesis 9:2155, 1988.
74. Gabrielson EW, Rosen GM, Grafstrom RC, Strauss KE, Harris CC: Cancer Res 48:822, 1988.
75. Yoakum GH, Lechner JF, Gabrielson E, Korba BE, Malan-Shibley L, Willey JC, Valerio MG, Shamsuddin AKM, Trump BF, Harris CC: Science 227:1174, 1985.
76. Reddel RR, Ke Y, Gerwin BI, McMenamin M, Lechner JF, Su RT, Brash DE, Park J-B, Rhim JS, Harris CC: Cancer Res 48:1904, 1988.
77. Amstad P, Reddel RR, Pfeifer A, Malan-Shibley L, Mark GE, Harris CC: Mol. Carcinogenesis 1:151, 1988.
78. Reddel RR, Amstad P, Malan-Shibley L, Pfeifer A, Ke Y, Gerwin BI, Mark GE, Rhim JS, Harris CC: Proc Am Assoc Cancer Res 29:449, 1988.
79. Pfeifer A, Mark GE, Reddel RR, Mann DL, Amstad P, Malan-Shibley L, Harris CC: Proc Am Assoc Cancer Res 29:170, 1988.

Growth Factors and Their Receptors: Genetic Control and Rational Application, pages 283–295

Forces Involved in Epidermal Growth Factor Receptor Binding

Kevin H. Mayo

Department of Chemistry, Temple University, Philadelphia, Pennsylvania 19122

Structure-function relations of epidermal growth factors (EGFs) and EGF-like peptides have become, in the last few years, an area of intense scientific interest. Since 1962, when Stanley Cohen published observations of early eyelid openings in neofetal mice which were injected with what we know today as murine EGF [1], many researchers pursued the study of how EGF induces biological activity and mitogenesis in cells. Mostly lacking in this research drive were efforts to answer the fundamental question, What is the molecular structure of EGF and how does that structure interact with the receptor to trigger such important biological activities? Limitations on amounts of isolatable material and on experimental methodology in structure research were the chief reasons which greatly hampered structural studies. With the advent of genetic engineering techniques and structural analyses via nuclear magnetic resonance (NMR) spectroscopy these limitations have been mostly removed. Our studies are currently focused on NMR structural analyses of rat and mouse EGF and physical properties of their interaction with cell receptors.

Amino acid sequences for mouse [2], rat [3], human [4], and guinea pig [3] EGFs are available and are given in Figure 1. As has been observed by many, these sequences contain a high degree of sequence homology. Since EGF from these species can compete with each other for binding to the same cell receptor in some cell lines, a receptor binding domain common to at least these EGF species is postulated to exist on the EGF protein surface. Poor anti-EGF antibody cross-reactivity among rat, mouse, and human EGFs [5], moreover, suggests that differences do exist in their structures. It has also been shown that certain virally transformed tumor cells produce an EGF-like peptide called transforming growth factor (TGF), which also competes with EGF for EGF receptors [6–10]; primary structures of rat [11] and human [12] TGFs are also given in Figure 1. Structural comparison of these EGFs and TGFs might aid in the identification of this proposed EGF receptor binding domain. Shown below these sequences (Fig. 1) is a bar graph which plots a residue frequency versus

Received April 1, 1988.

mouse EGF	NSYPGCPSSYDGYCLNGGVCMHIESLDSYTCNCVIGYSGDRCQTRDLRWWELR
rat EGF	NSNTGCPPSYDGYCLNGGVCMYVESVDRYVCNCVIGYIGERCQHRDLR
human EGF	NSDSECPLSHDGYCLHDGVCMYIEALDKYACNCVVGYIGERCQYRDLKWWELR
guinea pig EGF	QDAPGCPPSHDGYCLHGGVCMHIESLNTYACNCHIGYVGERCEHQDLDLWE
rat TGF	VVSHFNKCPDSHTQYCFH-GTCRFLVQEEKPACVCHSGYVGVRCEHADLLA
human TGF	VVSHFNDCPDSHTQFCFH-GTCRFLVQEDKPACVCHSGYVGARCEHADLLA

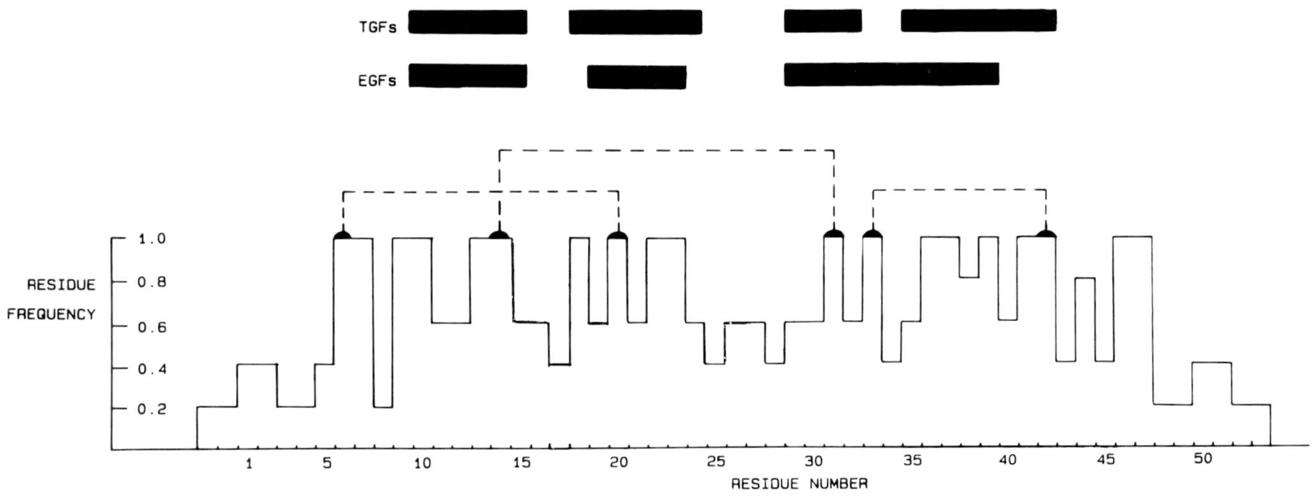

the residue number. The residue frequency, F, was calculated for each residue, ι, as:

$$F_\iota = \frac{n_\iota - 1}{N - 1} \tag{1}$$

where N equals the total number of sequences being compared, and n_ι equals the number of conserved residues for residue number "ι" relative to murine EGF. $F_\iota = 1$ means that a residue position is totally conserved in all sequences.

Among these growth factors, C- and N-terminal segments are least conserved, while about 40% of residues from position 6 to 47 are totally conserved. These F_ι values have been adjusted for some conserved amino acid types, i.e., all phenylalanine, tyrosine and histidine residues were treated equally, as were all leucine, isoleucine and valine residues. The six totally conserved cystines are indicated with half-filled circles at the top of the bar graph, and the conserved disulfide bridges are shown as dashed lines.

Conserved sequences generally relate to conserved higher order structures. By using the secondary structure predictive algorithm of Chou and Fasman [13], a high probability for β-structure is suggested for sequences shown by thick blackened lines for TGFs and EGFs (Fig. 1). On the whole, those regions of predicted β-structure are conserved in the sequences examined and run within the segment of residues 6 to 47.

Differences in amino acid C-terminal sequence length among these EGFs and TGFs might suggest that the C-terminus plays a less crutial role in receptor binding. Holladay et al. [14] state that in mouse EGF, the C-terminal pentapeptide is important to receptor binding since its removal reduces receptor binding tenfold; it is, however, possible that indirect effects, i.e., induced conformational changes, on peptide cleavage caused the observed differences in binding activity.

Reversed-phase HPLC of electrophoretically "pure" mouse [15,16] and rat [3] EGFs which had been initially purified by methods based on the original Savage and Cohen procedure [17] reveals that as many as four forms of the peptide actually exist; these have been identified as N-terminal cleavage products [3,16,18]. The terminology calls parent EGF as αEGF with subsequent cleavage products of one amino acid at a time as β, γ, δ, etc.... Matrisian et al. [19] claim that mouse αEGF is fourfold more potent than βEGF in stimulating DNA synthesis in quiescent Rat-1 cells, while others [15,18] state that no differences in activities can be observed among N-terminal derivatives. It is still unclear whether or not N-terminal or C-terminal sequence perturbations affect receptor binding or mitogenic activity.

In order to understand the forces involved in EGF receptor binding, one must have knowledge of the three-dimensional structure of EGF and of its receptor and have identified the receptor binding domain and the amino acid residues responsible for binding. Some progress is being made in this area, and we shall focus the remainder of this paper on recent NMR structural studies of EGFs.

Fig. 1. Amino acid sequences for EGFs and TGFs. Amino acid sequences for four species of EGF, i.e., mouse (2), rat (3), human (4), and guinea pig (3), and for two species of TGF, i.e., rat (11) and human (12), are aligned for comparison. The residue frequency as defined in the text is plotted in bar graph form versus the residue number. Chou-Fasman secondary structure algorithm predictions for β-structure for EGFs and TGFs are given as blackened bars.

STRUCTURE OF EGF

Early structural work on non-HPLC purified murine EGF by using circular dichroism (CD) spectroscopy [14] allowed estimation of about 26% β-structure and some possible C-terminal helical structure. CD, however, only allows one to have an average conformational picture of a protein without being able to specify, for example, inter-nuclear distances and specific, backbone folding patterns. Such structural analyses have normally been done by x-ray diffraction studies which are impotent in the case of EGF where suitable crystals have not been grown.

The first NMR studies on mouse EGF [20,21] clearly demonstrated the applicability of NMR spectroscopy to a structural analysis of EGF. Based primarily on observed nuclear Overhauser effect (NOE) cross-relaxation rates and preliminary two-dimensional (2D) NMR experiments, refinement of an approximate secondary structure model for mouse EGF [14] led to a proposed, partial tertiary structure containing antiparallel β-sheet segments and a clustered aromatic region, i.e., N-terminal domain [21,22]. This model resulted in a number of tentative resonance assignments of which few could be regarded as definitive in this early work; this tertiary model, therefore, should be viewed as a hypothetical model consistent with a limited set of observations. Many definite resonance assignments for mouse EGF were given in a later paper [16] and others are given herein.

During the last several years, methods which are based on two-dimensional, high-resolution ^{1}H-NMR experiments have been developed for complete structural analyses of low molecular weight, well-structured peptides. An overview of these methods has recently been published by Wüthrich [23]. Briefly, this approach involves 1) making sequence-specific, amino acid proton resonance assignments; 2) developing proximity relations among groups of protons and residues to deduce backbone folding patterns; and 3) structural modeling.

The first step in ^{1}H-NMR sequential resonance assignments is the identification of spin systems associated with individual protons of amino acid residues. A number of 2D-NMR experiments are normally performed to group resonances in a complex protein ^{1}H-NMR 2D-spectrum on the basis of thru-bond scalar coupling information, i.e., called COSY (correlated spectroscopy) type experiment. In any 2D-NMR experiment, the resulting data are usually displayed in a 2D-contour plot where the diagonal contains the normal 1D-NMR spectrum and the off-diagonal cross-peaks give various types of associations among or between proton resonances which lie on the diagonal. In the case of these COSY-like experiments, protons that are three bonds away from other protons and, therefore, covalently linked, can be identified and grouped into so-called "spin-systems." One spin-system is identified with one amino acid residue since the backbone carbonyl group decouples thru-bond proton scalar coupling to adjacent residues in the sequence.

Some spin systems are unique—i.e., Gly, Ala, Val, Ile, Leu and Thr—giving rise to a spectral pattern that is unambiguous. Other amino acid spin systems are generally nonunique, giving rise to several amino acid possibilities for one spectral pattern [23]. In the case of EGFs, the lack of lysine and phenylalanine residues helps by simplifying the spectral situation. Simple spin systems which contain only the NH-αCH-βCH_2 fragment, i.e., Ser, Cys, Asp, Asn, Tyr, and His, can be called "short" spin systems; and all other, nonunique ones, i.e., Glu, Gln, Met, Pro, and Arg, can be called "long"

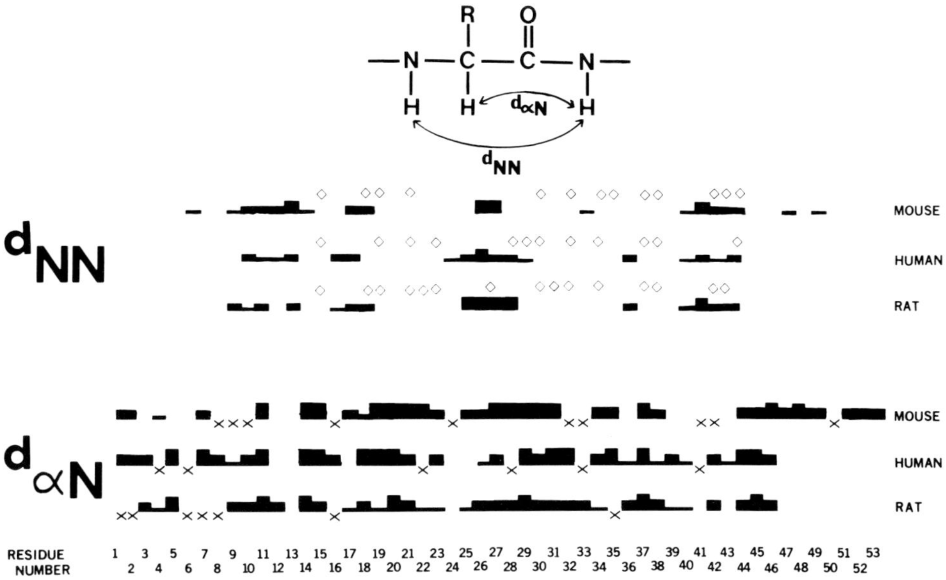

Fig. 2. Relative NOE magnitudes for EGFs. Backbone αCH to NH and NH to NH distances, i.e., $d_{\alpha N}$ and d_{NN}, respectively, are reflected in the relative NOE magnitudes and listed versus the residue number for mouse, human, and rat EGFs. The data are taken from [21] for human EGF, from [19] for rat EGF, and from [25] for mouse EGF. Long-lived backbone amide proteins are indicated as open-diamonds above the residue number.

spin systems. Hereby, many, if not all, proton resonances can be grouped and assigned to an amino acid or at least to one group of amino acids, i.e., "short" or "long."

Having grouped or assigned amino acid spin systems in this way, one can then sequentially assign them by using nuclear Overhauser effect spectroscopy (NOESY) which gives thru-space distance information. The NOESY experiment is sensitive to interproton distances which range from van der Waals contact distances to about 5 Å, depending on how the experiment is set up and on such variables as temperature and internal motions among protons. Normally, one can focus in on very close protons or open the NOE distance window to include slightly more distant protons. Relatively speaking, the larger the NOESY cross-peak, the closer the protons are. Regarding sequential residues, therefore, the NOESY experiment, in many cases, allows one to "jump" from one residue in the sequence to the next. For example, one can have a thru-space, NOESY connectivity between the αCH resonance of one residue to the backbone NH resonance of the neighaboring residue on the C-terminal side, or we can have a NOESY connectivity between backbone NH resonances of proximal residues.

By combining data from NOESY and COSY-like spectra, one can "walk" thru the peptide backbone and make sequence-specific (or sequential) assignments [23]. For example, we might start with the αCH to NH cross-peak of a unique valine spin-system in the COSY spectrum; then by following the αCH resonance line to the exact same position in the NOESY spectrum we should be able to identify the C-terminal neighbor NH resonance and by following its position back to the COSY spectrum we can identify its αCH resonance, perhaps of a glycine residue. This approach is repeated and a number of such di-, tri-, or polypeptide sequences can be thusly identi-

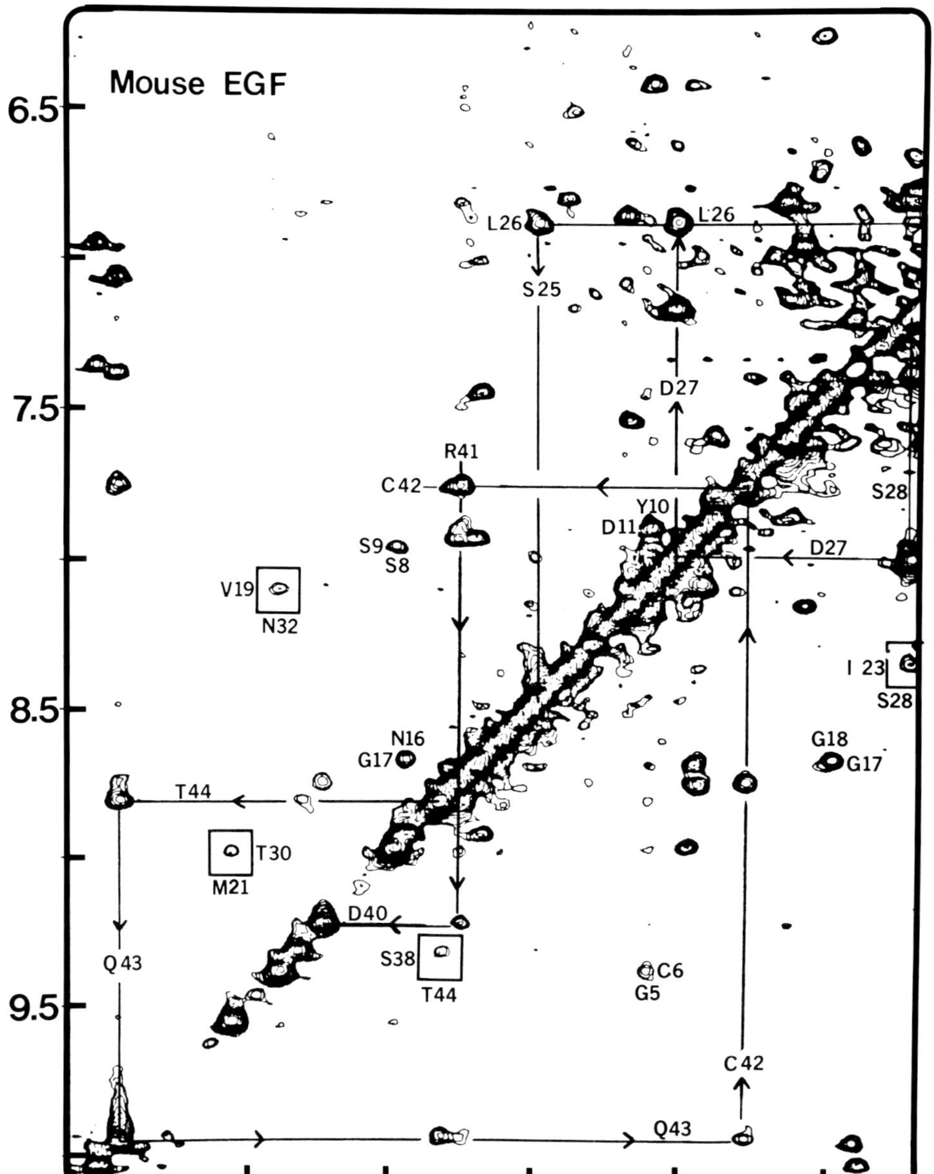

Fig. 3. Comparative NOESY contour plots for EGFs. The amide regions of 500 MHz NOESY spectra for mouse and rat αEGFs are shown for comparison. The data were acquired in 90% $1_{H_2O}/10\%\ 2_{H_2O}$ using pulse method solvent suppression; the protein concentration was approximately 2 mM in 20 mM potassium phosphate, pH 3 at 303 K. The mouse data are from Mayo and Burke [25] and the rat data are from [24]. Two chain reversal domains, i.e., residues 28–25 and 44–40 are indicated by solid lines. Interchain contacts are shown by boxed-in cross-peaks. Other NH to NH distances are also identified, as discussed in the text. Long-lived backbone amide protons are indicated as diamonds at respective residue positions. NOE magnitudes that could not be observed due to technical reasons are identified with an "X."

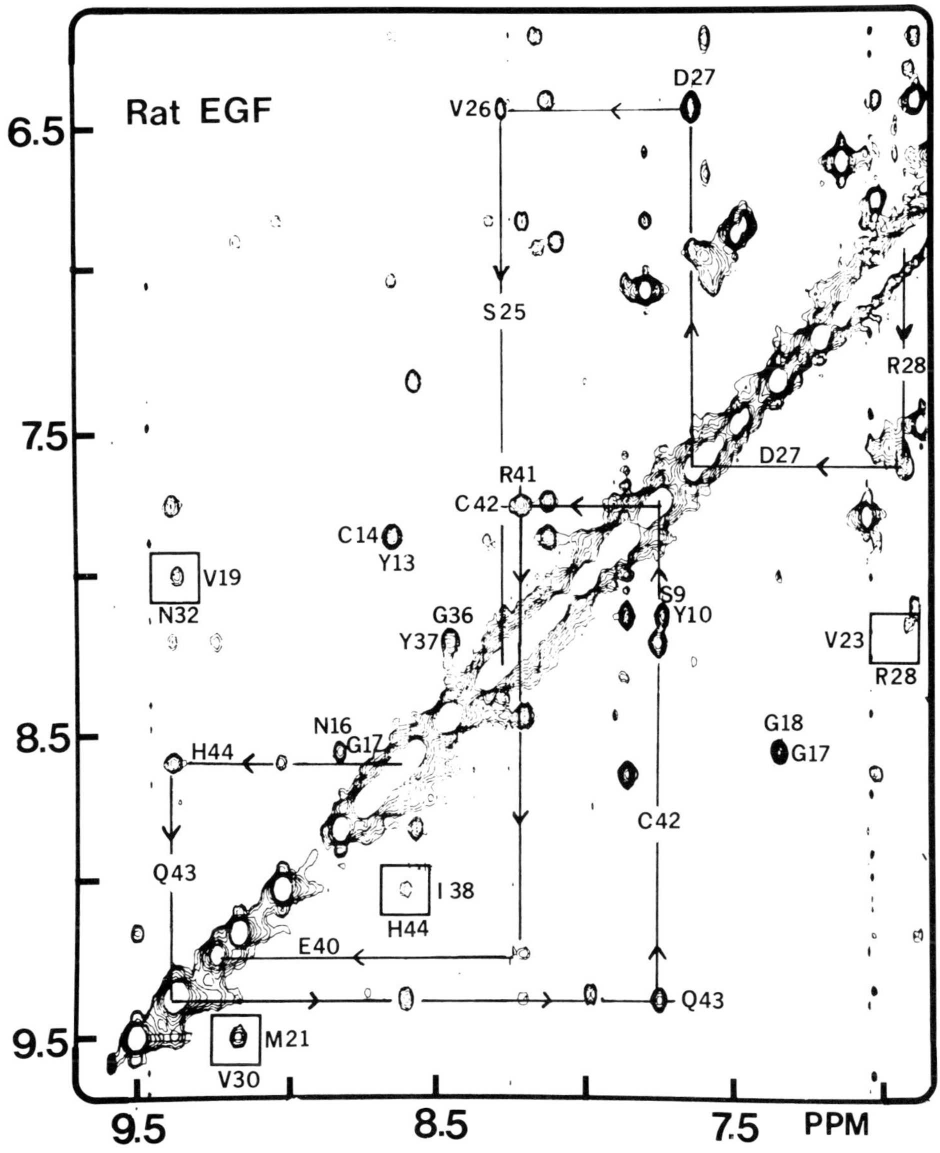

Figure 3

fied. In the case of our NH-Val-Gly-CO example, we can then search the known sequence for such a dipeptide and perhaps uniquely identify this sequence which gives us the sequence specific resonance assignments. Finally, unassigned peptides or peptide segments can be then assigned by using already uniquely assigned resonances as starting points. Other NMR data are then correlated with these to assure that all assignments are in fact unique [23].

Backbone conformations in terms of secondary structure may then be derived from NOESY data. Inter-residue backbone NH to NH and NH to αCH average distances along the chain sequence are reflected in the presence and magnitude of NOESY cross-peaks associated with resonances assigned to these protons. For example, a large NH to NH cross-peak connecting two sequentially proximal residues suggests that these two residues are perhaps in a helical or turn-like conformation where the NH to NH distance, designated d_{NN}, is normally smaller than in an extended or β-sheet conformation. On the other hand, a small αCH to NH distance, i.e., $d_{\alpha N}$, therefore a large NOE, would tend to suggest a β-sheet or extended conformation rather than a helical type structure. Usually, one piece of information does not surfice, and a composite picture of many pieces of NMR data are ultimately necessary to define structural types unambiguously. By compiling such NOESY/distance information, various structural features can usually be deduced. Figure 2 gives a list of such sequential NOEs observed in 2D-NMR structural studies on rat [24], mouse [25,26], and human [27] αEGFs. Overall, these data suggest a number of conserved secondary structures among these EGF species. For example, continuous segments of short $d_{\alpha N}$ distances suggest extended backbone conformations. In particular, residues 14–15; residues 18 or 19 to 23; residues 28–34; residues 37–38, and residues 44 to at least 47 have a high probability of existing in an extended backbone conformation. The N-terminus up to about residue 18 probably exists in a multiple bend structure, as evidenced by alternating short $d_{\alpha N}$ and d_{NN} distances. Chain reversals are indicated between residues 23 and 28, 34 and 37, and residues 39 and 43, where segments show continuous short d_{NN} distances, and $d_{\alpha N}$ distances are large, i.e., little or no apparent NOE.

As examples of our data, Figure 3 gives comparative NOESY spectra for mouse [25] and rat [24] αEGFs, showing many cross-peaks between backbone NH proton resonances. These spectral patterns and assignments are very similar for these two species of EGFs, suggesting the presence of similar structural domains. A number of NH resonance assignments are given in this figure. Continuous segments of short d_{NN} distances are also indicated with solid black-line vectors for two of the chain reversal domains mentioned above. Once again, notice the similarities in relative NOESY cross-peak magnitudes between rat and mouse αEGFs.

The NOESY experiment is also used to establish more distant inter-residue, proximity relations and backbone folding patterns. Boxed-in cross-peaks in Figure 3, for example, indicate interstand β-sheet NH-NH NOEs that reflect interproton distances of between 3 and 4 Å.

Figure 4 gives a general composite picture of backbone folding currently known to occur in mouse [16,21,22,26,28], human [27,29], and rat [24,30] αEGFs. Totally conserved residues in the four species of EGF given in Figure 1 have been circled in Figure 4. Position 35 is always an isoleucine or a valine residue, and position 40 is always a glutamic or aspartic acid residue; these residues might, therefore, also be added to this group. Interestingly, most of these residues do form a conserved structural domain which is constrained by the C33-C42 disulfide bridge. The short antiparallel β-sheet segment, V19-M21 and C31-N32, is, moreover, also as conserved and is structurally proximal to the C33-C42 double hairpin segment. Of these two conserved structures, the C33-C42 segment, is the more highly conserved when comparing them to TGFs (see Fig. 1) and might suggest a good candidate for a least part of the proposed receptor binding domain.

The N-terminus interacts weakly with the β-sheet segment containing residues 19–23 and is folded such that the Tyr-10/Tyr-13 segment is proximal to the Tyr-29

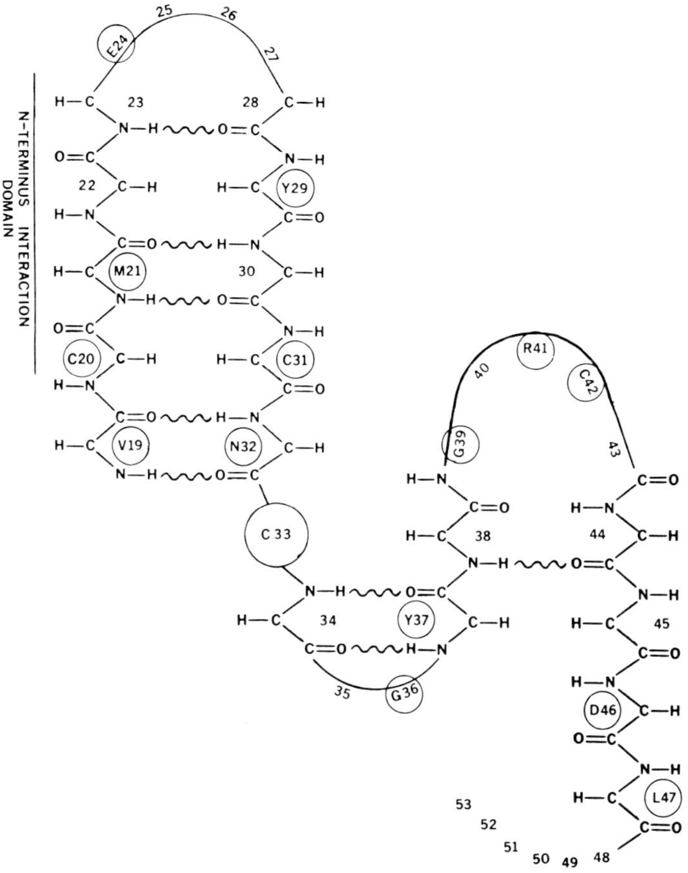

Fig. 4. Backbone folding patterns for EGFs. Well-defined structural domains in mouse, rat, and human EGFs are summarized. Proposed hydrogen bonds are shown as wavy lines. Totally conserved residues are circled.

β-sheet segment. In at least the mouse, rat, and human species of EGF, the N- and C-termini seem to be relatively highly mobile while the central portion of the peptide, i.e., residues 15 to about 44, seems to be rigidly constrained by the presence of three disulfide bonds and hydrogen-bonded antiparallel β-sheet structures [16,24,28,29]. This concept is supported by the presence of 12 (human) [27], 13 (mouse) [22], and 15 (rat) [24] long-lived, hydrogen-bonded amide protons, which are found within the central portion of any of these EGFs. These long-lived amide protons are indicated in Figure 2 (open-diamond symbols).

Eventually, complete structural, space filling models of these EGFs will be compared in order to more completely identify highly conserved surface domain(s). Such a conserved surface domain may be the receptor binding domain. The amino acid residues which compose this domain define part of the forces involved in EGF receptor interaction.

EGF RECEPTOR INTERACTION

Knowledge of the structure of EGF is only one side of the EGF protein-protein receptor interaction; the other side concerns understanding structural relationships in the membrane receptor itself. The EGF receptor is an approximately 170,000 dalton [31] glycoprotein [32,33] whose amino acid sequence is known [34]. This transmembrane protein consists of a single polypeptide chain [35] which contains several structural domains: 1) an extracellular portion containing the EGF binding site and various glycosylation sites, 2) a transmembrane region, 3) a tyrosine-specific kinase activity, and 4) autophosphorylation site(s). Several methods, e.g., antibody mapping, chemical cross-linking, and site-directed mutagenesis, are being used to identify these various receptor domains. No crystal structure of the receptor exists, and NMR structural methods like those discussed herein for EGF are inapplicable to the receptor for numerous reasons.

To complement our NMR structural work on EGF, EGF-receptor interaction studies via ^{125}I-labelled-EGF binding kinetics are being undertaken [36]. The specific, saturable binding of ^{125}I-EGF to target cells was first reported for human fibroblasts in 1973 [37], suggesting the existence of a specific cell surface receptor for EGF. Moreover, it is now known that EGF isolated from different species can compete at physiologic pH with each other for binding to the same receptor [5]. This suggests at least a partially conserved receptor binding domain. It is not known, however, if the binding domain on the extracellular receptor surface is exactly the same for each species of EGF; or if each species interacts at slightly different, although proximal, sites; or even if the amino acids which define the complementary protein hormone interaction domain can vary (and to what extent) and still explain competitive binding.

Our approach to gain insight into some of these questions has been to incubate ^{125}I-labelled rat and mouse αEGF with human fibroblasts in monolayer culture and monitor the kinetics of receptor binding as a function of pH from 5 to 9 and as a function of temperature from 4°C to 19°C. Varying these parameters and comparing different species of EGF with known structures allows us to probe some of the forces which may modulate binding activity. Figure 5 shows the time course of mouse ^{125}I-EGF receptor binding at pH 7.4 and 4°C. The lower temperature is used to suppress post-binding activities. A computer fit of a differential form of a simple rate equation to the kinetic binding data gives the broken line in Figure 5. The fit using this 2-state model is poor. EGF concentration dependence of initial binding rates allows one to estimate 2-state on and off rate constants ($1.2 \times 10^6\ M^{-1}\ s^{-1}$ and $5 \times 10^{-3}\ s^{-1}$, respectively). Application of a ternary complex model (36) which assumes complexation of the bound receptor with a cell surface interaction molecule, like a coated pit protein, gives a satisfactory fit (solid line in Figure 5). Biphasic ^{125}I-EGF receptor dissociation kinetic plots yield a fast component ($k = 1.7 \times 10^{-2}\ s^{-1}$) and a slow component ($k = 8 \times 10^{-5}\ s^{-1}$) which are consistent with ternary complex model parameters. Concave-up transient "pseudo-Scatchard" plots (positive cooperativity) taken at pre-equilibrium times are transformed to concave-down equilibrium Scatchard plots (negative cooperativity) strongly suggesting that simple receptor dimerization (39) can not explain the data, whereas bound receptor complexation with a coated pit protein can. At the very least, these data demonstrate that analysis of EGF receptor binding data using a simple 2-state model is inadequate and yields a smaller equilibrium dissociation binding constant than actually exists ($4 \times 10^{-9}\ M^{-1}$ relative to $1.5 \times 10^{-8}\ M^{-1}$

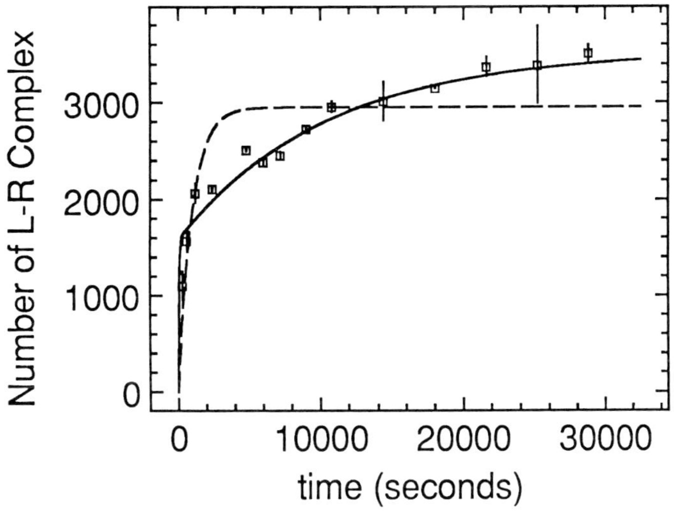

Fig. 5. Time Course of ^{125}I-labelled-EGF Receptor Binding. Human foreskin fibroblast cells were plated in 24 miniwells at pH 7.4 and allowed to grow into monolayers. ^{125}I-EGF at a concentration of 5×10^{-10} M was incubated with 2×10^5 cells (60,000 receptors/cell from Scatchard plot analysis) at 4°C. At various times after addition of ^{125}I-EGF, wells in duplicate were washed three times with 0.01 M PBS to remove free ^{125}I-EGF, and the cells were solubilized with sodium hydroxide and scintillation counted. All values were corrected for non-specific binding and are shown as ligand-receptor (L-R) complexes. Computer fit of the data using the simple 2-state model gives the broken line. Computer fit of the data using the ternary complex model yields the solid line.

from the ternary complex model). At present, other results are preliminary, but a few observations can be made: 1) EGF binding is pH dependent; as the pH is decreased from 9 to 5, apparent ligand receptor dissociation monotonically increases, suggesting a weakening of the forces which hold the complex together, and 2) rat EGF binds the receptor/complex more strongly than does mouse EGF.

Since cell viability and structure seem to be preserved in these fibroblasts during the time course of our binding experiments, observed differences should lie within the hormone-receptor binding domains, i.e., one on the receptor and one on EGF. The temperature trend for EGF receptor binding tends to reflect thermodynamic parameters generally associated with the "hydrophobic effect," [42] suggesting that mutual protein-protein interactions are at least partly due to nonpolar amino acid residues. Photo-CIDNP (chemically induced dynamic nuclear polarization) ^{1}H-NMR results on EGF-micelle interactions [40] have suggested, moreover, that EGF, in particular the C-terminal domain around Tyr-37, interacts somewhat selectively with SDS and DPC micelles in a partially hydrophobic manner. Many of the residues in this highly conserved structural domain are indeed of a hydrophobic nature.

Differences between rat and mouse αEGFs are enhanced at pH values below 7, suggesting that electrostatic interactions brought about by titration of ionizable groups on the hormone, or on the membrane receptor or on both are causal; His-22 in mouse EGF has a pK_a of 7.1 [21] and His-44 in rat EGF has a pK_a of 6.0 [30]. On the other hand, mouse EGF is known to undergo a pH-induced conformational transition around pH 7 [21,41], while the rat EGF structure seems to be more pH stable; this might play a role in variances in binding activities as well.

So far, these observations do not lend themselves to a clear explanation. One thing seems evident at least in that during endosome formation when the vesicular pH drops to about 4–5, the hormone-receptor complex is not totally dissociated and models for receptor recycling and hormone degradation should be analyzed with this in mind. Furthermore, differences in these binding properties between rat and mouse αEGFs also suggest that the proposed conserved receptor binding domain probably does contain some differences when comparing various species.

ACKNOWLEDGMENTS

This work was supported by the National Institutes of Health (GM-34662) and by the North Atlantic Treaty Organization (NATO) Scientific Affairs Division (RG 85/0424). These studies would not have been possible without the collaboration of Robert Kaptein, C. Richard Savage, Jr., and my graduate students Chris Cavalli, Carl Burke, and Marissa Nunez.

REFERENCES

1. Cohen S: J Biol Chem 237:1555–1562, 1962.
2. Savage CR, Jr, Inagami T, Cohen S: J Biol Chem 247:7612–6621, 1972.
3. Simpson RJ, Smith JA, Moritz RL, O'Hare MJ, Rudland PS, Morrison JR, Lloyde CJ, Grego B, Burgess AW, Nice EC: Eur J Biochem 153:629–637, 1985.
4. Gregory H: Nature 257:325–327, 1975.
5. Schandies P, Savage CR: Comp Biochem Physiol [A] 84:497–505, 1986.
6. De Larco JB, Todaro GJ: Proc Natl Acad Sci USA 75:4001–4004, 1978.
7. Todaro GJ, Fryling C, De Larco JE: Proc Natl Acad Sci USA 77:5258–5263, 1980.
8. Roberts AB, Lamb LC, Newton DL, Sporn MB, De Larco JE, Todaro GJ: Proc Natl Acad Sci USA 77:3493–3497, 1980.
9. Ozanne B, Fulton RJ, Kaplan PL: J Cell Physiol 105:163–172, 1980.
10. Sporn MB, Todaro GJ: N Engl J Med 303:878–900, 1980.
11. Marquardt H, Hunkapiller MW, Hood LE, Todaro GJ: Science 223:1079–1082, 1984.
12. Derynck R, Roberts AB, Winkler ME, Cohen EY, Goedel DV: Cell 38:287–297, 1984.
13. Chou PY, Fasman GD: Adv Enzymol 48:45–148, 1978.
14. Holladay LA, Savage CR, Jr, Cohen S, Puett D: Biochemistry 15:2624–2633, 1976.
15. Burgess AW, Knesel J, Sparrow LG, Nicola NA, Nice EC: Proc Natl Acad Sci USA 79:5753–5757, 1982.
16. Mayo KH, Burke C: Eur J Biochem 169:201–207, 1987.
17. Savage CR, Jr, Cohen S: J Biol Chem 247:7609–7611, 1972.
18. Di Augustine RP, Walker MP, Klapper DG, Grove RS, Willis WD, Harian DJ, Hernandez O: J Biol Chem 260:2807–2811, 1985.
19. Matrisian LM, Larsen BR, Finch JS, Magun BE: Anal Biochem 125:339–351, 1982.
20. DeMarco A, Menegatti E, Guarneri M: FEBS Lett 159:201–206, 1983.
21. Mayo KH: Biochemistry 23:3960–3973, 1984.
22. Mayo KH: Biochemistry 24:3783–3794, 1985.
23. Wüthrich K: "NMR of Proteins and Nucleic Acids." New York: Wiley.
24. Mayo KH, Cavalli RC, Peters T, Boelens R, Kaptein R: Biochem J 257:197–205, 1989.
25. Mayo KH, Burke C: (unpublished data).
26. Montelione GT, Wüthrich K, Nice EC, Burgess AW, Scheraga HA: Proc Natl Acad Sci USA 83:8594–8598, 1986.
27. Carver JA, Cooke RM, Esposito G, Campbell SD, Gregory H, Sheard B: FEBS Lett 205:77–81, 1986.
28. Montelione GT, Wüthrich K, Nice E, Burgess AW, Scheraga HA: Proc Natl Acad Sci USA 84:5226–5230, 1987.

29. Cooke RM, Wilkingons AJ, Baron M, Pastore A, Tappir MJ, Campbell ID, Gregory H, Sheard B: Nature 327:339–341, 1987.
30. Mayo KH, Schaudies P, Savage RC, De Marco A, Kaptein R: Biochem J 239:13–18, 1986.
31. Das M, Miyakawa T, Fox CF, Pruss RM, Ahronov A, Herschman HR: Proc Natl Acad Sci USA 74:2790–2794, 1977.
32. Carpenter G, Cohen S: Biochem Biophys Res Commun 79:545–552, 1977.
33. Hock RA, Hollenberg MD: J Biol Chem 255:10731–10736, 1980.
34. Ullrich A, Conssens L, Hayflick JS, Dull TJ, Gray A, Tam AW, Lee J, Yarden Y, Libermann TA, Schlessinger J, Downward J, Mayes ELV, Whittle N, Waterfield MD, Seeburg PH: Nature 309:418–425, 1984.
35. Buhrow AA, Cohen S, Garbers DL, Staros JV: J Biol Chem 258:7824–7827, 1983.
36. Mayo KH, Nunez M, Burke C, Starbuck C, Lauffenburger D, Savage RC: J Biol Chem (submitted).
37. Hollenberg MD, Cuatrecasas P: Proc Natl Acad Sci USA 70:2964–2968, 1973.
38. Pollet RJ, Standaert ML, Haase BH: J Biol Chem 252:5828–5834, 1977.
39. Yarden Y, Schlessinger J: Biochemistry 26:1434–1442, 1987.
40. Mayo KH, De Marco A, Menegatti E, Kaptein R: J Biol Chem 262:14899–14904, 1987.
41. Mayo KH, De Marco A, Kaptein R: Biochim Biophys Acta 874:181–186, 1986a.
42. Tanford C: "The Hydrophobic Effect: Formation of Micelles and Biological Membranes," 2nd Ed. New York: John Wiley & Sons.

Growth Factors and Their Receptors: Genetic
Control and Rational Application, pages 297–305

Biochemical Evidence for Multiple Classes of Platelet-Derived Growth Factor Receptor

Charles E. Hart, John W. Forstrom, James D. Kelly, Robert A. Smith, Russell Ross, Mark J. Murray, and Daniel F. Bowen-Pope

ZymoGenetics Inc., (C.E.H., J.W.F., J.D.K., R.A.S., M.J.M.), Seattle, Washington 98105; Department of Pathology (SM-30), University of Washington, Seattle, Washington 98195 (J.D.K., R.R., D.F.B.-P.)

Previous studies examining the biological activities associated with platelet-derived growth factor (PDGF) have been based upon the premise that a single receptor binds all three isoforms of the molecule (AA, AB, and BB). Through the use of purified preparations of all three isoforms of PDGF and the anti-PDGF receptor monoclonal antibody PR7212, we demonstrate that multiple classes of receptors exist for PDGF, which can be distinguished by their ligand-binding specificity. One class of receptor, designated the "A-receptor," binds and is downregulated by AA, AB, and BB, while the second class, designated the "B-receptor," binds and is down-regulated only by BB. The B-receptor appears to be expressed at sevenfold higher levels than the A-receptor on human dermal fibroblasts.

Key words: monoclonal, antibody, receptor, platelet-derived growth factor, PDGF

Platelet-derived growth factor (PDGF) is a potent mitogen for a variety of mesenchymal cells in culture [1]. It has been shown to stimulate a wide variety of processes, including chemotaxsis [2–4], phosphotidylinositol turnover, [5] and stimulation of tyrosine-specific phosphorylation [6]. PDGF is composed of a dimer of two chains, termed A-chain and B-chain [7]. PDGF can exist as a homodimer of two A-chains (AA) or two B-chains (BB) or as an A-chain B-chain heterodimer. PDGF purified from human platelets appears to be composed primarily of the latter form [8]. Previous studies involving PDGF have been based upon the premise that a single cell-surface receptor binds all three isoforms of PDGF (AA, BB, AB). With the multitude of events stimulated by PDGF and the demonstration that the A-chain and B-chain genes of PDGF appear to be independently regulated [9], it seems likely that the three isoforms of PDGF might stimulate effects unique to each isoform. This could either be through differential effects on a single receptor or interaction with different classes of

Received April 1, 1988.

PDGF receptor. To investigate these possibilities we examined the receptor-binding properties of each isoform of PDGF. In addition we examined the interactions for receptor binding between the three isoforms of PDGF and the anti-PDGF receptor monoclonal antibody PR7212 [10].

GENERATION OF REAGENTS

AA and BB homodimers were produced in a *Saccharomyces cerevisiae* expression system as previously described [11]. The A-chain and B-chain genes were constructed from synthetic oligonucleotides and encode 104 and 109 amino acid polypeptides, respectively. The homodimers have been purified from yeast culture media to greater than 95% as determined by SDS-polyacrylamide gel electrophoresis and aminoterminal sequence analysis. A mutant form of BB which has a tyrosine substituted for phenylalanine at position 23 has also been produced using the same yeast expression system [11]. This mutant form of BB allows iodination of the molecule using tyrosine-labeling techniques. AB heterodimer was purified from platelet-rich plasma by immunoaffinity chromatography using a monoclonal antibody, 121.6.1.1.1, which is specific for the B-chain of PDGF. This antibody recognizes both BB homodimer and AB heterodimer. Contaminating BB homodimer was removed from this preparation by immunodepletion using a second monoclonal antibody, 120.1.2.1.2, which specifically recognizes BB but not AB. The heterodimer material was further purified by reversed phase high-performance liquid chromatography to greater than 95%. Aminoterminal sequence analysis showed both the A- and B-chain sequences to be present in an equal molar ratio. The development and characterization of the monoclonal antibodies will be described elsewhere. The anti-PDGF receptor monoclonal antibody, PR7212, was purified from ascites fluid by protein-A chromatography as previously described [10].

AA, AB, and PR7212 were labeled with ^{125}I using Iodobeads as described by the manufacturer (Pierce Chemical Co. Rockford, IL). BB was labeled by Bolton-Hunter Reagent (DuPont-New England Nuclear Boston, MA). The tyrosine mutant form of BB was labeled with Iodobeads. The Bolton-Hunter labeled BB was used for the down-regulation and receptor competition binding experiments, while the iodinated tyrosine mutant form of BB was used for the saturation binding studies. The two labeled forms of BB have similar reactivities in both down-regulation and saturation binding procedures.

INTERACTION BETWEEN ANTIBODY PR7212 AND PDGF ISOFORMS FOR RECEPTOR BINDING

To determine the influence of receptor-bound ligand on subsequent PR7212 binding, monolayers of human dermal fibroblasts were incubated at 4°C with each of the purified isoforms of PDGF (AA, AB, BB). Following removal of the unbound ligand, the cells were subsequently incubated at 4°C with ^{125}I-labeled PR7212, or sequentially with unlabeled PR7212 and ^{125}I-labeled rabbit antimouse IgG (Table I). Preincubation of the cells with BB decreased ^{125}I-labeled PR7212 binding by 33%. This inhibition of binding was increased to 59% when PR7212 binding was detected using ^{125}I-labeled rabbit antimouse IgG. AB treatment inhibited PR7212 binding by approximately 10% under both the test conditions, while AA treatment had no effect.

TABLE I. Influence of Receptor-Bound Ligand on PR7212 Binding*

	% control binding (range)	
Ligand	^{125}I-PR7212	^{125}I-rabbit antimouse/PR7212
AA	110 (100–129)	106 (104–108)
BB	67 (54–75)	41 (40–42)
AB	89 (87–91)	93 (90–97)

*Monolayers of human dermal fibroblasts were incubated 2 h at 4°C with purified AA, AB, or BB (5 ng/ml). The cells were washed, then incubated 2 h at 4°C with either ^{125}I-labeled PR7212 or sequentially with PR7212 (2 μg/ml) followed by ^{125}I-labeled rabbit antimouse IgG. The cells were washed, harvested with solubilization buffer [16], and the cell-bound ^{125}I counts per minute (cpm) determined. The data given are the mean values for triplicate determinations from duplicate experiments and are presented as percent control binding occurring in the absence of ligand (range of mean values for the two experiments). Nonspecific binding for ^{125}I-labeled PR7212 was determined by the addition of 100-fold excess unlabeled antibody. Nonspecific binding for ^{125}I-labeled rabbit antimouse IgG was determined using a negative control monoclonal antibody in place of PR7212. Nonspecific binding was subtracted prior to calculating percent control binding.

TABLE II. Influence of PR7212 on ^{125}I-Labeled AA, BB, and AB Binding*

^{125}I-ligand	% Control binding
^{125}I-AA	103
^{125}I-BB	75
^{125}I-AB	90

*Monolayers of human dermal fibroblasts were incubated with 2 μg/ml antibody PR7212 for 2 h at 4°C. The unbound antibody was removed and the cells subsequently incubated with ^{125}I-labeled AA, BB, or AB for 2 h at 4°C. The cells were harvested with solubilization buffer [16] and the specific cell-bound ^{125}I cpm determined. The data shown are the mean values for triplicate determinations of duplicate experiments and are presented as percent control binding occurring in the absence of antibody. Nonspecific binding, determined for each of the iodinated ligands by the addition of 100-fold excess unlabeled ligand, was subtracted prior to calculating the results.

When human dermal fibroblasts were first incubated with PR7212, then subsequently with ^{125}I-labeled AA, AB, or BB, only ^{125}I-labeled BB had a significant decrease in receptor binding (Table II).

The overall results of these studies suggest that antibody PR7212 does not bind directly to the ligand-binding domain, but to a region of the receptor that is adjacent to this domain. The results also demonstrate that differences exist between the three isoforms of PDGF for their influence on receptor binding by PR7212. It is important to understand these influences when using PR7212 to monitor PDGF receptor expression following the addition of ligand to the receptor.

RECEPTOR-BINDING PROPERTIES OF THE THREE ISOFORMS OF PDGF

Binding of PDGF to cells at 37°C leads to depletion of cell-surface PDGF binding sites due to receptor internalization, a phenomenon known as "down-regulation" [12]. Down-regulation is the culmination of a complex series of events which include ligand binding, clustering of ligand-occupied receptors, and internalization of the

TABLE III. PDGF Receptor Down-Regulation*

	^{125}I-labeled probe (% control binding)			
Down-regulating ligand	AA	AB	BB	PR7212
AA	49	17	94	90
AB	54	9	87	88
BB	55	7	10	17

*Monolayers of human dermal fibroblasts were incubated for 2 h at 37°C with saturating levels of AA, AB, or BB (20 ng/ml). The cells were washed to remove unbound ligand, then incubated at 4°C for 2 h with ^{125}I-labeled AA, AB, BB, or PR7212. The cells were washed, harvested with solubilization buffer [16], and the total cell-associated ^{125}I cpm determined. The data are presented as mean binding as a percentage of control binding for determinations made from triplicate wells of a representative experiment. Nonspecific binding has not been subtracted. The inability of any ligand to decrease binding of ^{125}I-labeled AA below 49% reflects the high level of nonspecific binding for this preparation of iodinated AA. Variation between experiments was less than 10%. The specific activities for ^{125}I-labeled AA, AB, BB, and PR7212 were approximately 100 × 10^6, 65 × 10^6, 5 × 10^6, and 60 × 10^6 cpm/μg, respectively.

ligand-receptor complex. To study the stimulation of receptor down-regulation by the three isoforms of PDGF, monolayers of human dermal fibroblasts were incubated with either AA, AB, or BB at 37°C to allow receptor internalization to occur. PDGF receptors remaining at the cell surface were subsequently measured by binding at 4°C with ^{125}I-labeled AA, BB, AB, or PR7212 (Table III). All three isoforms of PDGF had equal effects in reducing subsequent binding of ^{125}I-labeled AA or AB. In contrast, ^{125}I-labeled BB and PR7212 binding was essentially unaffected by AA or AB.

The results from the down-regulation experiments suggest the existence of two classes of PDGF receptor. One class appears to bind and be down-regulated by all three isoforms of PDGF, while the second class appears to bind and be down-regulated only by BB. A second potential explanation for the down-regulation results could be the existence of a single class of receptor which binds all three isoforms, but which is not down-regulated following binding of AA or AB. In this situation the receptor would remain at the cell surface available for PR7212 binding. To investigate these two possibilities further we monitored the ability of AA and AB to stimulate receptor internalization, as determined by the loss of cell-associated ^{125}I-labeled AA and AB and the release into the culture media of trichloroacetic acid (TCA)-soluble, ^{125}I, which would occur following internalization and lysosomal degradation of the iodinated ligand. The results of these experiments demonstrated that ^{125}I-labeled AA and AB are internalized and degraded following receptor binding at 37°C with a half-time of approximately 45 min (Fig. 1). The addition of methylamine, a lysosomal inhibitor [12], to the culture media caused the rate of ligand degradation to decrease for both ^{125}I-labeled AA and AB, as demonstrated by the lower levels of TCA-soluble ^{125}I released into the media. In the presence of methylamine the half-time for ligand degradation was greater than 3 h. These results indicate that both ^{125}I-labeled AA and AB are internalized following receptor binding at 37°C and support the interpretation presented above that two binding classes of PDGF receptor exist which have different ligand binding specificity.

A second technique used to follow PDGF receptor internalization and to help delineate the existence of two classes of PDGF receptor was to monitor the internalization and degradation of receptor-bound ^{125}I-labeled PR7212 (Fig. 2). ^{125}I-labeled

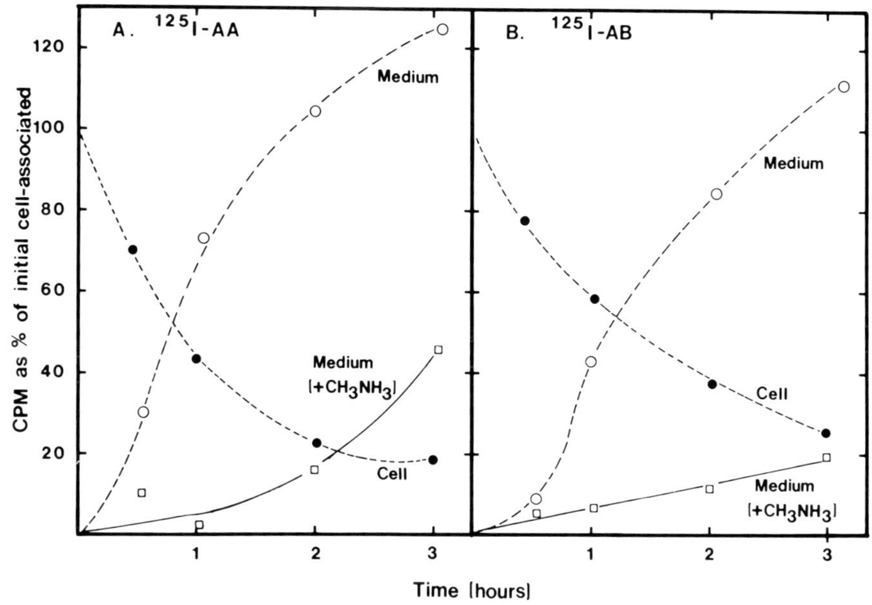

Fig. 1. Internalization and degradation of ^{125}I-labeled AA and AB. Monolayers of human dermal fibroblasts grown in 24-well culture plates were incubated at 4°C for 2 h with either ^{125}I-labeled (**A**) AA or (**B**) AB. The cells were washed to remove unbound ligand, then incubated for varying lengths of time at 37°C to allow the ligand-receptor complex to internalize and ligand degradation to occur. Conditioned media from the cells was collected and the amount of trichloroacetic acid (TCA)-soluble ^{125}I counts in the media determined (open circles). The cells were washed and cell-associated counts determined by extraction with solubilization buffer (closed circles). TCA-soluble counts released into the media were also determined for parallel cultures incubated with 25 mM methylamine, a lysosomal inhibitor [12], during the warming process (open squares). The data were calculated as the mean values for triplicate determinations from duplicate experiments and are plotted as the percentage of specific cpm initially bound to the cells at the time of warming. Data for the methylamine treatment are from a single experiment. Nonspecific binding, determined by the addition of 100-fold excess unlabeled ligand, was less than 10% of the total binding obtained for both ^{125}I-labeled AA and AB and has been subtracted.

PR7212 is internalized and degraded at 37°C with a half-time of approximately 3 h, consistent with the normal turnover rate of receptors not occupied by ligand (Fig. 2C) [8,13]. When cells were preincubated with ^{125}I-labeled PR7212 at 4°C, then subsequently incubated at 37°C in the presence of AA, no stimulation of antibody internalization was observed (Fig. 2A). In contrast, the addition of BB to the cells at 37°C stimulated rapid internalization and degradation of ^{125}I-labeled PR7212, demonstrated by the decrease in cell-associated counts and the release into the media of TCA-soluble ^{125}I (Fig. 2B). These results are consistent with the previously described down-regulation studies which showed that only BB stimulated significant receptor internalization, when monitored by ^{125}I-labeled PR7212 binding (Table III).

The capacity of AA and BB isoforms of PDGF to affect PDGF receptor internalization was further studied by immunoprecipitation of [^{35}S]methionine-labeled cell-surface PDGF receptor (Fig. 3). Preincubation of [^{35}S]methionine pulse-labeled human dermal fibroblasts with BB at 4°C (at which no ligand-induced PDGF receptor internalization occurs) resulted in a small decrease in the amount of cell-surface receptor immunoprecipitated by PR7212. This did not occur following preincubation

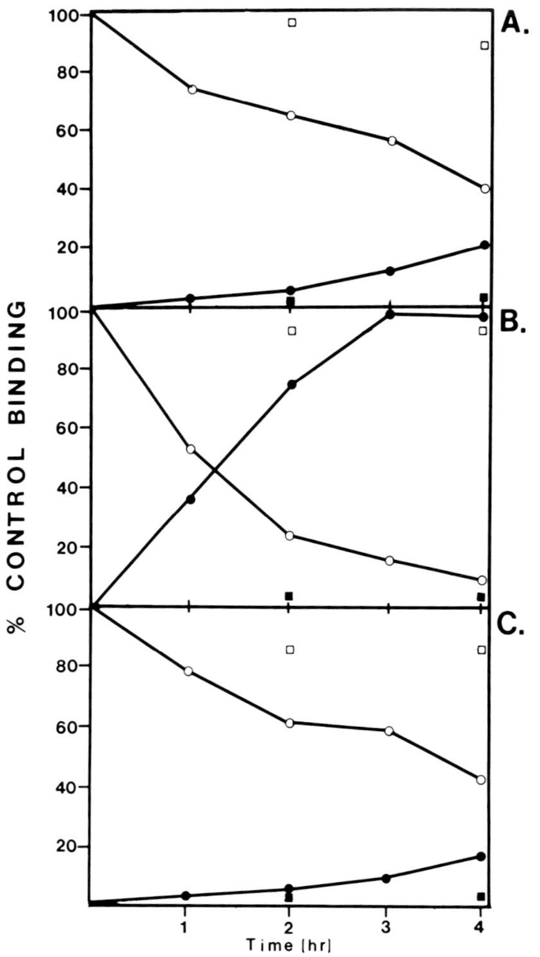

Fig. 2. Internalization and Degradation of ^{125}I-labeled PR7212 in the presence of AA and BB. Monolayers of human dermal fibroblasts were incubated with ^{125}I-labeled PR7212 for 2 h at 4°C. The cells were washed to remove unbound antibody, then incubated an additional 1 h at 4°C in the presence of either **A**) AA, **B**) BB, or **C**) binding media alone. The cells were then maintained at either 4°C (squares) or shifted to 37°C (circles) and incubated for increasing lengths of time. The conditioned media from the cells was collected and the TCA-soluble counts in the media determined (closed symbols). The cells were then washed and the cell-associated ^{125}I determined (open symbols). The data were calculated as the mean values for triplicate determinations and are plotted as the percentage of specific cpm initially bound to the cells at the time of warming. The nonspecific binding for ^{125}I-labeled PR7212 was less than 5% and has been subtracted.

with AA and probably reflects the limited competition for receptor binding between the antibody and BB (Table I). When the cultures were incubated at 37°C to allow ligand-induced receptor internalization, BB stimulated the loss from the cell surface of virtually all receptors recognized by PR7212. In contrast AA had essentially no effect on the level of receptors immunoprecipitated by PR7212.

To determine the number of binding sites for the two receptor populations, saturation binding studies were done with ^{125}I-labeled AB and BB on human dermal fibro-

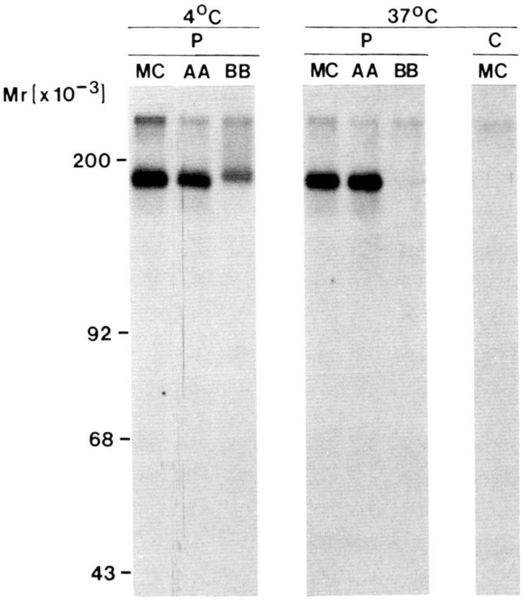

Fig. 3. Immunoprecipitation of [^{35}S]methionine-labeled cell-surface PDGF receptor. Human dermal fibroblasts grown in 25 cm^2 culture flasks were incubated with methionine-free media for 30 min, then pulse-labeled for 15 min with 0.15 mCi/ml [^{35}S]methionine (3,000 Ci/mmol, DuPont-New England Nuclear). The cells were washed, then incubated for 2 h with chase media containing 75 μg/ml methionine. The cells were then incubated for 1 h at either 4°C or 37°C with 10 ng/ml AA, BB, or media control (MC). The cells were washed to remove unbound ligand, then either antibody PR7212 (P) or a negative control monoclonal antibody (C) was added to the cells for an additional hour at 4°C. The cell membranes were then extracted by the addition of immunoprecipitation buffer, and the metabolically labeled receptors were immunoprecipitated as previously described [10]. The immunoprecipitated proteins were then analyzed by SDS-polyacrylamide gel electrophresis on an 8% gel in the presence of 5% beta-mercaptoethanol, followed by autoradiography. Molecular weight standards, shown on the left, include myosin (H-chain) 200,000, phosphorylase B 92,000, bovine serum albumin 68,000, and ovalbumin 43,000 (Bethesda Research Lab).

blasts. The results, shown in Figure 4, demonstrate saturable binding for both ^{125}I-labeled AB and BB with approximately 20,000 and 140,000 binding sites per cell, respectively. This corresponds to a ratio of 7:1 for BB to AB binding.

DISCUSSION

The data presented here suggests the presence of at least two classes of PDGF receptor which differ in ligand binding specificity for the three isoforms of PDGF. One class of receptor, which we have designated the "B-receptor," binds and is down-regulated only by the BB isoform of PDGF. The second class of receptor, which we have designated the "A-receptor," binds and is down-regulated by AA, AB, and BB. The identification of the two classes of receptor is most clearly demonstrated by the down-regulation experiments where each PDGF isoform was used both as a down-regulating agent and as a probe for detection of remaining receptor expressed on the cell surface. These results clearly show different patterns of receptor down-regulation

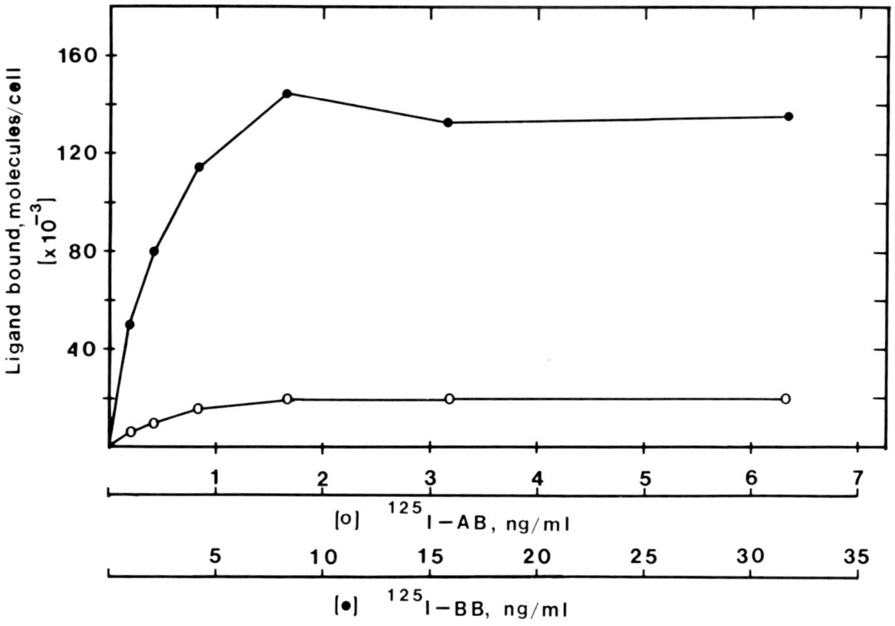

Fig. 4. Saturation binding with ^{125}I-labeled AB and BB. Monolayers of human dermal fibroblasts were incubated for 3 h at 4°C with increasing concentrations of ^{125}I-labeled AB (open circle) or BB (closed circle). The cells were washed to remove unbound ligand, harvested with solubilization buffer, and the total ^{125}I cpm bound to the cells determined. Nonspecific binding, determined by the addition of 100-fold unlabeled ligand, has been subtracted from the data. The results are mean values of triplicate determinations from a representative experiment and are expressed as molecules of ligand bound per cell. The specific activity for ^{125}I-labeled AB and BB were 94 × 10^6 and 15 × 10^6 cpm/μg, respectively.

for the three isoforms of PDGF and are consistent with the observation that BB stimulates internalization of receptor-bound ^{125}I-labeled PR7212 while AA treatment has no effect (Fig. 2). The distinction between the two classes of receptor is further supported by the studies examining immunoprecipitation of [^{35}S]methionine-labeled PDGF receptor (Fig. 3), where BB stimulated a loss of immunoprecipitable receptor from the cell surface, while AA treatment had minimal effect. The significance of these studies is supported by the observation that AA and AB can stimulate receptor internalization (Fig. 1). Thus, the lack of down-regulation activity stimulated by AA and AB, as detected by BB and PR7212 receptor binding, was not due to their inability to induce receptor internalization but due to their recognition of a limited subpopulation of PDGF receptors.

The ratio of the two classes of receptor on human dermal fibroblasts appears to be approximately 7:1 with the B-receptor being the predominant form. This ratio was calculated from the estimates of ligand binding sites per cell for ^{125}I-labeled BB and AB binding (Fig. 4). This ratio is supported by the results from the down-regulation studies, where AA and AB decreased ^{125}I-labeled BB and PR7212 binding by only 10–15%.

Previous studies using ^{125}I-labeled PDGF as a probe to monitor PDGF receptor expression have been done with PDGF iodinated by procedures that specifically label tyrosine residues. These labeling techniques would only iodinate the AA and AB forms of PDGF isolated from platelets, since only the A-chain of PDGF contains tyrosine. Thus, binding studies done with ^{125}I-labeled PDGF would detect the A-receptor class. Since BB would not be labeled by these techniques, the B-receptor would go undetected. The identification of multiple receptor classes has also been demonstrated for transforming growth factor-beta [14,15].

The origination of the two populations of PDGF receptor remains unclear. It is not known whether the two populations derive from two distinct gene products, modifications of a single gene product, or functional differences in a single molecule that result from different patterns of association with other cell proteins. Currently no functional differences have been assigned to the two classes of PDGF receptor. The recent availability of purified preparations of each of the isoforms of PDGF should allow us to rapidly address this question.

ACKNOWLEDGMENTS

This research was supported by NIH grants GM35501 and HL18645, by National Institute of Dental Research grant DEO7063-11, and a grant from the Life and Health Insurance Medical Research Fund.

REFERENCES

1. Ross R, Raines EW, Bowen-Pope DF: Cell 46:155–169, 1986.
2. Seppa H, Grotendorst G, Seppa S, Schiffmann E, Martin GR: J Cell Biol 92:584–588, 1982.
3. Grotendorst GR, Chang T, Seppa HEJ, Kleinman HK, Martin GR: J Cell Physiol 113:261–266, 1982.
4. Deuel TF, Senior RM, Huang JS, Griffin GL: J Clin Invest 69:1046–1049, 1982.
5. Berridge MJ, Heslop JP, Irvine RF, Brown KD: Biochem J 222:195–201, 1984.
6. Cooper JA, Bowen-Pope DF, Raines E, Ross R, Hunter T: Cell 31:263–273, 1982.
7. Johnsson A, Heldin CH, Westermark B, Wasteson A: Biochem Biophys Res Commun 104:66–74, 1982.
8. Hart CE, Forstrom JW, Bowen-Pope DF: unpublished observations.
9. Betsholtz C, Johnsson A, Heldin C-H, Westermark B, Lind P, Urdea MS, Eddy R, Shows TB, Philpott K, Mellor AL, Knott TJ, Scott J: Nature 320:695–699, 1986.
10. Hart CE, Seifert RA, Ross R, Bowen-Pope DF: J Biol Chem 262:10780–10785, 1987.
11. Kelly JD, Raines EW, Ross R, Murray MJ: EMBO J 4:3399–3405, 1985.
12. Heldin CH, Wasteson A, Westermark B: J Biol Chem 257:4216–4221, 1982.
13. Keating MT, Williams LT: J Biol Chem 262:7932–7937, 1987.
14. Cheifetz S, Weatherbee JA, Tsang ML-S, Anderson JK, Mole JE, Lucas R, Massague J: Cell 48:409–415, 1987.
15. Segarini PR, Roberts AB, Rosen DM, Seyedin SM: J Biol Chem 262:14655–14662, 1987.
16. Bowen-Pope DF, Ross R: Methods Enzymol 109:69–100, 1985.

Index